UNEQUAL JUSTICE

UNEQUAL JUSTICE

Lawyers and Social Change in Modern America

Jerold S. Auerbach

New York
Oxford University Press
1976

For
JEFFREY and PAMMY
        With Love

Preface

For every historian, answers from the past depend upon questions from the present, upon an often discordant conversation between past and present in which sounds from two eras compete for attention. Furthermore, a historian not only listens but speaks. Any pretense that he is aloof from his evidence may succeed as self-deception but it should not deceive readers. This book is a product of the interaction between historical past and historian's present. It proceeds from the assumption that a historian who is unaffected by the present can hardly claim sensitivity toward the past. Not only are historians selective toward the past; they are, like everyone else, selective toward the present. Experiences shape values which, in turn, filter judgments.

In 1957 I entered Columbia Law School as a first-year student. Although mine was the silent generation, and I embraced its values (silently), I nonetheless intended to become a civil liberties lawyer who would vindicate constitutional principles for embattled defendants who were noisier than I could be. I cannot account for this commitment. There were no lawyers in my family to emulate. Certainly I was ignorant of those great moments in American history when, as though by divine favor, a lawyer miraculously appeared to defend political pariahs: Andrew Hamilton for Zenger; John Adams for the British soldiers implicated in the Boston Massacre; Charles Evans Hughes for Socialists expelled from the

New York legislature. (I am equally certain that I was unaware of all those other moments, less conspicuous in our folklore if more characteristic of our national experience, when no such miraculous intervention had occurred.) I would like to think that my course was set by attorney Joseph Welch, whose televised indignation at the recklessness of Senator Joseph R. McCarthy during the Army-McCarthy hearings still remains vivid two decades later, or by Martin Luther King, Jr., whose visit to Oberlin College during the Montgomery bus boycott heightened my awareness of conflicts between law and justice. But I cannot be certain. Something about protesting against illegitimate authority drew me to law school.

Within a week disillusionment shattered my aspiration beyond repair. In my undergraduate innocence it had never occurred to me that legal education was the finest preparation available for a career in business. Quite the opposite: I entered law school in avoidance, not in pursuit, of that objective. But not only were Columbia Law School and Wall Street stations on the same subway line; they were stations on the same career line. The message was never explicitly conveyed but it was communicated through the curriculum we studied, the jackets and ties we wore, and the expected rewards for mastery of torts and contracts. Never was there a whisper of a suggestion that law related to choice, to history, to society, to justice. Its world was populated by appropriately anonymous A's, who fired bullets across B's land, wounding C's, who tumbled into D's well, after E's rescue efforts were thwarted by F's enraged bull. It was a year when all offers were rejected without consideration, legal or otherwise. I learned to demur, but I was never satisfied with *res ipsa loquitur* as an explanation for my doldrums.

Yet there was something quite compelling in legal education, despite the constricting shallowness of the case method and the arrogant callousness of some of its practitioners (who subsequently were memorialized in law reviews for their humane sensibilities). To apply a distinction quickly learned: the substance was a disappointment but the process, at best, was analytically rigorous and its corrosive skepticism was not without its virtues. After so many years spent learning what to think it was a relief to be told that I was learning how to think. The experience was paradoxical: the more I learned how to think like a lawyer the less I wanted to become one. Legal education was designed to evade precisely those questions which, in my naiveté, I believed that lawyers should contemplate: Is it just? Is it fair? If not, how can law be utilized to make it so?

In dismay, I left law school to begin graduate training in history. Although Clio was a less jealous mistress than Portia, fledgling historians, like fledgling lawyers, were trained to repress their own values in the service of professional austerity. First-year graduate students learned that Charles Beard's politics shaped his history and infused it with vitality, but there was no mention of the politics of professionalism or the disjunction it so often imposed between conscience and creation. We learned to be "objective" and "balanced," reflecting Sir Roger de Coverley's pronouncement in *The Spectator* that much may be said on both sides; that unless much *is* said on both sides the historian is dismissed as subjective and biased and his work is regarded as seriously flawed. No fair-minded assessment could ever be one-sided, since truth was defined as two-sided. The notion that all the evidence, when it is in, may compel a conclusion that is both fair and truthful because it is one-sided violated every conventional category of historical analysis. Not until long afterward did it become evident that this form of analysis was a mode of thought, not a categorical imperative—a mode of thought resting upon inherently political assumptions which professionalism disguised as value-free methodology.

My first sustained excursion into the past (my doctoral dissertation) was fortuitously deflected. In a marvelous twist of irony I was invited to spend a year in residence in a prestigious Wall Street law firm (by then relocated to Park Avenue) to prepare a biography of its senior partner. But I had unwittingly become the house historian for a law firm whose partners wanted hagiography, not biography. I searched in vain for documents that would clarify my subject's quite interesting political meanderings, only to discover that sheaves of personal papers relating to controversial public activities had been destroyed. I tried in vain to avoid the reconstruction of his professional career, which required immersion in incomprehensible case files, carted from warehouses by the truckload, whose subject matter had driven me from law school three years earlier. Worse yet, the patterns of deference and subservience which pervaded the firm stoked my antipathy toward my subject, who basked in their reassuring warmth. When I finally understood that historian's history was unacceptable to my employers, and that their self-celebratory history was unacceptable to me, I negotiated my release.

I retreated to a decade, the 1930's, and to a subject, civil liberties, whose allure I found unimpeachable. From civil liberties in the 1930's, the subject of my first book, I circled unerringly back to civil liberties lawyers. Their professional deviance intrigued me. I wanted to trace their emergence as a sub-group within the legal profession and to discover the historical and personal sources of their distinctiveness. Legitimate questions were then unanswered, and indeed unasked. What prompted certain lawyers to devote a substantial portion of their professional energies, often at high risk, to the defense of non-conformists? Why were a few lawyers preoccupied with justice when so many were only concerned with clients? To what extent did their commitment express ethnic or social marginality? What did the emergence of civil liberties lawyers reveal about the Bill of Rights in an urban industrial age? What were the implications of a specialized civil liberties bar for the legal profession and for American society?

As I began to read in legal history and in the sociology of law, questions multiplied and answers receded. In part this resulted from a vast chasm between the two disciplines. Legal history was still an arcane field of inquiry monopolized by antiquarians who were mesmerized by Blackstone. Sociology of law was exploding with exciting possibilities but the fixation of too many of its practitioners upon the present (or upon numbers) presupposed the creation of the world at the moment when the first questionnaire was distributed (and ignored, at their peril, the wisdom of law professor Thomas Reed Powell who declared that in the end counters didn't think and it was the thinkers who really counted). Several books decisively affected my own inquiry. Otto Kirchheimer's *Political Justice* shattered some traditional categories of time, place, and subject in its scrutiny of the legal process as a political instrument. Jerome Carlin's *Lawyers on Their Own* and Erwin Smigel's *The Wall Street Lawyer* destroyed the myth of a homogenized profession that lawyers nurtured to disguise gross disparities in the allocation and adequacy of legal services. Carlin's *Lawyers' Ethics* related professional conduct to social origins, professional status, and political purpose. And Willard Hurst's *The Growth of American Law* presented one of those rare forays into legal history that explored the social setting of law.

Amid the backing and filling that inevitably accompanies historical inquiry, two quite disparate sets of experiences fixed this book in its final form. The first resulted from an invitation to spend an academic year (1969-70) as a Fellow at Harvard Law School. As an observer, not a participant, I enjoyed obvious advantages upon my return to law school. At least I could differentiate what was unique to Columbia in 1957, or to Harvard in 1969, or to me at either time, from what was implicit in the ideology of legalism and in the structure of legal education.

Under pressure from a more demanding generation of students and a new agenda of public issues, law schools had diversified their curricula and relaxed the stringent hold of the case method. Yet in part because Harvard was where modern legal education began, but especially because I now had benchmarks of time and place for guidance, continuities were far more conspicuous than changes. To an outsider law school was still a church with its essential dogma preserved. Its teachers, only three generations removed from their patron saint (Langdell) and his apostles (Ames and Thayer), still inculcated acolytes with the conventional orthodoxy. To think like a lawyer at Harvard in 1969 was no different than at Columbia in 1957, or, as I would discover, at the beginning of the twentieth century when a new urban industrial society decisively transformed the American legal profession.

To a historian of that profession Harvard Law School simultaneously offered the perspectives of time and timelessness. The presence of the past was inescapable. Although I commuted between the archives, where I inhabited the world of the founding generation of teachers, and a course on the legal process taught superbly by a contemporary craftsman, it was evident how strong was the ideological affinity between them and how long was the methodological shadow of the founders, within whose professional enclave around the turn of the century the claims of law and society in a new era were received and transmitted across generations. These men, belonging to the legal profession yet detached from it, aloof from public life yet drawn to it, were "outside insiders" (a self-descriptive term later used by Felix Frankfurter) whose curious relationship to their profession and society threw both into sharp relief at a critical juncture in their history—and began to suggest the seamlessness of nearly a century of professional history.

The unity of that period became evident to me between 1968 and 1974, the relevant present for this book. I cannot pretend that my ideas about law and the legal profession were unaffected by the phantasmagoric events of those terrible years. Taken together (which was, after all, how they came), they depicted a society which, in one historian's apt phrase, was "coming apart" as legitimate authority was stripped from one institution after another—from university, government, presidency, military, police, prisons, courts, law.

Once historical and contemporary evidence converged, the evolution of this book—conceived in autobiography, grounded in history, and illuminated by contemporary events—ended. My conclusions about the past rest upon my reading of the historical record. The record is there, as my notes amply demonstrate. But so is *my* reading of it and the obvious fact that *when* I read is as important as *what* I read. If there is meaning to the concept of historical truth, and I believe that there is, it emerges from the historian's imaginative reconstruction of the past. The past, without the historian to order and interpret it, remains random chaos.

The historian, unbound to the past, writes fiction. My preferences are for order, not chaos; for history, not fiction.

I cannot, however, claim to be a neutral observer (nor, I believe, can anyone else). In any case, neutrality is no less partisan than engagement; it is merely an expression of different, but equally partisan, values. In writing the history of the legal profession, or of any social institution, it is possible to approach the subject functionally, to stress (and implicitly to praise) its adaptation to social values without examining either the values or the implications of adapting to them. But this would not be a neutral choice; it would only reflect a particular set of value judgments. It is not the choice I have made. I concede that to deplore professional limitations is ultimately to question some basic American values, which the legal profession has thoroughly absorbed. My commitment is not to those values, but to that equality of justice under law which those values persistently subvert. That is why I wanted to become a civil liberties lawyer. That is why I wrote this book.

J.S.A.

Belmont, Mass.
Jerusalem
1975

Contents

UNEQUAL JUSTICE

Introduction

The subject of this book is the response of elite lawyers to social change in the twentieth century. It explores the efforts of these lawyers to mold their profession to cope with the forces that have transformed our national life: industrial capitalism, urbanization, immigration, war, economic depression, and social ferment. For my purposes elite lawyers are more salient than others because their professional status and power afforded them special opportunities to articulate, and often to implement, solutions to the problems they perceived. I am most concerned with their explanations of change and with their efforts to restructure the legal profession to cope with change. This is not, therefore, a book about how lawyers practice law, nor is it an exegesis of legal doctrine. These subjects lie beyond my expertise. It is, instead, a social history of the professional elite. I am probing for attitudes and values which most fully reveal the legal profession as a social institution shaped by, and in turn shaping, society.

Recent students of the contemporary legal profession have described a highly stratified professional culture, not a monolithic profession. A lawyer is a lawyer—but only to a point. There are significant differences among lawyers which, the evidence suggests, are neither random nor necessarily meritocratic. These differences have profound implications for the presumed fairness of the adversary process, for the adequate provision of legal services to all who need them, and for the capacity of the legal process to provide equal justice. Their existence is the product of nearly a century of professional history, which lawyers (out of ignorance or self-interest) have ignored. Lawyers are understandably predisposed to belittle the idea that professional differentiation correlates with race, religion, class, color, sex, education, educational opportunity, and social origins—rather than with hard work, self-discipline, inductive reasoning skill, and academic achievement. Acknowledgment of such differentiation would cast serious doubts upon the fairness of the adversary process, the cornerstone of the American legal system. Yet the existence and persistence of these divisions is beyond dispute. How a stratified bar emerged in the twentieth century, what social values stratification reflected and perpetuated, and what the implications of stratification were for law and justice comprise a primary subject of this inquiry.

Stratification enabled relatively few lawyers, concentrated in professional associations, to legislate for the entire profession and to speak for the bar on issues of professional and public consequence. It once was written of New England Puritan ministers that they comprised a speaking aristocracy in the presence of a silent democracy; so it has been in the legal profession. Corporate lawyers and university law teachers emerged around the turn of the century as self-appointed guardians of professional interests as they defined them. Dominating major professional associations and institutions, they applied their leverage ostensibly but not always self-evidently in the public interest. They constituted a professional elite: a group able to define the terms of admission "to the circle of the . . . influential, terms which may include conformity to standards of wealth, social background, educational attainment and commitment to the elite's interests and ideology."[1]

A paramount objective of this elite was to structure the legal profession—its education, admissions, ethics, discipline, and services—to serve certain political preferences at a time when social change threatened the status and values of the groups to which elite lawyers belonged and whose interests they wished to protect.

They confronted choices between competing definitions of professional identity and obligation. Was their profession public or private? Should it be accessible or restricted? Were all citizens, regardless of social origins, politics, or income, entitled to adequate legal services? These questions provoked a prolonged struggle for power within the legal profession. Its outcome shaped the professional structure for decades to come.

The axes on which this struggle turned were ethnicity and class. The proportion of white Anglo-Saxon Protestants within the legal profession and American society was diminishing as changing immigration and demographic patterns swelled cities and the profession with the foreign-born and their children. Native American lawyers were determined to repel a dual challenge: to their ascendency in professional life and to the economic institutions of industrial capitalism which they served. First they consolidated their position within corporate law firms and professional associations. Once these enclaves were secure, they wielded their power to forge an identity between professional interest and their own political self-interest. Much evidence confirms the existence of opportunities for some members of ethnic minority groups to surmount obstacles to professional success. This evidence substantiates traditional assumptions about legal careers as avenues of social mobility. But it also begs important issues. It ignores the costs of success, for those who succeeded no less than for those who failed. It diverts attention from the fact that the recipients of the best legal services have remained consistently identifiable along class lines, even if the identity of those who serve them has changed. And it slights the tenacious hold, and pernicious effect, of elite values within the legal profession.

It was hardly coincidental that the structure of the modern legal profession was designed and built in the two decades between 1905 and 1925. It was then that the power of "malefactors of great wealth" within the new business system of industrial capitalism was vigorously asserted and sharply challenged. These were also the peak years of the "new" immigration from southern and eastern Europe; by 1925 the gates to mass immigration had been virtually closed by restriction laws. The tension between elitism and democracy within the legal profession was not unique to this era; it was, in some respects, a twentieth-century restatement of an earlier struggle during the age of Jackson. But the urban dimension was new and the economic stakes were high. The decisions made during this formative era for the modern legal profession still profoundly affect American law and society in the mid-1970's. How professional power was distributed, and the uses to which it was put, can hardly be ignored in a society committed in theory to equal justice under law and dependent upon the legal profession for its implementation.

Although the modern professional structure was in place by 1925, it is not my contention that nothing of consequence has occurred since. The depression of the 1930's contributed an important chapter to the social history of the American legal profession. Economic catastrophe produced severe social dislocation which momentarily weakened the power of the professional elite and the values that sustained it. An energetic reform administration in Washington opened new careers for lawyers whose social origins disqualified them from employment in the elite private sector. A temporary spirit of vitality in labor union organization and in civil rights litigation afforded additional opportunities for minority-group lawyers. A vigorous new professional organization challenged the American Bar Association, stronghold of the old elite. New conceptions of professional obligation to impoverished clients—and impoverished lawyers—emerged. By the beginning of World War II the structure of the legal profession no longer seemed to rest so securely upon a pre-Depression foundation.

Appearances were deceptive. A weakened professional elite, confronting massive political and economic pressure, had tacitly acceded to a quid pro quo: some outsiders finally would be admitted to elite positions within the professional structure in return for their loyalty to dominant professional values. Corporate clients still received the best legal services; only the identity of the providers changed. The nature of this compromise became evident during the Cold War when, with few exceptions, professional leaders not only permitted but encouraged the sacrifice of the rights of politically unpopular lawyers and defendants to public (and professional) hysteria. Confronting a choice between politics and professionalism, elite lawyers once again closed their ranks against deviance. Armed with disciplinary and disbarment powers and loyalty oaths, they ostracized dissidents who challenged their political or economic values. In the frigid climate of the Cold War the fragile shoots of professional democratization that survived from the depression years withered and died.

Amid the crisis politics of the last decade, the legal profession has become the subject of intensive scrutiny. With the erosion in legitimacy of one institution of law and justice after another, thoughtful scholars and critics even debated the question posed by the title of one recent volume of essays: Is law dead? A profusion of books about radical lawyers, relevant lawyers, people's lawyers, superlawyers, and mere lawyers testified to unprecedented interest in the legal profession and concern with its role in American society. Concern was justified, even before it became evident that the most callous disregard of the rule of law came from lawyers elected or appointed to the highest positions in government. As comforting as it surely would be to assume that individual venality, not a flawed structure, explains this crisis, there is abundant historical evidence to the contrary. Watergate raised the most profound and tormenting questions about the ethics and values of the legal profession. To those familiar with the past, Watergate served as a reminder that its own history is a nightmare from which the legal profession must try to awake.

Because this book was written over an extended period of time I enjoyed numerous opportunities to address portions of my argument, either in spoken or printed form, to critical audiences of legal scholars, law teachers, lawyers, and historians. Certain criticisms, especially from lawyers, have recurred with sufficient persistence for me to consider them briefly here. Since I hope this book will influence the way lawyers and law students think about their profession, I want to anticipate certain methodological inquiries that might, if unanswered, divert attention from the substantive thrust of my conclusions.

The first of these is primarily a matter of credentials: that because I am not a lawyer it is presumptuous, if not preposterous, for me to write intelligently about the legal profession. If law is a mysterious science then laymen surely are incompetent to comprehend the behavior of its practitioners. But because I believe that law and lawyers express policy preferences, I assume that it is necessary to examine which preferences (and whose) are expressed. Although the process of professionalization transforms many lawyers who experience it, and provides them with a special language, skills, and values, laymen are not incompetent to understand how lawyers have responded in their professional capacity to public issues. Indeed, laymen must insist—until lawyers venture from their professional cocoon—that law is social, not scientific; that legal history is a chapter of social history, not a self-contained entity. Even lawyers' craft, applied to the most technical professional tasks, can be studied as an expression of particular values in a particular culture at a particular time. It is incongruous, to say the least, that lawyers who pride themselves on being generalists equipped to resolve any social problem—a conceit which American society too eagerly encourages—should swiftly claim immunity as inscrutable specialists when outsiders poke into their professional activities.

Some critics have suggested that my research, which is "merely" historical, dwells unnecessarily on the failures of the past at a time when the organized bar is making constructive efforts to resolve contemporary problems. But past and present are not entirely separate. The historical appearance of a stratified legal profession, structured in its education, admissions, ethics, and services to implement certain ethnic and class values, created new problems in the process of alleviating old anxieties. These problems have not yet been resolved, nor can they be until their historical antecedents are understood—because they are products of that history. Legal education is still haunted by Langdell's ghost; too often it ignores precisely those value considerations that lawyers in their public policy role consistently confront. Ethical judgments still are framed within turn-of-the-century precepts as modified (slightly) by the recent Code of Professional Responsibility. The problem of minority-group admission has not disappeared; only the identity of the victims of discrimination has changed. The debate over federal legal services echoes a much earlier dialogue about legal aid. History, a law teacher has written, "is a systematic distortion of the past, designed to tell us something meaningful about the present."[2] One need not concur in that definition of lawyer's history to appreciate the conspicuous place of the past in the professional present.

My most insistent critics have asserted that the composition of the bar is too variegated and the tasks of its members too diverse to sustain generalizations about responses to social change. In part, this criticism expresses the lawyer's deep distrust of generalization, a product of that very process of professionalization that trains lawyers to focus on particular cases, to relentlessly spot issues, and to narrow holdings to their irreducible core. Thomas Reed Powell of Harvard once defined the legal mind as one that could think of something that was inextricably connected to something else without thinking about what it was connected to. Case analysis is a form of tunnel vision which doubtlessly facilitates certain tasks that lawyers perform. But its limitations need not be imposed upon historians, whose art is imaginative synthesis and re-creation from among the myriad fragments that are the residue of the past. Whatever the virtues of thinking like a lawyer may be for lawyers, historians must think like historians.

The charge of reductionism certainly would be justified if this book pretended to be a comprehensive history of the entire legal profession. But my primary interest, again, is the response of the professional elite to social change. The maldistribution of professional power makes it necessary to focus on bar leaders, who have exerted enormous influence within the profession and, as spokesman to the public, outside it. At elite levels the bar, as a matter of historical fact, has *not* been heterogeneous in its composition or purpose. It has represented identifiable interests and values which have guided the pursuit of certain objectives at the expense of others. To focus on the professional elite is not to distort reality but to illuminate part of it—the part with the broadest public implications. The words and deeds of bar leaders tell us things worth knowing about those who were empowered to speak for the legal profession and about a profession that permitted so few to speak for so many.

Within this framework my substantive conclusions have disturbed some people who suspect that my preoccupation with ethnicity is exaggerated. Beyond restating my belief that I have fairly represented the weight of the evidence left by a profession whose leaders often were preoccupied with the subject, I can only invite skeptics to go where I have gone, to the sources, and to see for themselves. Those whose knowledge of American society comes from the inside looking out (or from the top looking down) are likely to underestimate, perhaps to overlook, the depth of ethnic animosity in American history and its eruptive force in times of social stress. Ethnicity, to be sure, is related to class. But exploration of the complexity of that relationship in modern American history would carry me far beyond the boundaries of this book.

My point is not that lawyers have been more prejudiced than other Americans. It is, instead, that bias in the legal profession has had particularly serious consequences in a society that depends so heavily upon the legal profession to implement the principle of equal justice under law and, simultaneously, to harmonize law with social change. A legacy of prejudice poses fundamental questions for those who take professional obligations seriously. Is an adversary system of justice "just" when the adversary process is skewed by the social origins, ethnic identity, and financial resources of attorney or client? Is there equal justice when the legal profession, the primary instrument of its attainment, is structured to reflect and reinforce social inequality?

Not too long ago law was viewed as a "brooding omnipresence," a set of cosmic rules embodying right and truth that were unrelated to time, place, or political preference. Sociological jurisprudence and legal realism shattered that myth and compelled historians of legal institutions to relate law to society. Writing from the newer perspective, Willard Hurst, the foremost legal historian of our time, suggested that "in the interaction of law and American life the law was passive, acted upon by other social forces. . . . Law has been more the creature than the creator of events."[3] This sounds like a truism to sophisticated contemporaries, but the notion that law was a social institution, not a gift from the Anglo-Saxons or God, was a well-kept secret from legal historians until rather recently.

Every legal historian has by now internalized this truth. But legal history still reflects professional norms and the values of the legal culture. Since this is a matter of the historian's choice, not necessity, the costs of absorbing lawyers' norms should be carefully weighed. It is of the essence of the professionalization process to divorce law from politics, to elevate technique and craft over power, to search for "neutral principles," and to deny ideological purpose. Lawyers and laymen alike must be persuaded that law embodies reason, not will; that it is a mysterious science inaccessible to the uninitiated; that "thinking like a lawyer" is transcendent rather than time-bound. But historians need not accept those propositions at face value. Professionalization is not value-free; it is a process and an ideology which serves profoundly political objectives. Indeed, "it is the ideological character of professionalization that makes lawyer's history inevitably conservative," observed Morton J. Horwitz of Harvard Law School. Consequently, "an elitist and anti-democratic politics pervades most of the traditional writings on American legal history, just as it appears in virtually all of the rhetorical literature of the legal profession throughout American history."[4]

To write a social history of the legal profession that catches the dialogue between profession and society requires a conscious departure from these canons of conservative orthodoxy. Professional norms must be understood but they need not, and should not, be uncritically accepted. Not only must the legal profession be evaluated on its own terms; it must be assessed in terms that independently calculate its contribution to the well-being of society. "Over the years," Hurst has written, "the bar shared the prevailing religious, racial and national prejudices of middle-class Americans." But the conclusion does not follow that criticism of the bar has "unfairly, and with much hypocrisy, . . . assigned moral responsibility to the bar for conditions outside lawyers' control, or for which the community must share responsibility."[5]

Law is a mirror of social forces, yet different mirrors perform different functions. Some merely reverse the images they reflect. Others (those of which children are especially fond) alter and distort these images to accord with the conscious design of their planners. Law is both kinds of mirrors. It reflects what is in society, but often it exerts autonomous power to channel social problems and public issues into its own constricted framework of legitimacy and procedure. Then it may become detached from its social moorings. Consequently, the bar must be judged by two standards (but not by a double standard): its sensitivity to the values and mores of society; and its implementation of the obligation to provide equal justice under law. In theory, the bar assumes both transient and transcendent responsibilities: to serve clients and to serve justice. Although lawyers equate these, and assume that they do one when they do the other, their claim should be critically examined. In fact, a total commitment to the former may preclude any commitment to the latter.

In the United States justice has been distributed according to race, ethnicity, and wealth, rather than need. This is not equal justice. The professional elite bears a special responsibility for this maldistribution. Its members, absorbed with selective client-caretaking for a restricted clientele, have preserved social and economic inequality. Their efforts, in conjunction with the limitations of an adversary process largely dependent upon the ability to pay, have crippled the capacity of the legal profession to provide equal justice under law or to fulfill those paramount public responsibilities that alone can justify professional independence and self-regulating autonomy. This book presents a historical framework for comprehending the ominous gap between the services dispensed by the legal profession and equal justice. The past cannot be undone, but those who know it may enjoy the opportunity to escape its tenacious hold.

One

The Best Men
and the Best Opportunities

Alexis de Tocqueville, writing in the age of Jackson, told American lawyers what they most wanted to hear. In a sentence from his *Democracy in America* that would become a staple of bar association addresses for generations to come, he asserted: "If I were asked where I place the American aristocracy, I should reply without hesitation . . . that it occupies the judicial bench and the bar." Tocqueville's compliment was qualified by the knowledge that his aristocracy might be bridled by democratic constraints. American lawyers, he wrote, were hostile to the "unreflecting passions of the multitude," conservative and anti-democratic, and responsive to the lure of private interest and advantage. But if they exercised authority over the democracy they also derived their authority from it; comprising an aristocracy, they also belonged to the people; exerting disproportionate power within society, they were also responsible to it. Tocqueville's legal profession comprised the "friends of order" and the "opponents of innovation." Yet order and innovation were less antithetical than his polarity implied. As Tocqueville appreciated, a new order (shaped by lawyers) already was emerging from economic innovation.[1] The affinity between elite lawyers—his aristocracy—and nascent corporate capitalism was there, and Tocqueville saw it. Even in Andrew Jackson's day, as one perceptive historian has suggested, "there seems strong evidence that an identifiable, self-conscious, and dominant portion of the profession was found consistently in collusion with the advance guard of commercial and industrial capitalism."[2] Tocqueville's aristocrats and their successors stood boldly astride the portals of the legal profession as the United States entered the modern era.

Daniel Webster, whose ascendancy in law and politics accompanied his service to the emergent commercial-industrial elite in New England, probably was the lawyer whose career inspired Tocqueville. But the country lawyer, a quite different professional man, touched even deeper chords in American hearts. Practicing alone in a small town, he prepared for his profession by reading Blackstone and Kent and by apprenticing himself to an established practitioner for whom he opened and cleaned the office, copied documents, and delivered papers. The commercial bustle of the city was another world. Whether he rode circuit or lounged around the local courthouse, he absorbed the camaraderie of his profession and cherished the respect of his neighbors. An independent generalist, he served all comers, with no large fees to turn his head toward a favored few. He moved easily between his casual, cluttered office, where informality (it was assumed) nurtured trust and loyalty, and the courtroom, where skill as an advocate earned him local renown. Self-reliant and persevering, he was the common man's lawyer in a pre-urban, pre-industrial society.

In time, especially after he was pushed to the edges of an urban industrial society, the country lawyer assumed heroic dimensions. After the martyrdom of Abraham Lincoln his apotheosis was complete. Antebellum America receded, but images remained vivid of that child of the American frontier, self-taught in a Kentucky log cabin, the circuit-riding country lawyer in Illinois who became President to save the Union and died to make men free. Lawyers cherished Lincoln's memory as fondly as schoolboys treasured tales of Old Hickory and Davey Crockett. Never again, a lawyer from the land of Lincoln concluded at the century's turn, could his profession recapture the popular affection once extended to it during the ascendancy of country lawyers. Modern attorneys sought results, he complained; country lawyers followed principle. The luster of the country lawyer, a Missourian observed, grew "pale and anemic in the shadows cast by the huge bulk of a city's commerce." Even some of the newest professional institutions were obeisant to the past. A correspondence school —a phenomenon of the urban age—advertised for students with a picture showing a young man standing in a barn orating to farm animals. The caption read: "Henry Clay began his illustrious career as statesman, orator, and lawyer by declaiming to the chickens and cattle in the barn."[3]

The country lawyer and the Tocquevillian aristocrat were distinct, even rival, professional types. But as the antebellum era of Tocqueville's aristocrat and Lincoln's country lawyer receded into national memory and the homogeneous rural society that had molded them disappeared, lawyers struggled to recapture those elusive moments when aristocrat and democrat had coexisted harmoniously in a noble profession within a stable society. In professional lore they created a fragile synthesis from both figures. The country lawyer assured equal opportunity, social mobility, and professional respectability for the man of humblest origins, thereby preserving the democratic flank of the profession. The aristocrat promised wealth and stature for those who reached the top, thereby enhancing professional elitism among those who served corporate business interests. Most important, the society in which lawyers were praised as aristocrats yet honored for their common virtues became etched in memory as the good society. And the age when this happy condition existed became known, in retrospect, as the Golden Age. Wrenching social changes might destroy its foundations, but its luster brightened the further it receded into memory. Few lawyers, then or subsequently, seemed troubled by the fact that their golden age coincided with an era of rambunctious democracy, generally considered to have been the nadir of professional life in America, or by the observations of David Dudley Field, an esteemed lawyer writing not long after Tocqueville, who looked at the same profession only to complain that the bar was already "crowded with bustling and restless men" who lacked the manners and wisdom of their refined forebears.[4] The lines between myth and fact, between perception and reality, were easily blurred amid the stress of changing professional roles in a rapidly changing society.

It was perilous for lawyers to believe, as too many did, that their profession had reached perfection in an Eden uninhabited by the serpents of industrialization, urbanization, and immigration. By drinking so deeply from the cup of nostalgia lawyers impaired their ability to cope with social change. A restorationist impulse infused their professional culture during the early decades of the twentieth century, years when the accelerating pace of change profoundly affected legal institutions. Lawyers were not uniquely susceptible to romantic hindsight. Others shared their myopia; indeed the American experience itself seemed to confirm man's ability to leap backward from a sinful present and land with both feet in paradise. The American Adam haunts our national consciousness, reappearing whenever change is too swift, or its shock too sudden, for men to comprehend or tolerate. But in a society where political issues were invariably framed in legal terms, where lawyers operated the machinery of government, and where entry into the legal profession provided access to political power, lawyers who were too preoccupied with their professional past to cope with contemporary social problems severely impeded social change.

As the United States moved uneasily but irrevocably from its homogeneous, rural, agrarian past to its heterogeneous, urban, industrial future, its legal institutions were afflicted with cultural lag. In impassioned language which captured the cadences of turn-of-the-century professional rhetoric, Dean Roscoe Pound of Harvard Law School described in 1912 a legal system already shackled by principles suitable only to small-town life. The principles endured, but the society that once had been comfortably governed by them no longer existed. It was virtually impossible, Pound wrote, to apply traditional legal doctrine "in a heterogeneous community, divided into classes with divergent interests, which understand each other none too well, containing elements hostile to government and order, elements ignorant of our institutions. . . ." This was especially true of a community "where the defective, the degenerate of decadent stocks, and the ignorant or enfeebled victim of severe economic pressure are exposed to temptations and afforded opportunities beyond anything our fathers could have conceived. . . ."[5]

The professional implications were ominous. Like American society, the legal profession abided by nineteenth-century values; indeed soon after the twentieth century began, it vigorously reasserted these in its Canons of Ethics. Like that society, it had become more heterogeneous and stratified; hence its reverence for antebellum simplicity and its effort to recapture it. If social tranquility was undercut by corporate power and by a proliferating *lumpen* class of immigrants, professional harmony was likewise destroyed by the emergence of corporate law firms and by a professional underclass drawn from the immigrant communities. As society confronted conflict and disorder, members of the legal profession voiced unease over the declining force of law—and the declining stature of lawyers. Profession and society: each wrenched by change; each crippled in its choice of means to cope with change by its tenacious hold on the past. Until lawyers could discard their myths and memories, their profession would teeter precariously between the world that was lost and the world that was becoming—unable to relinquish one; unable, therefore, to enter the other.

Even before the nineteenth century ended, the legal profession seemed far removed from any golden age. Lord Bryce, another aristocratic visitor to the United States, found a weakened sense of professional dignity and "a latent and sometimes an open hostility between the better kind of lawyers and the impulses of the masses." In contrast to its stature in an earlier day, he concluded in 1888, the bar "counts for less as a guiding and restraining power." At one extreme, the growth of a moneyed class had diminished the prestige of professional men; at the other, mass education had narrowed the distance between the multitude and the profession.[6] Tocqueville's America had disappeared; worse yet, during the last decade of the century social cataclysm seemed imminent. American society was rent by a severe economic depression, agrarian and labor unrest, attacks on corporate power and private wealth, a vigorous third-party challenge, labor-management conflict, and a violent defense of the new industrial order waged by private and public armies in behalf of business interests. Legal instruments of social control, ranging from injunctions to the police, were pressed into service. The judiciary vigorously defended private property against the regulatory efforts of legislative majorities. Supreme Court decisions voiding the federal income tax law and upholding an injunction issued against workers during the Pullman strike in 1894 precipitated a "conservative crisis" which split the legal profession. Bench and bar divided over the Court's sweeping assertion of judicial prerogatives and its application of law as an instrument of class advantage.[7]

With McKinley's victory over Bryan in 1896 the national crisis subsided. But the crisis in the legal profession had barely begun. Fed by converging currents of ethnic and economic conflict, it swept through professional life during the first three decades of the twentieth century—until the Great Depression jolted the profession into new concerns. Ethnic homogeneity had been one of the salient characteristics of the legal profession during most of the nineteenth century. Daniel Webster, defending the interests of the Second Bank of the United States before the Supreme Court, might share little in common in his daily professional life with the rambunctious country practitioner so evocatively depicted in Joseph Baldwin's *Sketches of the Flush Times of Alabama*. But as native American Protestant lawyers, they not only shared a heritage defined by centuries of Anglo-Saxon legal development but a common national cultural experience. Embodied in social and legal institutions, it was—in a distinctly proprietary sense—*theirs* to honor and to perpetuate. It was the cement that held otherwise disparate attorneys fast to each other. Stratified by education, wealth, power, and style, Tocqueville's aristocrat and the country lawyer nonetheless belonged to one society, one culture, one past. But by 1900, lawyers no longer could inhabit such a homogeneous national and professional culture because it no longer existed.

With accelerating momentum in the second half of the nineteenth century, the traditional, cohesive social structure was threatened with disintegration. Industrialization, urbanization, and immigration were the forces weakening the foundations of agrarian, rural, Protestant America. There were varied and complex responses to this process of social and cultural subversion. Some Americans vigorously reasserted traditional values. Others experimented with new instruments of social control and order. Still others created privileged sanctuaries for the "best" people and their progeny. Legal institutions and the legal profession were profoundly affected by these developments.

The city and the immigrant frightened respectable, middle-class, American-born professional and business people. The city was the sordid home of criminality, corruption, vice, disease, and delinquency; the immigrant was the presumed carrier of these contagious social germs. Both needed cleansing with the values of rural Protestantism. New controls over deviance were devised to assuage xenophobic fears. Incarceration was one response. Prisons, reformatories, asylums, and almshouses became "a dumping ground for social undesirables." They were filled with a disproportionate number of inmates drawn from the foreign-born and urban lower-class population—the "dangerous classes." Children from urban slums were "saved" by removing them to rural reformatories where thrift, discipline, and hard work were taught. Delinquency was "invented" and child-saving was practiced to preserve parental authority, family stability, and the work ethic as defined by native-born middle-class Americans. Here, perhaps, were the most extreme nativist responses to foreign ideologies. But even "progressive" reforms were tinged with xenophobia: women's suffrage might counteract the immigrant male vote; prohibition would circumscribe immigrant pleasures; good government could defeat immigrant bosses and restore the "best men" to political power; vice control imposed censorship in the name of rural innocence; and education reform socialized the children of the immigrant poor.[8] Quite aggressively, native Americans attempted to retain the values of a pristine, arcadian past destroyed by immigrants in an urban industrial society.

Besieged Americans not only created coercive institutions for recalcitrant outsiders; they built sanctuaries from which they could assert their own values against the infidels, and retreated to them. Capital eased the task for some. As mass immigration and urbanization inundated the dominant Anglo-Saxon culture, the fortunate few moved to the safety of selected social institutions— Eastern schools, for example, and careers in business and finance —which could protect, or extend, their power and status.[9] As immigrants gained political control in cities, native Americans took refuge in business careers. Big business served as "a new preserve of the older Americans, where their status and influence could continue and flourish. . . . The social patterns established within Big Business bureaucracies at the turn of the century helped to close off key areas of the economy and to keep them virtually impenetrable to even the most gifted outsiders."[10]

The emergence and proliferation of corporation law firms at the turn of the century provided those lawyers who possessed appropriate social, religious, and ethnic credentials with an opportunity to secure personal power and to shape the future of their profession. These lawyers were not conspirators who subverted either law or society. They were propelled by the same social and economic forces that were transforming American civilization. They capitalized upon historical circumstance to hitch professional values, which they were advantageously located to define, to the service of social stratification and corporate profit. The corporate law firm was their fortress. Its priorities—more precisely, the priorities of its clientele—shaped professional education, career patterns, ethics, mobility, and the availability and distribution of legal services—indeed, the very meaning of law and justice. It functioned as a prism, refracting social change upon the professional culture and back again to the larger society.

Only lawyers who possessed "considerable *social* capital" could inhabit the corporate law firm world. Born in the East to old American families of British lineage, they were college graduates (a distinct rarity) who followed their fathers into business and professional careers.[11] They molded the law firm to resemble the corporation; both restricted access to those who presented proper ethnic and social credentials. According to folklore, the doors of access to the legal profession always swung open to anyone stung by ambition; lawyers might prefer a restricted guild, but democratic realities required them to settle for less. But this is a half-truth, which conceals the fact that doors to particular legal careers required keys that were distributed according to race, religion, sex, and ethnicity. Myths notwithstanding, mobility was not a ladder whose rungs all could climb; for outsiders a career in a Wall Street, State Street, Market Street, or LaSalle Street firm was "more like scaling a wall than climbing a ladder."[12] Immigrant and farm boys who became corporation lawyers in prestigious firms, like those who became business leaders, "have always been more conspicuous in American history books than in American history."[13]

By the turn of the century corporate law firms were edging to the pinnacle of professional aspiration and power. Structurally the legal profession still was, and would long remain, "mostly a cottage industry of single practitioners."[14] But the emergence, rapid proliferation, and growth of corporate law firms, their impact upon patterns of recruitment and styles of practice, and their appeal to ambitious young attorneys invested them with significance (and their partners with professional power) that far exceeded their number and size. As early as 1900 the corporate firm expressed the palpable stratification of professional life and the application of professional expertise to the service of particular values and interests in an urban industrial society. It is necessary to understand the values it defended, the services it provided, the opportunities it created, the power it accumulated—and the concern that it elicited.

Corporate law firms were creatures of an age of organization. They reflected the needs of the most powerful institutions within the new urban industrial society. With business enterprise growing in size and complexity, and government by legislation and administration accelerating in tempo, the resolution of disputes required the preventive techniques of the counselor who spoke to the future rather than the forensic skill of the advocate who litigated the mistakes of the past. Business corporations needed efficient organizations to service them, a need which nineteenth-century law firms could not meet. Few in number, the older firms were indistinguishable in organization and competence from the traditional two-man law office. Dimly lit, lacking telephones and typewriters, staffed at most by a stenographer and a clerk, they were suited to unhurried times and to small claims. Corporate law firms—organized, departmentalized, and routinized—filled the vacuum. To contemporaries they were as appropriate (and, to some, as menacing) an expression of the age as their most conspicuous clients. In an era whose dominant impulses were rationalization, specialization, and professionalization they epitomized the search for order in a complex society.[15]

Corporate lawyers became the new professional men of the new century. Their "flair for figures, passion for facts and more facts, and insistence on realistic economic analysis rather than polished rhetoric, literary allusion or poetical quotation" set them apart.[16] Renowned for their mastery of the intricacies of corporate finance, they earned their reputations for work done at their desks and in conference rooms, not in court. Attuned to the accelerated pace of business activity, they were described even by admirers as "highstrung, tense, and driving personalities."[17] The prototype of the new breed was Paul D. Cravath. As clerk to Walter S. Carter, an early titan of the corporate bar, Cravath had grasped the responsibilities and opportunities of law practice in a new era. When Cravath left Carter to join the Seward firm in 1899, he instituted the "Cravath System," which quickly spread as the model for other firms.

Until Cravath arrived, the Seward partners worked independently of each other; the firm, less than the sum of its parts, lacked members with sufficient business acumen to serve its corporate clients adequately. By molding the firm into a "cohesive team," and by providing for an ever-replenished source of talented specialists in newer fields like securities, taxation, and reorganization, Cravath built a dynamic, efficient firm from a sputtering partnership. Systematic recruitment and training formed the lynchpin of the Cravath system. Cravath learned from Carter the advantages of hiring recent law-school graduates. As Carter explained: "I thought the best way to get a good lawyer was to make him to order, instead of getting him ready-made."[18] No longer did friendship with a client or a partner automatically qualify a lawyer for firm membership. New recruits followed a carefully prescribed path: college (perhaps Phi Beta Kappa); Harvard, Yale, or Columbia law school; preferably a law review editorship. In addition to academic credentials they were expected to possess "warmth and force of personality" and "physical stamina."

Once inside the firm, Cravath men were trained methodically as generalists before being entrusted with the responsibilities of specialty practice. According to the firm historian, they were "not thrown into deep water and told to swim; rather, they are taken into shallow water and carefully taught strokes." Those who swam with demonstrated competence were rewarded with responsibility until, after approximately five years, they earned the ultimate reward of partnership. Cravath's insistence upon selecting partners from within the firm reflected his conviction "that the office and its clients would get the best service from men confident of unimpeded opportunity for advancement"—and, doubtlessly, his belief in the efficacy of strenuous competition. Cravath men were made to understand that the practice of law must be their "primary interest" and that all business transacted in the office was Cravath business. The nature of this business was clear: "The practice of the office is essentially a civil business practice. Cravath desired a staff equipped to serve corporate and banking clients in any of their legal problems."[19] By 1910 the Cravath system, widely emulated, dominated the expanding world of corporate law firms.

Underlying the Cravath system was the assumption that business practice within a law firm would attract and hold the ablest graduates of leading law schools. The assumption was valid. Young lawyers responded with alacrity to the challenge and income that awaited them in metropolitan corporate practice. The rewards were ample. Within five years after their graduation from law school lawyers in the larger cities, where corporate work was concentrated, were earning considerably more than their less urban brethren; an alumnus of Harvard reported that the lawyer who derived satisfaction *"from merely being in touch with big things and big business"* had no excuse for unhappiness in New York practice.[20] There was always the possibility, Elihu Root told the graduating class at Yale Law School in 1904, that the truly successful lawyer might win "some prize of business life" and become a corporation president.[21] Even those who remained in corporate practice could derive solace from reports that William D. Guthrie earned a million-dollar fee—at a time when two-thirds of the lawyers in New York City earned less than $3,000 annually and two thousand Chicago lawyers had incomes lower than union brickmasons.[22] As Charles Evans Hughes observed of the New York bar at the close of the nineteenth century: "These highly privileged firms seem to hold in an enduring grasp the best professional opportunities. . . ."

Only the best men, however, were permitted to seize the best opportunities. There was, as Hughes also noted, "little room for young aspirants outside the favored groups."[23] Barriers to access became more formidable as the desirability of access increased. Young men, doubtlessly nurtured on the belief that success rewarded perseverance and integrity, learned that professional opportunity depended upon ethnic, social, religious, and educational credentials. A Columbia graduate (later an appellate division judge) realized that "the doors of most New York law offices were closed, with rare exceptions, to a young Jewish lawyer." Even John Foster Dulles, who in time would qualify for partnership in Sullivan & Cromwell, vainly applied for a position with eminent New York firms only to discover that a law degree from Harvard or Columbia was preferred to his own from George

Washington University. Impoverished applicants were advised to avoid combining day work with night law classes because only university law schools offered access to desirable professional positions.[24] Certain areas of practice receded into the professional shadows cast by corporate work. "It is a reproach in our profession and in the community," wrote Samuel Untermeyer, "for a man practicing law in the city of New York to be regarded as essentially a criminal lawyer. . . . Unlike the custom of former days no man with large corporate interests to protect would think of going to a man however eminent and learned in the law, whose reputation had been achieved as a criminal lawyer."[25] The point was succinctly made by a Cravath partner, who told a neophyte interested in litigation not to come to New York: "The business connected with corporations and general office practice is much more profitable and satisfactory and you will find that the better class of men at our Bar prefer work in that line."[26] Another attorney described the stratification of the New York bar into constitutional lawyers, corporation lawyers, and collection lawyers. Cromwells and Cravaths rose to the top; "Hebrews" sank to the bottom.[27]

There were, of course, conspicuous examples of nonconformity. Felix Frankfurter spurned private practice after brief exposure to it because it meant nothing more than "putting one's time to put money in other people's pockets"; he complained that "the intellectual process involved does not sufficiently appeal to me to make me forget the ultimate end."[28] Similarly, Henry L. Stimson grew disenchanted with the business transactions that consumed his energy in Elihu Root's firm. When the accelerated pace of federal regulation under Theodore Roosevelt's Square Deal expanded the need for government lawyers, Stimson left the firm to become a United States attorney. "The profession of law was never thoroughly satisfactory to me," he told his Yale classmates in 1908, "because the life of the ordinary New York lawyer is primarily and essentially devoted to the making of money. . . ."[29] New York practice doubtlessly exacerbated such feelings, but the phenomenon knew no geographical bounds. Progressive reformer

Frederic C. Howe, partner in a Cleveland firm whose clients were drawn primarily from railroads, banks, and manufacturing companies, deserted law practice entirely; he wrote in his autobiography, "I never overcame my dislike of the profession and got little enjoyment out of such success as I achieved."[30]

Yet those who left, no less than those who remained, testified to the growing professional dominance of the corporate firm. When Stimson left the United States attorney's office he was reassured that "a high stand in private business is worth more than too much public office," and he was reminded that "more happiness and money" surely awaited him in practice.[31] A law student's magazine, commenting upon James Beck's departure from the Justice Department for a Wall Street firm, casually dismissed notions of public service: "Most any man can hold a government position. . . . If the government wants good men let it pay for them. That is business."[32] One of Frankfurter's Harvard professors reminded his eccentric pupil that "most young men have a right to have in mind the pecuniary needs of later life," which government work could not satisfy. But a friend, taking the opposite tack, urged Frankfurter to remain in government at least temporarily—and justified his suggestion with the prescient observation that public service might translate itself into a larger income upon return to private practice.[33]

The Anglo-Saxon Protestant retreat to corporate enclaves was facilitated by changing patterns of legal education, which enabled corporate firms to camouflage their prejudices under the cover of academic achievement. By the turn of the century university law schools were assuming a substantial share of the burden of legal education. Law teaching was professionalized, and a standardized curriculum, emphasizing the staples of business practice, was instituted. Most significantly, corporate practitioners accepted student certification and ranking by law teachers—especially after grades became convertible into law review experience. Once Harvard blazed the trail in 1887, the proliferation of law reviews helped to certify a professional meritocracy and to systematize hitherto random hiring practices.[34] In special circumstances lawyers like Stimson and Frankfurter communicated directly with law teachers to obtain the pick of well-stocked university law school talent pools. A strong nexus developed between law firms and law schools. An inexorable channeling process directed those lawyers designated as "best" into law firms that served the new corporate elite. Lawyers, by serving that elite, joined it.

If academic merit alone had defined "best," the result would have been a truly democratic meritocracy in which elites circulated to reflect the shifting social base of American society. But this hypothetical model did not accurately depict reality, although most lawyers doubtlessly believed that it did. None defended it more vigorously than Felix Frankfurter when he reminisced about Harvard Law School, where he was a student between 1903 and 1906, a law review editor, and then a member of the faculty for a quarter of a century. Frankfurter sang paeans to "the democratic spirit" of the institution and to its ranking process:

> . . . rich man, poor man were just irrelevant. . . . The thing that mattered was what you did professionally. . . . The very good men were defined by the fact that they got on the *Harvard Law Review*. . . . All this big talk about "leadership" and character, and all the other things that are non-ascertainable, but usually are high-fallutin' expressions for personal likes and dislikes, or class, or color, or religious partialities or antipathies—they were all out. . . . There was never a problem whether a Jew or a Negro should get on the *Law Review*. If they excelled academically, they would just go on automatically.[35]

By restricting his scrutiny to the academic basis of selection to the *Harvard Law Review*, however, Frankfurter overlooked *social* filters which functioned at antecedent and subsequent stages to deplete drastically the pool from which the *Law Review* meritocracy, and law firm partners, were chosen.

Frankfurter described Harvard law students with alluring simplicity: "They were rich and poor."[36] Yet, ironically, Harvard was the worst example Frankfurter could have chosen to make his point because it was, until 1916, the only law school to require a college degree for admission. Among Frankfurter's American contemporaries, born between 1885 and 1889 and likely to enter law school between 1905 and 1910, fewer than 4 percent finished high school *and* college (while approximately 20 percent finished high school).[37] Harvard's admission requirement effectively eliminated 96 percent of the eligible population from *consideration;* admissions policies at less exclusive university law schools eliminated 80 percent. Restrictions based upon sex, race, ethnicity, class, and family background permeated the admissions process. The school excluded women. The financial expense of undergraduate and legal education, in addition to the substantial loss of income during the seven years required to earn two degrees, eliminated the most impoverished, among whom racial and ethnic minority group members were disproportionately concentrated. In theory, professional education was open to all who qualified. In fact, rich and poor enjoyed "Anatole France equity" (both being forbidden to beg or sleep in public parks). University law schools, like their undergraduate branches, remained "especially open . . . to children of northern European origin whose fathers did not work with their hands."[38] Frankfurter, the Austrian-born son of Jewish immigrant parents, was the conspicuous exception, not the prototypical product of a democratic meritocracy. His analysis was correct as far as it went, but it stopped too soon. The Jew or black who excelled academically *would* become a law review editor. But the critical question was whether a Jew, a black, a woman, or the Polish Catholic son of a day laborer could first qualify for admission to the school.

Even assuming that equal opportunity for a law review editorship existed—a false proposition—possession of the coveted honor was not legal tender for every holder. Frankfurter overlooked the law firm selection process just as he ignored the law school admission process. Again, with some conspicuous exceptions to the contrary, Jewish law review editors were excluded from partnerships in the prestigious corporate law firms until after World War II; blacks and women were outsiders until their token entry in the late 1960's and early 1970's; and other ethnic minority group members have barely begun to gain entry. Consequently, Protestant partners in these firms comprised the professional elite; comprising it, they defined it; defining it, they excluded nonwhites, non-males, and non-Christians. Academic achievement was necessary, but insufficient, for entry. Social origins, together with racial, sexual, and ethnic identity, determined both the possibility of academic achievement and the opportunity to reap its rewards. The Wall Street firm, and its counterpart in other cities, was the crucial link between corporate capitalism and social elitism within the legal profession.

Harvard Law School and legal education were neither synonymous nor coextensive—although the Harvard model dominated university law training. Similarly, there was more to the legal profession than corporate lawyers; indeed, only "a handful" of lawyers held large-firm partnerships around the turn of the century, and nearly half a century *later* the bar remained "a profession of highly individualistic practitioners," organized in 1940 as it had been organized in 1840. But the newer corporate firms, as legal historian Willard Hurst has observed, "symbolized a new role of the bar. They reflected the demands of big business clients."[39] If their numbers were small, their power—economically and professionally—was considerable. As yet, they barely resembled the Wall Street firms that would dominate corporate practice in later years. The dangers of retrospective projections of size, organization, and efficiency are obvious. In 1903, when James Beck left the government for Shearman and Sterling, which became a giant among giants, he described the advantages of a partnership in a "small firm."[40] That same year Samuel Untermeyer complained of the chaos in Guggenheimer, Untermeyer & Marshall. With a trial pending one afternoon he discovered "absolutely no preparation"—no statement of facts, no brief, no list of witnesses. "I find to my amazement," he told his partner, "that there is not a soul in the office who knows anything about this case. . . ." Yet measured by the experiences and expectations of contemporaries, these firms and their partners represented a sharp break with the past and an omen for the future. In Richmond,

Virginia, where only one firm in 1901 had as many as three members, the organization of a four-man firm elicited comparison with "the larger New York law concerns that are equipped for handling all legal matters."[41] Evidence of their small size, occasional disorganization, or minuscule numerical beachhead within the profession could not eradicate concern that one age had ended and a newer, menacing era of concentrated economic power had begun.

The invective directed against corporate lawyers and their firms was embellished by rhetoric that doubtlessly exaggerated their presence. But the fear that generated the rhetoric was deeply rooted in the realities of twentieth-century life. The size of law firms did not matter; it was the service they provided, and the clientele they served, that elicited distress. The looming presence of the metropolitan firm made many lawyers uneasy because no other institution so accurately reflected the altered contours of professional and economic life in the new century. With nostalgic fervor they harked back to an earlier day of presumed professional esteem—a day invariably located in the pre-urban, pre-industrial past. The independent small-town lawyer, whose clients were his friends, acquaintances, and townsmen, could not comprehend the mores of the austere partner who counseled corporations, not people. Even some city lawyers echoed that discomfort. A Detroit attorney advised the graduates of Michigan Law School in 1902 to seek employment in a country lawyer's office. "You will learn more there in a year than you will in a city office in five years." City practice, he cautioned, was a "great mistake."[42] A New York judge insisted that for training and self-reliance the old-time lawyer "is as far superior to the average modern lawyer as the self-sufficient pioneer is above the dweller in the city."[43] In a revealing metaphor the modern firm was described as "a money-making mechanism, inelastic, rigorous, unsympathetic; into which the young man, just from his studies, fits . . . like a fresh adjusted cog into a well-oiled machine."[44]

Corporate lawyers presented the most visible target for critics within and outside the profession. As the symbol of change this new legal elite drew the fire of those whose notions of professional propriety were shaped by images, whether real or imagined, derived from an earlier era of professionalism. Critics used the legal profession as a surrogate for society. By attacking corporate lawyers and a commercialized profession they could displace some of the anger, fear, and resentment stirred by their perception of the declining quality of life in an urban industrial age. Once the corporation became the object of public scrutiny and then the target of public hostility, as it increasingly did after the turn of the century, the new professional elite was vulnerable. The private corporation, a legal "person" entitled not to be deprived of its liberty or property without due process of law, owed its legal existence and therefore much of its social power to the innovative skills of lawyers and judges. Their reworking of the Fourteenth Amendment into a shield for corporate power had met the wishes of a society whose brief concern for freed slaves had yielded to a more enduring concern for free enterprise. But lawyers became identified in the public mind with the corporate clients whose interests they served. It was hardly possible to pillory malefactors of great wealth yet ignore those who counseled the malefactors.

The lawyer as surrogate for the corporation presented an alluring target. Thorstein Veblen, writing in 1899, concluded that the profession of law was nothing more than a form of employment "immediately subservient to ownership and financiering. . . . The lawyer is exclusively occupied with the details of predatory fraud, either in achieving or in checkmating chicane, and success in the profession is therefore accepted as marking a large endowment of that barbarian astuteness which has always commanded men's respect and fear."[45] Journalist Herbert Croly, describing the betrayed promise of American life, insisted that the lawyer was no longer qualified to interpret and guide American constitutional democracy because he had abdicated his role as representative citizen to defend special interests.[46] Perhaps the unkindest cut came from President Theodore Roosevelt, once a student at Columbia Law School, who delivered an address at Harvard in 1905 sharply critical of the new professional elite. "Many of the most influential and most highly remunerated members of the bar in every centre of wealth," Roosevelt charged, "make it their special task to work out bold and ingenious schemes by which their very wealthy clients, individual or corporate, can evade the laws which are made to regulate in the interest of the public the use of great wealth." These lawyers, Roosevelt concluded, were encouraging the growth of "a spirit of dumb anger against all laws and of disbelief in their efficacy."[47]

An acute sense of loss pervaded the legal profession during the early years of the new century. In a representative complaint, John R. Dos Passos described "a transformation from a profession to a business."[48] For Dos Passos the Civil War divided the old order from the new. The postbellum years represented not only an era of change but of "intellectual decadence" in the bar. The lawyer's "aristocratic and social prestige" had disappeared; his "moral and intellectual standard has been lowered"; dignity, learning, and influence had declined.[49] Once the lawyer had been a cultured man who was treated with deference, observed Theron G. Strong, member of a Connecticut family of lawyers dating back to the Revolution. But "the incursion of the money-making power" had robbed the lawyer of his stature and self-respect. Furthermore, "many of the best-equipped lawyers of the present day are to all intents and purposes owned by the great corporate and individual interests they represent, and while enormous fees result they are dearly earned by the surrender of individual independence."[50] James Hamilton Lewis, who mourned "the end of lawyers," observed sadly that "the lawyer who is but a lawyer, however talented, learned, and refined, must take second place beside the director of the company for which he is counsel or beside the client who is rich."[51]

With independence undermined, influence must deteriorate. Lawyers were richer, corporate attorney Edward M. Shepard told the New Hampshire Bar Association in 1906, but their public influence had diminished. In the nineteenth century, Shepard claimed, lawyers had wisely refused to commit their professional skills to the interests of any single client or type of client. But corporate needs imposed irresistible pressures upon the profession. The more lawyers counseled corporations the less public esteem the bar retained.[52] Harlan F. Stone, delivering the Hewitt lectures at Columbia University, referred to "a deterioration of our bar both in its personnel, its corporate morale, and, consequently, in the public influence wielded by it; and . . . this deterioration has been very considerably accelerated during the present generation." Stone expressed unease over the leadership of the business lawyer, skillful and resourceful at his best, but at his worst "the mere hired man of corporations."[53] Woodrow Wilson, speaking to the American Bar Association in 1910, contrasted the golden past with the sordid present. Principled lawyers had once offered disinterested service to the community. But American society, Wilson observed, "has lost something or is losing it. . . ." The constitutional advocate, once the pride of the profession, had virtually disappeared. In his place stood "lawyers who have been sucked into the maelstrom of the new business system of the country. . . . They do not practice law. They do not handle the general, miscellaneous interests of society. They are not general counsellors of right and obligation." The modern lawyer, Wilson declared, counseled individuals, not the community. "He does not play the part he used to play. He does not show the spirit in affairs he used to show. He does not do what he ought to do." Worse yet, lawyers did what they ought not to do: as corporate counsel they applied their skills to the destruction of the old order. They were "intimate counsel in all that has been going on. The country holds them largely responsible for it. It distrusts every 'corporation lawyer.' " Wilson hoped to "recall lawyers to the service of the nation as a whole," but his remarks offered scant solace to those who may have shared his wish.[54]

No lawyer articulated this sense of concern with greater precision than Louis D. Brandeis. In an address to Harvard undergraduates in the spring of 1905, he attributed declining popular esteem for the bar to the lawyer's fall from independence. "Instead of holding a position of independence, between the wealthy and the people, prepared to curb the excesses of either, able lawyers have, to a great extent, allowed themselves to become adjuncts of great corporations and have neglected their obligation to use their powers for the protection of the people. We hear much of the 'corporation lawyer,' and far too little of the 'people's lawyer.' " For nearly a generation, Brandeis claimed, leaders of the bar had opposed constructive legislative proposals in the public interest, while failing to oppose legislation in behalf of "selfish interests."[55]

Wilson and Brandeis: moralists in politics and in law. Both men had launched public careers whose rallying cry would be a new freedom based upon the restoration of old values. Community cohesion, stability, and fixed principles of "right and obligation" characterized the old order, which the "new business system" had undermined. The destructive power of corporate capitalism fed their unease and drew their fire. Corporation lawyers incurred mistrust, not because they were inherently venal, but because, in contemporary reform rhetoric, corporate interests were *ipso facto* antithetical to social interests. By definition, lawyers who counseled corporations were derelict in their social obligations. The language varied, but the meaning remained despairingly constant. Tocqueville's aristocrat, to say nothing of the country lawyer, was unrecognizable as Stone's "hired man," or as Brandeis' "adjunct." The omnipresent corporation seemed to hold the lawyer's professional soul in its grasp; he no longer commanded the respect that accompanied professional detachment. In the heyday of the new corporate lawyer, when James C. Carter, William D. Guthrie, and Paul Cravath served as models for aspiring neophytes, an old-timer sadly complained that "the American Plutarch will find little material for his pen in the lives of modern lawyers, but among the lives of bankers, manufacturers, pioneers, railroad men and other men of business, material of the richest sort awaits him."[56]

Corporate lawyers were not indifferent to these jeremiads, but neither were they persuaded by them. In their struggle for professional acceptance they enjoyed important advantages. Above all, they rode the rising tide of national economic development. In addition, their corporate clients, facing the common enemy of governmental regulation, demanded the efficient services that co-hesive law firms were best able to provide. Furthermore, just as they shared bonds of social origin, education, religious affiliation, and club membership with corporation managers, their relation-ship with university law schools assured a perpetual flow of talent to meet the needs of their clients. Finally, bar associations, in which they wielded power disproportionate to their professional numbers, provided an organizational base for their interests, a forum for their views, and leverage for the implementation of their programs. Unwilling to be confined to nineteenth-century molds, unafraid of corporate power, and eager to confront the challenges of organization and control in modern society, they committed their professional energies to the service of industrial capitalism. At bar association meetings and in professional jour-nals they flung the gauntlet of criticism back at their detractors.

James B. Dill, a prototype of the new corporate lawyer (he enjoyed a lucrative New York practice, drafted the New Jersey holding company and incorporation statutes, and advised Theo-dore Roosevelt on trust policy), asserted that the successful mod-ern lawyer no longer was the last resort of a businessman facing destruction, but his constant consultant at every stage of business enterprise. Consequently, "the more nearly the lawyer brings his profession into touch with business methods the greater will be his success, and the profession is to-day [1903] beginning to real-ize the fact and to act upon it." The corporation lawyer had evolved as the necessary counterpart of the corporation manager. "Yes," Dill concluded, "law is a business, and if the young man wants to practice it, the sooner he makes up his mind to do so with an eye single to some particular branch of it, the better lawyer will he become."[57] Defenders of the old order might chastise Dill for deflecting aspiring lawyers from traditional professional ideals, but Dill's call for specialization and business methods found many echoes in professional circles.[58]

In one city after another, corporate lawyers repudiated the no-tion of professional decline, asserting that never before had tal-ented practitioners enjoyed such abundant opportunities. None of them disputed the *fact* of change; their disagreement was over the *quality* of change. "Let the flamboyant orator and the sensational newspaper writer say what they will," declared Henry Wollman of New York, "there never was a time when lawyers, as a class, were so genuinely respected and looked up to and trusted as they are now." A commercial age inevitably produced commercial lawyers; instead of lamenting this development the bar might point with pride to lawyer-presidents of the Union Pacific Railroad, United States Steel, United States Rubber, and the International Paper Company and conclude, with Wollman, that "there never has been a time when so many lawyers have been called . . . to fill so many places of enormous responsibility."[59] Levy Mayer of Chicago asked a familiar question: "Is the lawyer of today a mere material tool of commercialism, narrow in mind, lacking in logic, and deficient in legal acumen?" His answer, like Wollman's, was vigorously negative. The modern lawyer was indeed more businesslike than his mid-nineteenth-century counterpart, but he also was better educated, better equipped, and more skilled.[60]

Many attorneys stood the argument of the Cassandras on its head, asserting that the role of the lawyer in business affairs reflected his *ascending* position and power. They saw new opportunities for the lawyer who could cope with modern conditions. Charles W. Needham, assistant solicitor for the Interstate Commerce Commission, observed that "new conditions in social and political life demand new types of men, or at least, men of special training and equipment."[61] A New York lawyer reassured Cornell law students that the change from advocacy to counseling did not signify professional deterioration; rather it meant that lawyers had learned to adjust their skills to changing social conditions.[62] Edward P. White, another New Yorker, spoke unselfconsciously about the "law business," asserting that lawyers had much to learn from businessmen. The commercial spirit, he declared euphorically, "transforms barren abstractions into fruitful realities. It is essentially just . . . and it is inspiring because it deals with things and not theories." White implored his professional brethren to "get in touch with business men and in sympathy with business methods so that we can be of real assistance in carrying on business enterprises." Few lawyers were as indecorous as White, who bluntly asserted that "the profits of the business [man] are our most practical concern."[63] But many, aware of changes in the lawyer's role, refused to equate change with decline.

Knowledge of opportunities in corporate practice spread quickly through the profession. Within a decade, declared Dean Harry S. Richards of the Wisconsin Law School in 1914, "an entirely new field of usefulness for the lawyer has been created, and its boundaries are not yet fixed." Professional opportunity was greater than ever, he concluded, even if "the character of the business" had changed rapidly.[64] The generational appeal of corporate practice was evident. "It is significant of the development of the Bar of our generation," wrote Julius Henry Cohen, "that the successful lawyers—the men who have attained supremacy—are men who combine business skill with the training of the law." The modern office lawyer, Cohen observed, must be both administrator and executive; in sum, "a business manager as well as an adviser to business men."[65] In a society whose economic and political life increasingly reflected the power and values of corporate enterprise, the professional dominance of corporate lawyers was assured. So, too, was the evolving diversity of their roles: they were attorneys who were also business counselors, lobbyists, and public relations men. "Their new importance by the turn of the century," concludes one historian, "reflects the growth of bureaucratic managements typically in need of help in navigating legal and political labyrinths and in conciliating public groups often made hostile by the results."[66]

The ascendancy of corporate lawyers marked a critical turning point in the emergence of the modern American legal profession. Capitalizing upon advantages conferred by corporate growth and ethnic stratification, they exerted professional leverage quite disproportionate to their numbers. As a pyramidal social structure based upon the ethnic identity of lawyers and the class interests of clients emerged within the legal profession, corporate lawyers, at the apex, associated professional responsibility with group interest. Responsive to the economic opportunities of the new age, they thrust themselves into professional leadership as the legal profession experienced the growing pains of modernization. Once in the ascendancy, the best men began to set their professional house in order and to set themselves resolutely against challenges to their own professional authority or to the political and economic power of their clients. Especially at the metropolitan bar, where corporate lawyers were concentrated, traditional folkways were unsuited to the changing demographic patterns, accelerating commercial pace, and shifting values of an urban industrial society. Within a generation the profession pulled away from the old moorings. Bar associations expressed an impulse toward professional cohesion. Bar admissions standards were tightened, and ethical norms were promulgated to define and deter deviance. University legal education, especially the case method, elevated academic excellence above practical experience and encouraged the professionalization of law teaching. Systematized recruitment patterns channeled the talent flow to corporate firms which provided comprehensive services to a restricted clientele. The transformation in structure and values was completed as the new elite guided the profession through the disruptive fluidity of the late nineteenth century to the uncertain stability of the early twentieth century and forged a new professional identity that would profoundly affect law and the administration of justice for decades to come.

Two

A Stratified Profession

Early in the twentieth century lawyers began to contemplate the impact of social change upon their profession. "Commercialization" was the protean term of invective that most vividly expressed their unease. At one extreme, it expressed anxiety about concentrated economic power and hostility toward corporation lawyers, whose rapid professional ascendance and apparent subservience to businessmen were deeply unsettling. At the other, as an indication of concern with immigrant ghettos and urban poverty, it demonstrated antagonism toward lawyers from ethnic minority groups—the profession's new and growing underclass. Both the Wall Street lawyer and the ambulance chaser threatened models of professional independence and esteem associated with a homogeneous society and an arcadian past.

Lawyers who clung to older values launched a crusade for professional purification. The immediate impetus was provided by President Theodore Roosevelt who, in his Harvard address in the spring of 1905, sharply rebuked corporate lawyers for aiding their clients in evading regulatory legislation. Stung by the President's speech, Henry St. George Tucker, president of the American Bar Association and scion of a venerable Virginia family of lawyers, urged his associates to inquire whether "the ethics of our profession rise to the high standards which its position of influence in the country demands."[1] The association authorized a committee to consider the wisdom of a code of professional ethics. It reported back a year later with a vigorous affirmative recommendation. In soaring rhetoric it implored the bar to return to its golden age of virtue. "Our profession is necessarily the keystone of the republican arch of government," it declared. "Weaken this keystone by allowing it to be increasingly subject to the corroding and demoralizing influence of those who are controlled by graft, greed and gain, or other unworthy motive, and sooner or later the arch must fall." An influx of new lawyers who pursued an "eager quest for lucre" had diverted the bar from its traditional standards and from its "pristine glory." A code of ethics, drawing upon the wisdom of the past, "should provide a beacon light on the mountain of high resolve to lead the young practitioner safely through the snares and pitfalls of his early practice up to and along the straight and narrow path of high and honorable professional achievement."[2] Although Roosevelt had denounced those who counseled the malefactors of wealth, and the cry of "commercialization" bracketed lawyers in the highest and lowest professional strata, the committee report was more selective. Indeed, once bar associations exerted control over the campaign for canons of ethics, corporate lawyers, who were disproportionately represented in their councils, shifted the onus to "ambulance chasers" and "shysters," who were disproportionately excluded.

The new canons drew heavily upon George Sharswood's *Essay on Professional Ethics*, published in 1854. Sharswood's *Essay* was, at best, antiquated; at worst, irrelevant. He had addressed it to a generation accustomed to moral exhortation and confident that its own definitions of character, honor, and duty were eternal verities. Warning even in the 1850's of "a horde of pettifogging, barratrous, money-making lawyers," Sharswood had urged "high moral principle" as the bedrock of professional dignity. Passivity and patience were his cardinal virtues. Like young maidens awaiting suitors, aspiring lawyers must await clients. "Let business seek the young attorney," Sharswood insisted. It might come too slowly for profit or fame (or never come at all), but if the lawyer cultivated "habits of neatness, accuracy, punctuality, and despatch, candor toward his client, and strict honor toward his adversary, it may be safely prophesied that his business will grow as fast as it is good for him that it should grow."[3] Sharswood's safe prophecy may have comforted a young nineteenth-century attorney in a homogeneous small town, apprenticed to an established practitioner, known in his community, and without many competitors. It could hardly reassure his twentieth-century counterpart, the new-immigrant neophyte in a large city where restricted firms monopolized the most lucrative business and thousands of attorneys scrambled for a share of the remainder. He could draw scant comfort from Sharswood's confident assertion that some preordained rule determined that his practice grew no faster than was good for him. He either hustled or starved.

Yet the new ABA Canons, which were quickly adopted at the state level, measured the social texture of twentieth-century urban practice against antebellum memories. Emphasis on reputation as the key to professional dignity and success—understandable in the tight and comfortable social structure of Sharswood's Philadelphia —presupposed the vanished homogeneous community whose lawyers were known, visible, and accessible and whose citizens recognized their own legal problems and knew where to turn for assistance. In twentieth-century cities, ironically, these conditions flourished only in the subculture inhabited by corporate law firms and their business clients. Here, too, was a small homogeneous community whose members enjoyed shared values, ease of communication, and a network of mutually reinforcing educational, religious, and social ties. The Canons, reflecting values appropriate to a small town, were easily adaptable to an equally homogeneous upper-class metropolitan constituency, where they served as a club against lawyers whose clients were excluded from that culture: especially the urban poor, new immigrants, and blue-collar workers.[4] These lawyers confronted problems of client procurement which an established corporate practitioner did not experience.

A cluster of canons pertaining to acquiring an interest in litigation, stirring litigation, and division of fees almost exclusively affected the activities of struggling metropolitan solo lawyers. They did not apply to the conduct of the firm members or securely established practitioners who formulated them. The canon that uniformly prohibited lawyers from indicating publicly that they engaged in a specialty practice equated the firm known to serve a roster of corporate clients with the solitary lawyer, especially the newcomer, who specialized in negligence work and depended upon a constant client turnover for economic survival. The prohibition against advertising instructed lawyers that success flowed from their "character and conduct," not from aggressive solicitation. It thereby rewarded the lawyer whose law-firm partners and social contacts made advertising unnecessary at the same time that it attributed inferior character and unethical behavior to attorneys who could not afford to sit passively in their offices awaiting clients; it thus penalized both them and their potential clients, who might not know whether they had a valid legal claim or where, if they did, to obtain legal assistance. The canon prohibiting solicitation discriminated against those in personal injury practice, who bore the pejorative label "ambulance chasers." And although the fee-fixing canon reminded lawyers that their profession was "a branch of the administration of justice and not a mere money-getting trade," only the contingent fee (the hallmark of negligence lawyers) was explicitly subjected to judicial scrutiny. That distinction suggested that the decisive question was who earned the fee, not the size of the fee earned.

In tandem, these canons condemned the acquisitive urge (especially among lawyers who earned least), consigned the lawyer to his office to await a client who wandered by with a case that assured fame and fortune, and attributed success (hardly unrelated in American society to monetary accumulation) to good character. The lower the fee a lawyer earned, and the less discreet he was in pursuit of it, the more likely it was that his "money-getting" activities would be scrutinized and criticized.[5] The Canons especially impeded those lawyers who worked in a highly competitive urban market with a transient clientele. Little wonder that they made more sense to an attorney in Sullivan & Cromwell or in Fairfax County, Virginia, than to a personal injury lawyer on the Lower East Side of Manhattan.

The class and ethnic biases that appeared in the Canons were nowhere more evident than in the special treatment reserved for contingent fees. Few other issues cut so deeply into social mores and professional concerns in an urban industrial age. An alarming proliferation of work and transportation accidents, most often borne by those least able to afford lawyers' fees, generated human tragedies which a profit economy and its legal doctrines exacerbated. Accident victims—and the surviving members of their families—were compelled to bear the full burden for the risks inherent in dangerous work. Corporate profit was the primary social value. Legal doctrine impeded the opportunity of an accident victim to recover damages; furthermore, legal services were available only to those who could afford to purchase them.

The consequences were starkly exposed in Crystal Eastman's study *Work-Accidents and the Law*, conducted for the Russell Sage Foundation and published in 1910 as part of the landmark Pittsburgh Survey. In more than half of all work-accident fatalities in Allegheny County, widows and children bore the entire income loss. In fewer than one-third of these cases did an employer pay as much as five hundred dollars—the equivalent of a single year's income for the lowest-paid workers. Similarly, more than half of all injured workers received *no* compensation; only 5 percent were fully compensated for their lost working time while disabled. The result, Eastman concluded, was "not hardship alone, but hardship an outcome of injustice."[6] Hardship was evident; injustice was buried in a thicket of legal doctrine that exonerated an employer from liability even when he was demonstrably at fault. From a well-stocked arsenal of defense weapons even the negligent employer might claim that his maimed or dead employee was contributorily negligent, or that he had assumed the risk, or that a co-worker was at fault. Liability for fault retained its doctrinal potency until 1910, when New York enacted the first workman's compensation law (solely for dangerous occupations), which the state court of appeals overturned as being "plainly revolutionary." Not until well into the twentieth century in most jurisdictions could personal injury victims bypass doctrinal impediments to recovery—assuming they could afford the cost of counsel.

The contingent fee was therefore a necessary financial inducement for the provision of legal services to personal injury victims. This gambler's chance, which assured profit to the counselor and counsel to the needy, offered the victim his only hope of recompense—minus at least one-third of his recovery sum as the attorney's fee. In this diabolical game a negligence lawyer could claim a sufficiently high percentage from his successful suits to compensate for his losses. Nevertheless the contingent fee arrangement did enable some workers to secure otherwise unattainable legal services. But the costs to the legal system were high: enormous pressure to litigate, with predictable expense and delay which served the employer's advantage; sustained public criticism of these delays and costs; and allegations of widespread ethical improprieties once the contingent fee was denigrated to the level of original sin and negligence lawyers were denounced by the professional patriciate as inferior ambulance chasers or shysters.[7]

Nothing plunged the professional elite deeper into despair than contingent fees and the proliferation of negligence lawyers whose practice depended upon them. Rather than cure the doctrinal disease, and adversely affect their corporate clients, bar leaders and authors of treatises on legal ethics preferred to denounce the symptoms: contingent fees and ambulance chasers. The legality of the contingent fee—assumed to be "beyond legitimate controversy" by the Supreme Court in 1877—was unquestioned.[8] Its morality was another matter. Sharswood had insisted that it created a "new and dangerous relation" between client and lawyer, deposing the latter from his position as officer of the court and reducing him to a partner in the claim who was "tempted to make success, at all hazards and by all means, the sole end of his exertions." The fee, by encouraging litigation, transformed law into a "lottery" and lawyers into "higglers" with their clients.[9] By the early twentieth century the evils attributed to the contingent fee included declining professional spirit, loss of professional independence, commercialization, and homogenization of lawyers into "one indistinguishable crowd"—with a commensurate diminution in status for elite practitioners tainted by association with their ambulance-chasing brethren.[10]

Consequently when bar associations considered the adoption of canons of ethics the propriety of contingent fees occupied a conspicuous place in their deliberations. Members of the American Bar Association heard a virtually unprecedented floor debate over the contingent fee canon. In its original version the canon permitted such fees but cited their "many abuses" as justification for court supervision. Thomas J. Walsh, the future Progressive senator from Montana, vigorously protested against such efforts to distribute professional status according to a lawyer's willingness to contract for contingent fees. No self-respecting attorney, he suggested, would tolerate judicial scrutiny of his fee schedules. Yet the proposed canon imposed precisely this requirement upon negligence lawyers. Only contingent fees, Walsh observed, enabled the victims of work and transportation accidents—especially indigent immigrants—to obtain some measure of economic recovery. "It is altogether quixotic," he added, "to imagine that members of the Bar of high standing are going to come forward to offer gratuitously to prosecute such claims." Walsh condemned the double standard which subjected the fees of negligence lawyers to court scrutiny but ignored the attorney who capitalized upon his social contacts to procure far more lucrative corporate retainers.[11]

To most ABA members, however, such distinctions were appropriate. Court supervision was justified on behalf of the personal injury victim, who presumably needed protection from his attorney more than he needed monetary damages for his injury. Indeed the ethics committee, prodded by expressions of dissatisfaction with the broad tolerance for contingent fees under the proposed canon, offered a more restrictive substitute which the members adopted. Contingent fees, it stated, *"where sanctioned by law*, should be under the supervision of the Court in order that clients may be protected from unjust charges."[12] The critical change was to couple judicial scrutiny with a states-rights position that limited ABA acquiescence to those jurisdictions that permitted contingent-fee arrangements. In one of these, Massachusetts, both the state and Boston bar associations adopted a more stringent provision than the ABA canon. Lawyers were urged not to stipulate a fixed percentage of the recovery as their fee, and they were advised to accept contingent fees only when the client had a meritorious cause of action. The promotion of "groundless and vexatious suits" was condemned—a veiled reference to one class of lawyers and the type of litigation they stirred. Any social value of stirring litigation was slighted: that if a lawyer took action, legitimate claims might be pressed; that legal services would be provided to the needy; and that the assertion of rights would be channeled into the legal process.[13] The provision of legal services to a working-class clientele was distinctly secondary to concern for the public image of the legal profession and for corporate profits.

Insistence upon court supervision of contingent fees to "protect" clients from their lawyers was highly incongruous at a time when no other attorneys' fees were supervised and when no principle was more tenaciously cherished by corporation lawyers (and increasingly by courts) than liberty of contract. This legal fiction required terms of employment to be contracted free of such legislative infringement as minimum wage or maximum hours guarantees. To justify judicial nullification of state and federal welfare legislation, liberty of contract established a steel worker as a bargaining partner equivalent with the president of the United States Steel Corporation.[14] Yet, as the debate over contingent fees demonstrated, reverence for liberty of contract could vanish as quickly as a magician's rabbit. Rather than permit personal injury victims to contract freely with negligence lawyers for a contingent fee, bar associations "protected" these victims by singling out their fee arrangements for special attention and supervision. The presumption of bargaining equality, the core of the doctrine when miserable working conditions at minuscule wages were in question, disappeared when a negligence lawyer negotiated a contingent fee with his client. Attorney and client, the argument ran, were unequal partners in the bargaining process. The client, in his ignorance or physical distress, would acquiesce in an unfair fee arrangement. Therefore, courts must intervene to protect those who could not protect themselves from "unconscionable bargains" extracted by greedy lawyers.[15]

The outpouring of concern over contingent fees accompanied the development of stricter standards of corporate liability for activities resulting in injury or death. Once the social costs of corporate immunity were perceived as exorbitant, corporations were assessed damages with greater frequency. Contingent fees provided an inducement for such litigation, increased pressure for a substantial settlement, and, thereby, threatened corporate profits. Any limitation upon contingent fees promoted the financial interest of corporations in avoiding settlements with accident victims. In addition to assuring counsel for those who could not otherwise afford it, contingent fees furnished cases for lawyers who lacked social connections and law-firm ties that attracted business retainers. Understandably, therefore, elite practitioners invested much energy and emotion in the contingent fee issue. Not only were the financial interests of their own corporate clients at stake; additionally, their own professional self-esteem rose as they cast aspersions upon negligence lawyers, whom they consigned to the lower professional depths.[16]

Ambulance chasers became the scapegoats in a heterogeneous profession increasingly populated by foreign-born lawyers. As Boston attorney Reginald Heber Smith concluded, the contingent fee system might be the logical outcome of the existing maldistribution of legal services but it nonetheless constituted "the great blot on the history of the American Bar"—primarily because the lure of money attracted "undesirable persons" to the profession.[17] Years later the problem still would be perceived as personal rather than systemic, the responsibility of mercenary lawyers rather than the result of institutional inequity. In 1929, after a lengthy, publicized investigation in New York into the evils of ambulance chasing, resulting in recommendations of disciplinary proceedings against seventy-four lawyers, the chief counsel pointedly observed that some attorneys who had testified "could not speak the King's English correctly. . . . These men by character, by background, by environment, by education were unfitted to be lawyers." The only remedy, he suggested, was a character examination, prior to law-school admission, to eliminate those who lacked proper antecedents, home environment, education, and social contacts. If such an examination created a legal aristocracy, he told applauding members of the New York State Bar Association, so be it.[18]

By attributing responsibility to inferior character, lawyers deflected criticism from a social system which accepted uncompensated injury and death to a substantial number of workers as a tolerable consequence of economic growth and private profit. Once the consequences finally were seen as intolerable, maximum and minimum solutions were available: alteration of legal doctrine and provision of legal services. The doctrinal approach was far-reaching, and therefore belated. As an interim measure it was easier to permit (and criticize) contingent fees than to alter assumptions that sustained profits. But the costs were always high: plaintiffs sacrificed a large portion of their recovery; negligence lawyers paid with their professional stature.

"Ambulance chasing" was never precisely defined. As a term of art it ostracized plaintiffs' lawyers who, representing outsiders to the economic system, solicited certain types of business. Once fee-hungry ambulance chasers were isolated, they could be excluded from professional respectability by a series of discriminatory ethical judgments. Their methods of solicitation were condemned, but nothing was said about company claim agents who visited hospitalized workers to urge a quick and inexpensive settlement. Their fees were isolated for judicial scrutiny, but larger corporate retainers were ignored by professional associations. Their pecuniary interest in the outcome of litigation was criticized, but the pecuniary interest of most lawyers in their cases (even when not ascertainable in advance) was disregarded.[19] Not only were they criticized for professional malfeasance; their speech was mocked (many were recent immigrants) and their perseverance was denigrated as aggressiveness (many were Jewish). Commercialization, speculation, solicitation, and excessive litigation were decried, but there was no mention of the contribution of contingent fees to the enforcement of legitimate claims otherwise denied by the victim's poverty. Few lawyers complained about the Hobson's choice imposed upon an accident victim, who could either relinquish all hopes of recovery or merely relinquish (to his attorney) one-third of what he might recover if he overcame doctrinal impediments. Aggressive solicitation of personal injury litigation through a contingent fee arrangement doubtlessly produced its share of abuses. But when the sole alternative was the waiver of legitimate claims, one is compelled to agree with the conclusion that "the social advantage seems clearly on the side of the contingent fee."[20]

The ethical crusade that produced the Canons concealed class and ethnic hostility. Jewish and Catholic new-immigrant lawyers of lower-class origin were concentrated among the urban solo practitioners whose behavior was unethical because established Protestant lawyers said it was. There were serious "social, economic and occupational differences between the rulemakers and others subject to the rules. . . ."[21] Elite practitioners insulated themselves from "ethically contaminating influences" while they compelled lawyers whose low status was attributable to their ethnic and class origins to bear the brunt of such pressures. Rules of ethical deviance were neither universal nor timeless. They were applied by particular lawyers to enhance their own status and prestige. Deviance was less an attribute of an act than a judgment by one group of lawyers about the inferiority of another.[22]

The Canons were justified in quite different terms. Their purpose, declared one member of the American Bar Association, was "to elevate the standing of the profession. . . . The public and the press expect that we shall formally promulgate some decalogue. The time is opportune, if we wish to maintain the traditional honor and dignity of our profession."[23] The spirit of the decalogue was very much in evidence. Canon 32, the peroration, called upon lawyers to render service in "exact compliance with the strictest principles of moral law."[24] Professional and lay reactions played variations on this theme. David J. Brewer, an associate justice of the Supreme Court and a member of the committee that drafted the Canons, had noted in their support that the ideal lawyer "looks above the golden calf and the shouting crowd, and ever sees on the lofty summits of Sinai the tables of stone chiseled with imperishable truth by the finger of God."[25] One newspaper hailed the "moral possibilities" of the ABA crusade; the Canons, according to another, will move the nation along "the higher path of moral progress." The Chicago *Tribune* praised the Canons as "a reaffirmation of right principles." And the Philadelphia *Press*, although observing that the Canons merely restated familiar ethical principles from the past, lauded them for this very reason, declaring: "It is a good thing to have a revival . . . for the tendency to backslide is universal."[26]

The Canons reflected and reinforced an increasingly stratified profession. Particularly in larger cities, declared *Bench and Bar*, "where the practice of law has become more nearly than ever a mere commercial pursuit, in which the wiles and deceits of the horse-trader are deemed good form by many, it is necessary to set up, so that all may see, some of the more elementary ethical precepts of our professional forebears of the 'old school.' "[27] (Professional forebears, of course, had registered the same complaint: back in 1879 the president of the New York State Bar Association complained that "men are seen in almost all our courts slovenly in dress, uncouth in manners and habits, ignorant even of the English language, jostling, crowding, vulgarizing the profession.")[28] The Canons embodied these "old-school" precepts. Ironically, however, it was the low-status lawyer, the target of the Canons, whose generalized practice and range of human contacts most closely approximated the traditional professional ideal of the accessible generalist. He had become the Abraham Lincoln "gone urban," whose ethnic origins and urban habitat destroyed any resemblance to his idealized rural ancestors and made him professionally vulnerable.[29]

As the Canons demonstrated, the faster the old order changed the more tenaciously its defenders at the bar clung to it, and the more resolutely they attempted to build a professional structure that would be resistant to social change. The Canons represented a counter-revolutionary thrust within the legal profession. The professional prototype—the independent country lawyer—had been rudely elbowed aside both by corporate attorneys and by solo practitioners from ethnic minority groups. Corporate lawyers might taint the profession by serving business so blatantly, but at least they shared the common ethnic origins of the "best" men and they were allied with wealth and power. Solo lawyers adhered most closely to the traditional professional model, but their scramble for clients and cases threw the onus of unethical behavior upon the entire profession, and, as ethnic outsiders, they were expendable. Ethical admonitions offered protection on both flanks. As a reaffirmation of traditional values they provided reassurance that corporate clients and high fees had not, as so many lawyers feared, transformed law from a profession to a business. More important in the years to come, as the children of new immigrants entered the legal profession in unprecedented numbers, the Canons separated the bar's best men, Protestants of English and northern European origins, from the "unfit"—Jews and Catholics from eastern and southern European backgrounds. In the end, ethnic bonds mitigated fears that corporation lawyers had wrought destructive dislocations upon the profession. Old-style practitioners could cooperate with them in a united front to preserve the legal profession—or, at least, its elite stratum—as an Anglo-Saxon Protestant enclave.

The Canons of Ethics measured the depths of elite discomfort in an urban industrial age. But no mere reassertion of honor and dignity sufficed to mute criticism of lawyers and legal institutions. Public dissatisfaction with the administration of justice ran too deep for that. During the early years of the new century it was increasingly evident that justice was available only to those who could pay its price; even the contingent fee was, at best, a half-way compromise. But the bar responded indifferently, or dilatorily. The boundaries to its energy, and the sources of its belated recognition that a potentially explosive social problem existed, are illuminated by the history of the legal aid movement.

According to professional legend the dedication of the bar to the promotion of justice seldom was more vividly demonstrated than by the establishment of legal aid societies around the turn of the century. Lawyers, it was said, responding with personal compassion and professional altruism, promoted legal reform in the interest of the disadvantaged, whose need for low-cost legal services was as evident as the supply was lacking. In thus serving the poor, lawyers met their professional obligation to provide equal justice under law.[30]

The history of legal aid lends slight credence to this view. The pioneers were not members of a professional group but German-Americans who, in 1876, organized a legal committee, the *Deutscher Rechts-Schutz Verein*, within the German Society of New York to protect new German immigrants from sharpers who preyed upon them. A statement of purpose issued by its anglicized offspring, the New York Legal Aid Society, revealed its narrow objectives: "to render legal aid and assistance, gratuitously, to those of German birth, who may appear worthy thereof, but who from poverty are unable to procure it." Other cities responded laggardly to the New York example. In 1886 a society was formed in Chicago, primarily to protect young women who were lured into seduction under the guise of employment offers. Federal legislation furnished additional impetus. In 1892 federal judges were authorized to assign attorneys to poor people with meritorious claims (an authorization expanded in 1910 from civil to criminal actions and from trial to appellate proceedings).[31] By the turn of the century legal aid societies existed in six cities, but not until 1909 did a bar association evince sufficient interest in legal aid to sponsor a society.

Such limited success as legal aid enjoyed during its early years was almost entirely attributable to the unbounded energy and commitment of Arthur von Briesen, who came from a proud German family that traced its lineage back to the ninth century. Von Briesen's father, a retired Prussian army officer, had come to New York in 1856; Arthur followed him two years later at the age of fifteen. Serving in the Union army as a sergeant in the First New York Voluntary Engineers, he suffered his most searing wartime experience when he returned home during the terrible anti-draft rioting and devastation of 1863. Fire destroyed the family drug business, and a mob injured his father for coming to the aid of a wounded army officer. Young von Briesen developed an abiding fear of violent upheaval. After the war he worked as a patent draftsman, graduated from New York University Law School, and opened a patent law office in 1874 where he prospered throughout his professional career. He joined the Legal Aid Society in 1884, became a director in 1889 and president the following year—a position he held until anti-German hysteria compelled his resignation in 1916.[32]

Von Briesen's leadership nurtured legal aid from its parochial origins as a German immigrant aid society to its maturity as an institutionalized national reform and also made the movement into an extension of the man. By social origin and professional stature von Briesen belonged to that group often characterized (and self-described) as the "best men"—metropolitan business and professional people of British and northern European origins, whose perceptions of rapid social change and fear of social upheaval in an urban industrial society impelled their active participation in public life.[33] Von Briesen's horror of incipient class warfare shaped his approach to legal aid, his posture toward its recipients, and his justification of its social utility. He envisioned legal aid above all as a bulwark of social cohesion. By convincing impoverished immigrants, its primary constituency, that justice was within their grasp it would deflect them from anarchy, socialism, and bolshevism and would strengthen their loyalty to American institutions.

Von Briesen, who expressed a Junker's uneasiness toward the dislocating encroachments of an urban industrial society, warned that poor people deprived of legal redress for wrongs inflicted upon them were "ripe to listen to those social agitators and disturbers who are only too prevalent." Legal aid was vital because it "keeps the poor satisfied, because it establishes and protects their rights; it produces better workingmen and better workingwomen, better house servants; it antagonizes the tendency toward communism; it is the best argument against the socialist who cries that the poor have no rights which the rich are bound to respect." Protected by legal aid, "a weak and helpless person . . . is very apt to become a staunch supporter of the social organization of that community and a very poor listener to the preachers of discord and discontent."[34] Von Briesen often struck a hyperbolic note; doubtlessly salesmanship mingled with conviction as he attempted to win converts to his cause. His idealism was as genuine as his passion for stability. Revolution was unlikely, with or without legal aid. But fear of social disorder, von Briesen sensed, would encourage the growth of the legal aid movement.

Notwithstanding von Briesen's warnings, legal aid limped along until turn-of-the-century rumblings of dissatisfaction with the administration of justice, combined with apprehensions of social conflict and cataclysm, propelled it to the front rank of social control institutions. The problem was not new: back in 1888 an officer of the Chicago Legal Aid Society had warned that among the poor, "hundreds, perhaps thousands of people . . . have lost confidence in the law and its administration. . . ."[35] But perception of its dangers increased as boatloads of immigrants spilled into rapidly growing cities, as radical movements flourished, and as labor-management discord intensified. In 1901, at the twenty-fifth anniversary dinner of the New York Legal Aid Society, editor Lyman Abbott warned that if the poor could not obtain justice, "the firebrand of revolution will be lighted and put into the hands of men" and Vice-President Theodore Roosevelt described legal aid as a necessary bulwark against "chaos" and "violent revolution."[36]

The problem of providing justice to the poor was compounded by the attitudes of the Society toward work and poverty and by its attentiveness to the sensibilities of the bar. Although von Brie-sen insisted that legal aid must furnish justice, not charity, only the "worthy" poor, according to the Society's constitution, could receive legal aid. "Whoever receives our attention," von Briesen wrote, "must show that he *has* rendered some service, that he has done some work, and that he is entitled to a corresponding consid-eration, which, being denied, we enforce in his behalf." The "un-worthy" poor—those without jobs—were disregarded; the chosen beneficiaries were not people "who are always poor but only peo-ple who are made poor for the time being by the wrongful acts of others." With the assistance of legal aid they would be converted from "dissatisfied grumblers into self-satisfied citizens . . . [who] will promptly join the ranks of those who are the most ardent supporters of our institutions."[37]

Other self-imposed restrictions limited the effectiveness of legal aid societies. Not only did legal aid confine its energies to the deserving poor but, fearful lest it be accused of competing with the bar, it restricted itself to claims so small that no practicing lawyer would consider handling them. Consequently even the "worthiest" poor had to sacrifice some of their legitimate claims to the best interests of the bar. There was no guarantee that a claim too large for legal aid would in fact be large enough to tempt a private lawyer. Furthermore, until well into the twentieth century legal aid societies took only civil, not criminal, cases. Criminal defense, observed a Boston legal aid attorney, would diminish the reputation of the society before the courts—a primary consideration. Legal aid spurned the endless supply of personal injury cases, which were left to negligence lawyers in private practice to handle on a contingent fee basis. In addition, many so-cieties declined divorce and separation cases in deference both to the bar and to prevailing mores regarding the sanctity of marriage and the family—at least for those who could not afford high legal fees. Societies compounded the problems generated by their own selectivity by refusing to refer aggrieved persons to lawyers who *might* handle their claims—again lest the wrath of the bar descend upon them for selective referrals. Thus a "deserving" poor person whose claim legal aid rejected was no closer to access to a lawyer than when his search began. The obligation to help the poor seldom was permitted to overcome deference to the bar.[38]

Rarely was such concern reciprocated by the legal profession. Most societies were unincorporated private charities dependent upon community largesse for their survival. Bar associations were indifferent to their plight; law schools were minimally involved in their work. To a limited extent the federal statutes of 1892 and 1910, which established the legal right to assistance in certain circumstances, marked an administrative departure from the charitable tradition of legal aid. This partial compromise with laissez-faire theory provided some protection for the rights of the poor without imposing direct obligations upon the state. In the end, however, the "gratuitous services" of the legal profession were insufficient.[39] Only a smattering of lawyers displayed occasional interest. The Boston legal aid society, the fourth to be established, existed between 1900 and 1910 as an extension of the practice of various Boston law firms, but it is difficult to determine whether this relationship demonstrated the solicitude of lawyers or the impotence of the society. In Chicago, where the society's president pleaded with lawyers to strengthen their contacts with legal aid, Dean John H. Wigmore of Northwestern Law School was instrumental in persuading twenty firms to contribute $250 annually.[40] Such support was rare; generally the societies fended for themselves beyond professional borders. Von Briesen, who contributed $5000 annually to their work, consequently expended considerable energy trying to enlist the support of "the best people of the City." But the societies, inhibited by their decision (consistent with the Canons of Ethics) not to advertise, and constrained by professional mores, suffered from perpetually precarious finances.[41]

Working conditions for legal aid attorneys created additional impediments. Von Briesen hoped that young lawyers would "earn their spurs" in legal aid, which he envisioned as "a stepping stone to success to all those who faithfully served the poor in the beginning of their careers." Citing educational benefits, especially courtroom experience and exposure to a wide range of legal problems, he suggested that legal aid work was the appropriate analogue to a physician's internship.[42] But virtue often went unrewarded and reality seldom measured up to von Briesen's expectations. Louis Stoiber, chief attorney in the New York office for many years, described its tasks as "deadening, routine work, which would kill any sensible, ambitious man in two months." Only the chief in each branch escaped professional tedium; for the others the pay was miserably low, and most of the work, Stoiber wrote, "can be done by an office boy."[43] A society rule prohibited lawyers from handling private cases during their tenure at legal aid. This restriction, reinforced by prevailing patterns of professional channeling and stratification, made it unlikely that a society lawyer could utilize legal aid as a "stepping-stone to success." The longer he served the society the less likely he was to escape the confines of its work. More likely, considering the nature of their constituency and the structure of the profession, legal aid societies offered temporary employment to minority-group lawyers whose access to desirable sectors of private practice was strewn with obstacles.

Handicapped by the magnitude of the task of providing even limited legal services to the poor, by its fear of offending the bar, by woefully inadequate financial resources, and by staff frustration and instability, legal aid could have been expected to founder early in the twentieth century. Its subsequent vigor, not its early impotence, requires explanation. Between 1900 and 1910 the number of societies tripled (from 5 to 15); by 1920 it had nearly retripled (to 41); in 1923, 61 societies handled 150,000 cases annually.[44] Studies of legal aid work were published; the National Association of Legal Aid Organizations was formed; bar leaders occasionally referred to the problem of class injustice; and the American Bar Association even appointed a standing committee on legal aid.

Fear of social unrest accounted for this growth. Immigration peaked between 1905 and 1914, with southern and eastern Europeans comprising the bulk of the newcomers. Their deviant cultural patterns directed attention to the virtues of the melting pot. Here was a vast unassimilated mass concentrated in urban ghettos, generating concern about lawlessness and disorder. In these circumstances legal aid assumed new significance. Its defenders were galvanized into a renewed appreciation of the importance of their Americanizing mission. For the first time they won recruits among the professional elite, whose members fretted over mounting evidence of public discontent with the legal system—especially its delays and costs. Additional impetus came from the Bolshevik Revolution and the postwar Red Scare. With social disintegration threatening, there was all the more reason to strengthen legal aid societies as ameliorative and control institutions.

In 1919, against this ominous social backdrop, appeared Reginald Heber Smith's *Justice and the Poor,* an indictment of class injustice as a pervasive fact of American law and life. Smith, a partner in the Boston law firm of Hale and Dorr who served as general counsel to the Boston legal aid society, had drafted a preliminary report which documented the handicaps of poor people before the law. The Carnegie Foundation then subsidized a national study to determine whether legal aid merited support. Smith concluded that it did, although his evidence demonstrated that forty years of legal aid had not changed a legal system in which justice was available only to those who could pay for it. Court costs and fees, the expense of retaining counsel, and delay contributed to the denial of justice by transforming the issue from the merits of a claim to the availability of financial resources. Both bench and bar, Smith charged, displayed "an ignorance of, or indifference to, the disadvantages under which the poor have struggled."[45]

These disadvantages troubled Smith precisely because he was convinced that immigrants should "be assimilated and taught respect for our institutions." His goal of assimilation was undercut by the persistence of class injustice. But Smith could not believe that the conditions responsible for the demonstrably massive breakdown in the administration of justice for the poor were inherent in American institutions. Rather, he said, they accompanied rapid social change: immigration, urbanization, and the growth of a wage-earning class. Legal aid societies had struggled to reverse the tide and to promote good citizenship. Each year they proved to their clients "the integrity and fairness of our institutions," which Smith believed to be "virile and sound of heart." Yet despite his boundless faith in American institutions, Smith was impelled to a stark conclusion: "The administration of American justice is not impartial, the rich and the poor do not stand on an equality before the law, the traditional method of providing justice has operated to close the doors of the courts to the poor, and has caused a gross denial of justice in all parts of the country to millions of persons."[46]

Smith's message was unequivocal, but 1919-20 was not a propitious time to win a sympathetic audience for it. Within the legal profession the mood of the Red Scare distorted Smith's ill tidings: his theme of injustice for the poor was repressed; the need for rapid Americanization of immigrants was emphasized. In a Foreword to *Justice and the Poor* Elihu Root cited the value of the book to "the multitude of Americans who are interested in the Americanization of the millions of foreigners who have immigrated to this country, and who fail to understand or who misunderstand American institutions."[47] At its 1920 convention the American Bar Association devoted a session to legal aid in which Charles Evans Hughes blunted Smith's conclusion by referring to the "false notion that our judicial establishment is only the mechanism of privilege. . . . To spread that notion is to open a broad road to Bolshevism." For Hughes the crux of the problem was "the question of Americanization." The best remedial institution, therefore, was the legal aid society. Organized and efficient, it provided a service that law firms could not (and hardly cared to) duplicate. Indeed, Hughes concluded, urban lawyers in private practice could best discharge their obligation to the poor by supporting legal aid, not by handling claims which (in his revealing phrase) were "foreign" to their experience.[48]

Here—for the professional elite—was the ideal solution to the Pandora's box opened by Smith. If class injustice was not the problem, but Americanization of foreigners was, lawyers were absolved of direct responsibility. The happiest resolution was for them to continue to confine themselves to the familiar claims of their corporate clients, while encouraging legal aid to bear a burden which, according to Smith's own evidence, had already overwhelmed it. Within a few years, after the postwar hysteria over Americanization had subsided, lawyers would celebrate legal aid for its contributions to national harmony. They could safely ignore the problem of injustice for the poor. In his Foreword to a semi-centennial history of the Legal Aid Society, William D. Guthrie concluded that legal aid promoted respect for law, belief in the impartial administration of justice, confidence in the fairness of American institutions, and loyalty among "the unfortunate poor, ignorant and helpless." Without such accomplishments, "a drift toward communism, revolution, and anarchy would have been inevitable. . . . Failure of the legal aid movement might have spelled ultimate national disaster."[49] But it was less the success of legal aid than the end of the Red Scare and enactment of draconian immigration restriction laws that accounted for Guthrie's revival of confidence in American institutions.

Preservation of a stable legal order amid a rapidly changing society was the dominant theme in the early history of legal aid. Even Smith could not fully accept the implications of his own evidence: that much of the legal system, with or without legal aid, was designed to control poor people. Two years after his book was published he referred to the maladministration of justice as a product of conflict between the honest and the dishonest, not between rich and poor. The public image of legal aid and the bar took precedence over the substantive reality of injustice for the poor. Smith urged Chief Justice Taft to accept designation as honorary president of the national legal aid organization so that legal aid attorneys could cite his interest to reassure clients that American institutions were "fundamentally sound." Similarly, he referred to the overriding need of the bar, not the poor, for a vigorous legal aid movement. Public respect for lawyers would increase "if we can get into the public mind the fact that the bar really champions and directs the legal aid work for poor persons in this country."[50]

But in a society that dispensed justice for a fee (which many of its members could not afford), a society frightened by immigration and rapid social change, legal aid could not succeed either as a socializing institution or as a remedial agency. Injustice to the poor, not Americanization, was the basic problem. It required a subsequent generation, liberated from an obsession with cultural assimilation and more sensitive to the persistence of poverty and its relation to injustice, to acknowledge what Reginald Heber Smith had discovered, to point to the inadequacy of legal aid, and to devise new institutions and new instruments that might bring equal justice under law to those for whom law was, in Smith's words, "an impotent sham." But in the early decades of the twentieth century the energy invested by lawyers in the promulgation of ethical norms and in the fashioning of law firms to serve corporate clients was conspicuously absent when the problem was inadequate provision of legal services to poor people. Ethnic and class animosity, even when softened by *noblesse oblige*, left a legacy of injustice. Legal aid, tenuously grafted to the principle of service for a fee, might reassure elite lawyers, but it did not eliminate injustice for the poor.

In the years before World War I the structural transformation of the legal profession neared completion. White Anglo-Saxon corporation lawyers were concentrated at the professional apex, and new-immigrant metropolitan solo practitioners were restricted to the professional base. Ethical norms were promulgated; whether enforced or not they distributed status and morality according to social origin and type of professional practice. Legal services were available according to economic resources, not need. Yet professional stability was only tentative, never final. Recurrent threats required attention, and bar associations began to serve as caretakers of elite interests.

The bar association movement was a characteristic feature of the decades surrounding 1900. Lawyers, in common with doctors, social workers, teachers, and engineers, flocked into professional associations, whose growth—the number of bar associations jumped from 16 in 1880 to more than 600 by 1916—expressed the impulse for professional cohesion in a fragmented society undergoing rapid change. Their local revival after the Civil War was attributable to unease over urban corruption. Nationally, the American Bar Association, organized in 1878, had more diffuse purposes: to promote the administration of justice, to advance jurisprudence, to uphold professional honor, and to encourage social intercourse among lawyers. But the ABA exuded the genial tone of a social club, set by its predominantly Southern members who came to Saratoga Springs each year to escape the summer heat. The "benefit of the waters," one member declared, rivaled in importance the professional business of the association.[51] Simeon Baldwin, the moving spirit behind the association, labored to confine membership "to leading men or those of high promise. . . ."[52] Local associations often were similarly exclusive. The Boston Bar Association seemed to exist solely for the benefit of State Street and Federal Street lawyers. The Chicago Bar Association, founded (in the words of one of its presidents) to bring "the better and the best elements of the profession together," charged high admission fees and annual dues to achieve its purpose. The strongest pillars of the Association of the Bar of the City of New York were Yale, Harvard, and Protestantism.[53]

Bar associations were not the exclusive preserve of corporate lawyers, but lawyers whose practices provided them with a sufficient margin of wealth and leisure to pay fees, attend conventions, and participate in committee work were bound to predominate. Leadership patterns within the American Bar Association measured the ascent of the corporate lawyer. During the 1890's association members still looked for leadership to prominent courtroom advocates like James C. Carter and Joseph Choate. Between 1899 and 1905, a transition period, more specialized railroad lawyers were elected president. In 1909 the "lawyer as businessman" reached the ABA pinnacle.[54] State associations were more diversified; even in the most heavily urbanized and industrialized states general practitioners retained a strong voice. In fact, in states like New York, Massachusetts, and Illinois, metropolitan and state associations, representing respectively the urban and rural wings of the profession, engaged in constant internecine sniping. Whether corporate or country lawyers predominated, however, the "best men" used bar associations as a lever of control over professional ethics, educational qualifications, and bar admission. Claiming the right to represent and to police the entire profession, they discriminated against an increasingly substantial number of urban practitioners from ethnic minority groups.

Understandably, bar associations defended stability, order, and control. Confronting social turmoil endemic to a society undergoing rapid change, lawyers sought firm bedrock. They found it in natural law, Adam Smith economics, Tocquevillian assurances of high status, social Darwinism, and constitutional certitudes defended by a judiciary that would stand as Gibraltar against regulatory legislation. In 1915 the president of the American Bar Association asserted that law was "as omniscient and omnipotent as God because it is an attribute of God, and its home is the bosom of God." But a younger lawyer, less certain of legal divinity, accused his professional colleagues of using bar associations "as a broom to sweep back the waters of . . . modernism." Lawyers, he suggested, should not permit their "rapt contemplation of the past" to cripple their ability to cope with the present.[55] Janus-like, professional associations plunged into the twentieth century beset by nineteenth-century yearnings.

Bar associations did venture timidly into the shallower waters of law reform, but they usually skirted the dangerous shoals of substantive charge. In 1912 ABA president Stephen Gregory declared that professional associations were "the chief instrumentality of constructive legal reform."[56] Rarely, however, did their concern extend to such problems as the provision of legal services. At best, they preoccupied themselves with the most technical, professional aspects of legal issues—for example, the ethical proprieties of contingent fees rather than the social and individual costs of lives broken in industrial accidents. The result was that law reform served as "a banner of rectitude waved in the public eye," a shield to deflect public criticism. The law reform movement, sponsored by bar associations after the turn of the century, became the plaything of the legal specialist, whose job "was not to turn society on its head, but to refine and embroider his own special product. . . . Law was supposed to be socially relevant; but the areas chosen for reform were not the areas of law most socially relevant."[57]

During the second decade of the twentieth century the American Bar Association began to assert itself aggressively as a professional protective organization. Its purpose was twofold: to preserve its own exclusiveness (and the status that accompanied its preservation) and to exert professional leverage upon the political process. Two prewar episodes provided a test of its strength and scope: the admission of black lawyers and the nomination of Louis D. Brandeis to the Supreme Court.

In 1912 the executive committee of the American Bar Association unknowingly admitted three black lawyers to membership. Informed of its carelessness, it quickly passed a resolution rescinding the admission and—"since the settled practice of the Association has been to elect only white men as members"—referring the matter for determination by the entire association.[58] Attorney General George W. Wickersham protested (one of the contested members, a Harvard Law School graduate, was his assistant in the Department of Justice)—not from any commitment to racial equality but from disgust with procedural irregularities that violated association by-laws. He was assured by the association's secretary that the recision resolution had been adopted only with "a sincere purpose to do what seemed . . . to be right and just. . . ." And he was sternly chastised for his "discourteous and dogmatic" criticism, a display of pique unbecoming an association member.[59] But Moorfield Storey, a past president of the bar association and the first president of the National Association for the Advancement of Colored People, was incensed. "It is a monstrous thing," he complained, "that we should undertake to draw a color line in the Bar Association." Storey repudiated the notion that blacks were excluded by association policy, although he conceded that none had ever been admitted.[60] The association was in a quandary. Claiming to be a national organization, it fuctioned as a restricted social club. The admission of blacks, in the words of its membership chairman, posed "a question of keeping pure the Anglo-Saxon race." A compromise resolution precluded future associational miscegenation. Prodded by Storey, members permitted the three duly elected black lawyers to remain but provided that all future applicants must identify themselves by race.[61] The association thereby committed itself to lily-white membership for the next half-century. It had elevated racism above professionalism.

Professionalism converged with politics in the Brandeis donnybrook. The first of several dramatic twentieth-century Court nomination controversies, it brought into sharp focus the public implications of professional parochialism. More was at stake than a judicial seat, although a place on the Supreme Court was hardly inconsequential at a time when the judiciary was praised or blamed as the most reliable defender of vested property interests against public regulation. On the surface the division seemed clear. Brandeis' opponents, drawn largely from State Street law firms and from the American Bar Association, could plausibly view the Boston people's attorney as a threat to their restricted professional world. They spoke of law as a bulwark of private property; Brandeis, who would not have disagreed, had often used it as an instrument of social change to make property owners more responsible to the public. They devoted their careers to counseling private interests; Brandeis committed much of his to public service. Their law was a "brooding omnipresence"; his was shaped by contemporary social needs. They defined themselves as counselors to corporations; Brandeis, an opponent claimed, "acts the part of a judge toward his clients instead of being his clients' lawyer."[62] They were Protestant; he was the first Jewish nominee to the Supreme Court.

These differences masked some striking similarities between Brandeis and his critics: his commitment to efficiency and order;

his application of business values to the operation of his law firm; his admiration for the great New York firms; his fear of radical challenges to American institutions; and his insistence that only lawyers were competent to criticize and remedy defects in the administration of law and justice.[63] But the differences were crucial. They determined that the challenge to Brandeis would cut across every major professional concern of the day: ethnicity; the social function of law; the role of lawyers; and standards of professional character, conduct, and ethics. As "an outsider, successful, and a Jew," Brandeis was suspect.[64] His confirmation fight was a symbolic crusade, pitting the newest defenders of the established professional order against the outsider who was especially dangerous because he shared so many of their attributes yet put them to such different use. It was precisely because Brandeis' credentials were so impeccable—a brilliant record at Harvard Law School and a lucrative corporate practice—that the opposition to his appointment was so revealing. Even the most qualified of outsiders—qualified according to professional terms set by the insiders—encountered a wall of antipathy from the elite. Their resistance, and their defeat, exposed both the sources of their professional power and its limits.

Brandeis remains an elusive figure because his passion for personal privacy still makes access to the inner man difficult. Yet there is sufficient evidence to conclude that ambivalence toward modern America pervaded his life and career. It has been suggested that Brandeis, whose distinguished public career fell entirely in the twentieth century, possessed "one of the finest minds of the nineteenth century."[65] He was nearing fifty when the century turned, and his most memorable public service commenced with his *Muller* brief in 1908. Brandeis labored incessantly to reconcile an older morality with novel conditions. In the end, ambivalence triumphed. Brandeis, far more than all but a few of his contemporaries, mastered techniques for coping with social change within the liberal reform tradition. His point of reference, however, always was the past, specifically the Jeffersonian vision of an independent yeoman or artisan living within social institutions commensurate with the "wee individual." Understandably, Denmark and a Jewish state in Palestine would appeal to him as models. Yet the United States could not be either of these—that was the very point of the social trends that consumed Brandeis' public energies. In his private life Brandeis could spurn automobiles and telephones, those most modern and characteristically American appurtenances; but in public life it was more difficult to escape the present. In 1906 Brandeis wrote that "true conservatism involves progress"; as for so many Progressives, the destination of his reform odyssey was the past by way of the future.[66]

Only in his law practice, suggestively, was Brandeis unabashedly the modern man. Yet here, too, paradox abounds. Making it, for this young lawyer in 1879, had meant achievement "unassisted by the fortuitous circumstances of family influence or social position."[67] Yet just six weeks earlier, while exploring the possibilities of a partnership with Samuel D. Warren, he had withheld a decision until Warren was more certain of his own business prospects, based upon "social and financial position."[68] And, though he later invariably advised his acolytes to return to their local communities, the young Brandeis had himself departed for Boston at his earliest opportunity, becoming the prototypical lawyer on the make: searching for the proper partner, joining the proper clubs, fashioning a proper firm, and jealously guarding professional prerogatives. Similarly this passionate denunciator of the curse of bigness defended size and impersonality in his own law firm, justifying both by the test of efficiency for clients and pecuniary gain and status for members. The "great New York firms" were his model; the pleasure of clients was his aim. Hardly incongruous —except for a lawyer who earned his reputation (and considerable enmity) for standing in judgment of his business clients and for preaching individual identity through decentralized institutions. Brandeis transformed independence into a moral verity; so one can only wonder how the young lawyer William H. Dunbar reacted when Brandeis assured him that, despite his dissatisfaction as an anonymous associate in Warren & Brandeis, he would be "less happy standing alone." Brandeis reminded Dunbar that large specialized modern law offices were "the most effective means of doing the law work of this century—so far as clients are concerned." The proof was "the success of the great New York firms —the pecuniary success and the professional success or reputation of the individual members."[69] Finally, Brandeis described courts and lawyers as "a strong reactionary eddy" against the Progressive tide; yet when confronted with mounting evidence of popular dissatisfaction with the administration of justice he insisted that the privilege of criticism and change was reserved for lawyers— thereby insuring that there would be little criticism and virtually no change.[70]

Brandeis' credentials might entitle him to admission in the Union Club, a citadel of proper Boston society, but they could not provide access to the legal elite—especially after he had crossed swords with its members in litigation that pitted corporate interests against his conception of the public interest. Yet Brandeis' marginality as the Jewish reformer, the source of so much opposition to him in 1916, may also have enabled him to view his professional culture with uncommon perspective. In his famous Phillips Brooks House address Brandeis had confronted many of the issues that tormented his professional brethren. He conceded that the lawyer had become an inextricable part of the business world, but he rejected the notion that the closer lawyers moved toward business the further they departed from professionalism. He tried to convince his audience that the legal profession still afforded "unusual opportunities for usefulness." According to Brandeis, lawyers had assumed their prominent role in guiding industrial and financial affairs because they possessed "particular mental attributes and attainments" which must be utilized if these affairs were to be properly handled. In modern times, he claimed, such matters had passed beyond the realm of private activities to become essentially issues of statesmanship. They required "the exercise of the highest diplomacy. The magnitude, difficulty, and importance of the questions involved are often as great as in the matters of state with which lawyers were formerly frequently associated. The questions appear in a different guise but they are similar." Lawyers in business merely applied traditional skills to novel situations. Their legal training was in demand in the business world, Brandeis observed, "because business had become largely professionalized." By reversing the standard formulation of the problem—the commercialization of the legal profession— Brandeis tried to underscore continuity of function amid changed circumstances. It enabled him to assert that the lawyer's influence had not receded at all; rather, "it is simply a question of how that influence is to be exerted."

But Brandeis conceded halfway through his address (in what virtually amounted to the beginning of a new speech) that the legal profession and the nation had paid a high price for the lawyer whose social vision was restricted by his corporate counseling. First came the loss of independence of those lawyers who "allowed themselves to become adjuncts of great corporations." Then followed the social consequences of dependence: "For nearly a generation the leaders of the bar with few exceptions have not only failed to take part in any constructive legislation designed to solve in the interest of the people our great social, economic and industrial problems, they have failed likewise to oppose legislation prompted by selfish interests." The next generation, Brandeis predicted, would witness "a continuing and ever-increasing contest between those who have and those who have not." For precisely this reason he urged his Harvard listeners to study and to practice law, for "there is a call upon the legal profession to do a great work for this country."[71]

Brandeis' address reflected his own ambivalence, and that of his professional culture, toward change. He could argue that lawyers had lost their independence but retained their influence; that the bar was not commercialized but that business had become professionalized. These tenuous distinctions mirrored the split in Brandeis' own career between corporate counseling and public service. Yet Brandeis' plea for professional responsibility and public service struck a responsive chord. Felix Frankfurter, then a law student at Harvard, never forgot the speech; another young lawyer recalled, years later, that Brandeis was "the first great lawyer I ever knew who had a social conscience. . . . Look at the lawyers who were our leaders. . . . None of them made any contribution to the social growth of America. Brandeis did and they hated him for it."[72]

The opportunity for retribution came in 1916, when Wilson nominated Brandeis to the Supreme Court. Brandeis' opponents staked their claim on the ground of ethics and character. Moorfield Storey, the venerable Boston mugwump and past president of the American Bar Association (who had opposed Brandeis in important railroad litigation), testified to Brandeis' reputation as "an able lawyer, very energetic, ruthless in the attainment of his objects, not scrupulous in the methods he adopts, and not to be trusted." A spokesman for the Boston Bar Association concurred: Brandeis was a lawyer "of great ability, but not straightforward." The "Brief on Behalf of the Opposition," submitted to the Senate, rested on Brandeis' "defective standard of professional ethics." Prepared by Austen G. Fox, a wealthy Wall Street lawyer, it accused Brandeis of acting unethically—with references to his "duplicity" and "sharp practice."[73] At the instigation of Storey and ABA president Elihu Root, six former association presidents petitioned the Senate Judiciary Committee, declaring that Brandeis was "not a fit person" to sit on the Supreme Court.[74] None of the signers was more incensed than William Howard Taft, the former President and future Chief Justice, who nurtured strong judicial ambitions of his own and an equally strong dislike of Brandeis. Dipping his pen in vitriol, he dispatched letter after letter of calumny to friends and family, berating Brandeis for his ethics, politics, and religion.[75]

By resting their public opposition on ethical and character defects, opponents of the nomination avoided a direct confrontation on the grounds of religion or reform. Storey, for example, vigorously denied that criticism of Brandeis was attributable to anti-Semitism or to politics. Lawyers, he insisted, objected to Brandeis solely "on the ground of his character."[76] But "character" already had become a term of art in the legal profession, applied unnerringly to those lawyers--and only to those—whose religion, national origin, or politics threatened the professional status quo. Certainly it is impossible to know (and unnecessary to establish) what distressed his opponents more: his Jewishness, his public service, his successful practice, his outspoken opposition to corporate arrogance, his social approach to legal problems, or his judgments upon the justness of a client's case. Success aside, these traits made Brandeis a professional outsider—reason enough to contest his nomination.[77]

The insiders themselves were a motley group. Moorfield Storey spoke for the established Boston bar; Simeon Baldwin and Francis Rawle, venerable leaders of the American Bar Association, expressed the forebodings of the old-line professional elite; Austen Fox and Elihu Root voiced the dismay of the new breed of corporation lawyers. The anti-Brandeis alignment suggests the extent to which hitherto antagonistic professional groups had buried their differences by 1916. Corporate lawyers, only a few years earlier considered subverters of the professional status quo, had become its staunchest defenders. With their professional primacy established, they pledged themselves to the defense of tradition and stood shoulder to shoulder with spokesmen for the golden age of professional purity. The newest members of the court, they were more royalist than the king. This was a congenial arrangement both for the traditionalists and for the corporate lawyers which freed old timers to accept the ascendancy of corporate lawyers and assured the newcomers of heightened status in exchange for their defense of tradition—which they were quite willing to assert since they now belonged to it.

Ethnicity was the cement that solidified this professional alliance. Though *arriviste* corporate lawyers were an unsettling force in the profession, the accelerating dangers to American social and legal institutions posed by those who lacked Anglo-Saxon credentials seemed far graver, especially when corporate lawyers were demonstrably eager to defend the stratified professional culture in which they were securely ensconced. In an era of rapid social change and strident social protest corporate lawyers, once the most potent agents of change, tried to slam shut the doors of professional mobility after they themselves achieved elite status. With their opportunity and their success assured by their ethnic origins they joined hands with the traditionalists in an effort to prevent lawyers who lacked the necessary ethnic credentials (and desired political commitments) from gaining access to their firms or to the professional associations which they had come to dominate—or to the Supreme Court.

The Brandeis confirmation struggle exposed both the depths of elite exclusiveness and the limits of its reach. Within the profession the new elite was virtually free to design its structure, impose its values, and anathematize its enemies. But its power diminished at the profession's edge. It might entreat the United States Senate, or state legislatures, but it could not command. Once elite lawyers required the affirmation of political institutions for their professional designs they were no longer rulers, but competitors. And even within professional confines all was not yet secure. A new challenge, issued by law teachers, demanded attention.

Three

Scientific Expertise: The Triumph of the New Professoriat

Upset by changes in American society, the legal profession was simultaneously wrenched by internal stress. The stratification of practice was accompanied by the professionalization of law teaching. Both the corporate firm lawyer and the university law teacher were new men of power in a new age. Twin offspring of modernization and specialization in an urban industrial society, they displayed sibling rivalries and loyalties. The growth and influence of university law schools paralleled the expansion and power of corporate law firms. A law-school hierarchy emerged in which Harvard Law School and its emulators trained aspirants to the professional elite, while night law schools prepared members of ethnic minority groups for careers in business, politics, and in the professional underclass.[1] The university law teacher and the corporate lawyer shared many values, but their temporarily divergent views regarding the social function of law and the professional role of teachers were sources of continuing friction early in the twentieth century. Consequently they maintained an ambivalent relationship which vacillated between enmity and amity.

In 1873 James Barr Ames received a five-year appointment as assistant professor of law at Harvard Law School. Ames was a recent graduate of the school, without experience in practice. "What a venture that was," President Eliot of Harvard recalled two decades later. "This School had never done it; no school had ever done it; it was an absolutely new departure in our country in the teaching of law." The implications, for Eliot, were momentous: "In due course . . . there will be produced in this country a body of men learned in the law, who have never been on the bench or at the bar, but who nevertheless hold positions of great weight and influence as teachers of the law, as expounders, systematizers and historians. This, I venture to predict, is one of the most far-reaching changes in the organization of the profession that has ever been made in our country."[2]

By the turn of the century the accuracy of Eliot's prediction was beyond dispute. In the thirty years since Dean Christopher Columbus Langdell had introduced the case method at Harvard, law teaching had evolved into a new profession. Prior to 1870, teachers were indistinguishable from practicing lawyers. The road to the classroom began in practice and, for elderly members of bench and bar, in retirement. As long as lectures provided the staple of legal education, teaching was an avocation rather than a profession; to qualify, an aspirant displayed credentials earned outside the law school. The case method was both a symptom and a cause of change. Its stringent demands upon the instructor excluded the part-timer who restricted his preparation to dusting off old lecture notes. Especially in its early years, when its defenders engaged in virtually a holy war for supremacy, it imposed demands that neither busy practitioners nor retired gentlemen could meet. Few practitioners could spare the time required by Langdell's innovation; and venerable attorneys with leisure time lacked the intellectual flexibility to cope with the analytical, inductive process that comprised the core of the case method.[3]

The assumptions underlying the case method, even more than the rigors implicit in its teaching, created what Ames was to call "the vocation of the law professor."[4] To Langdell, law was a science, the library was its laboratory, and cases were its natural elements. "If it be a science," he asserted, "it will scarcely be disputed that it is one of the greatest and most difficult of sciences, and that it needs all the light that the most enlightened seat of learning can throw upon it." Since law was not "a species of handicraft," learned by serving a practical apprenticeship, the student must be taught "by teachers who have travelled the same road before him."[5] That road was the case method; as Langdell's disciples traversed it they spread the gospel to their professional brethren.

Their commitment to law as a science implied a parallel commitment to teaching as a profession. As James Bradley Thayer told the American Bar Association in 1895: "You cannot have thorough and first-rate training in law, any more than in physical science, unless you have a body of learned teachers; and you cannot have a learned faculty of law unless . . . they give their lives to their work."[6] A few years later Ames commented on the vigorous growth of law schools—the number of law students tripled between 1889 and 1899—and found its significance to lie in "the opening of a new career in the legal profession, the career of the law professor."[7] When Ames spoke, only one-quarter of American law professors were full-time teachers, but he accurately predicted the reversal of these proportions within a generation. In proposing "the sound general rule that a law professorship should be regarded as a vocation and not as an avocation," Ames set a standard to which virtually every university law school would aspire.[8]

The spirit of scientific expertise inspired the generation of law teachers who spanned the Progressive era after the turn of the century. In an age of reform, teachers perceived law as an instrument of social engineering, with broad public implications that transcended client-caretaking.[9] They must, Roscoe Pound insisted, create "a true sociological jurisprudence"; the proper source of legal doctrines could only be "a scientific apprehension of the relations of law to society and of the needs and interests and opinions of society of today."[10] The marriage of science to reform, at a time of heightened self-consciousness among law teachers, was a source of boundless energy and enthusiasm. A Chicago professor, citing the growing impact of law schools within the profession, spoke with a measure of awe about his new professional responsibilities: "Contemplate . . . what it means to be able to train the men who directly and indirectly are to exercise the most potent influence over the growth and development of our law." He referred to law schools as laboratories, where new ideas would be "tested, compared, analyzed and reported upon."[11] Law teachers, to conclude the metaphor, were the scientists who gathered, sorted, and evaluated data to produce legal solutions for social malfunctions.

Teachers who asserted the scientific expertise of the professoriat enhanced their power in public affairs. By claiming to speak with the authority of disinterested scientists on the reform side of public issues, they could reduce practicing lawyers to the status of special pleaders for the parochial interests of their clients. No one more acutely perceived how law teachers might apply the scientific spirit to public life than Felix Frankfurter, who had studied under Ames and revered Thayer while developing "a quasi-religious feeling" about the Harvard Law School.[12] In 1913, after assisting Henry Stimson in the United States Attorney's Office and serving in the War Department, Frankfurter was invited to return to the Law School as a member of the faculty. He did not see the decision as a choice between public service and an academic career; teaching law was, by Frankfurter's definition, involvement in public life. Traditional scholarship—remote and cloistered—had no appeal. "On the other hand," Frankfurter told his mentor Stimson, "I do feel very deeply the need of organized scientific thinking in the modern state and, particularly, in a legalistic democracy like ours, the need of a definitely conceived jurisprudence coordinating sociology and economics. In other words, I am struck with the big public aspect of what should be done by our law schools."[13] Private practice in New York was unappealing; there, he concluded, he would "peg away most of my vital activities in matters not of dominant appeal to me." But Cambridge, especially the Law School, was "the center of the liberal movement." There, Frankfurter told Stimson (and convinced himself), the work "*is* public work—our Universities increasingly should *be* in politics."[14] Within a year after his appointment, in an address before the American Bar Association, Frankfurter delineated the responsibility of law schools and their faculties. The swift pace of change since the Civil War had rendered obsolete the legal doctrines that were suitable for a homogeneous agrarian nation but thoroughly inadequate for a heterogeneous industrial society. Neither bench nor bar could assume the entire burden of doctrinal modernization. "What we need are . . . men who labor steadily upon law as an organic whole, who should produce tentative working hypotheses to be tested, revised and modified as the actualities of controversy require." Teachers must demonstrate to students that law was "an instrument and not an end of organized humanity." And law schools must be "fit for . . . the work not merely of training practitioners but of helping to develop the law, of participating in a great state service."[15]

From Langdell through Thayer and Ames, to Pound and Frankfurter, wound the strands of professional self-consciousness. The bar association addresses of Thayer in 1895 and Frankfurter in 1915 bracketed the formative decades for professors in American law schools. As Ames foresaw, teaching had indeed become a vocation. The laying on of hands in Cambridge was characteristic of national trends. A new profession was emerging, comprising a group of men who, in Richard Hofstadter's apt description, were "keepers of the professional conscience."[16]

It is debatable whether something inherent in the case method, which provided an initial impulse toward professionalization, also predetermined a reform function for teachers. Certainly the earlier practice of law lecturing implied that teachers dispensed fixed, authoritative principles; students dutifully learned "that law was a body of pre-existing rules which they had no power or authority to alter. . . ."[17] But the case method, at the very least, encouraged student skepticism toward judicial reasoning, even if it did so only in the narrowest terms of craftsmanship rather than on the broader ground of doctrinal substance. In comparison with their practicing brethren, the generation of teachers nurtured on the case method who came of age after the turn of the century was bent toward reform. (Law was not the only profession functionally divided into staid practitioners and crusading teachers, each struggling to establish its separate identity.) Especially in the Progressive era, a time of reform ferment when Americans first tried to grapple with the problems that accompanied industrialization and urbanization, law teachers were distinguished by their sensitivity to the sociolegal implications of these problems. Their status as teachers, and their relationship to universities, may have strengthened this impulse. Law teachers inhabited two worlds but were marginal to both. They were simultaneously lawyers who were functionally distinct from practitioners, and teachers with transacademic responsibilities not shared by their liberal arts colleagues. With due regard for individual variations, it seems likely that the conjunction of two developments—the professionalization of law teaching and the nationalization of public issues within a reform setting—provided law teachers with their distinctive identity.

The Langdell-Ames contribution was necessary for this—since it divided the legal profession along functional lines—but it was not sufficient. Without the particular rhythms characteristic of early twentieth-century life, law teachers might have remained merely the "expounders, systematizers, and historians" described by Eliot—rather than the reformers so many of them in fact became. Such, indeed, was the experience in England, where the narrowness of legal education was as directly attributable to historical accident as was its breadth in the United States. When university legal education was revived at Oxford and Cambridge in the mid-nineteenth century, the new academics found themselves imprisoned by the parochialism of their institutions. Consequently the impact of academic law was minimal and the profession of law teaching foundered.[18] The delayed birth of university legal education in the United States, and its coincidence with national reform, was fortuitously decisive. Had it emerged before 1890, or after 1915, the profession of law teaching probably would have been conspicuously different.

Yet as liberating as the case method was in its day, its emphasis upon rigorous doctrinal analysis set constricting limits to Langdell's "science." Preparing students only for mastery of judge-made law, it deluded its practitioners into believing that law was science, not policy, and that other scholarly disciplines, and even practical experience, had nothing to offer. As early as 1900 an English legal historian observed that "the final triumph of the Harvard professoriate" was their ability to convince bar leaders that law "is a science, that it rests on valid grounds of reason, which can be so explained by men who have mastered its principles. . . ." Harvard demanded compliance with its precepts. When Chicago approached a Harvard faculty member to serve as dean of its new law school, Ames insisted upon "a School like ours, that is a School with a curriculum of pure law, with a Faculty made up exclusively of professors who are lawyers, . . . believing in Harvard standards and Harvard methods." The prospective dean warned Chicago that no curricular deviation was permitted. There could be "no serious work outside the strictly legal subjects we teach," nor could the length of courses vary since law was "most naturally" taught that way. Indeed, he observed, it was Columbia's heretical willingness to include political science courses in its law curriculum that accounted for its "striking failure" to achieve institutional greatness. "Pure law," unsullied by the impinging forces of society and transmitted through the rigors of the case method, was the new professional orthodoxy. Law would be taught as a science, but as long as its teachers elevated logic and doctrine above practical social policy and human experience, process would submerge the substantive needs of clients and society. In time, sharp challenges to the case method would come from those wishing to move beyond appellate opinions to analysis of the social forces that shaped law and to the practical experiences that molded lawyers.[19]

In the years after 1900, however, the novelty of the case method in university legal education and the special skills that its teaching required contributed to the widening gulf between law teachers and practitioners, many of whom preferred doctrinal affirmation to critical analysis. The legal profession, and ultimately American society, paid a price for the happy accident of timing that transformed law teachers into legal reformers. Although it surely was necessary for the profession to assume a share of responsibility for the modernization of law and legal institutions, it was hazardous for teachers alone to shoulder the burden of reform. This division of labor drove a wedge deep into the profession, separating teacher from practitioner and eliciting mutual suspicion. The more fervently teachers proclaimed their sense of public responsibility the wider and more menacing this split became.

In the two decades preceding World War I a sense of public responsibility and an identification with political reform provided law teachers with their special identity.[20] Their mission was to redeem the profession and reform the nation. Amid widespread allegations of unethical practice, delays in the administration of justice, ambulance chasing, and commercialization, improved legal education was hailed as the first step toward the restoration of professional integrity. Courses in legal ethics, the panacea for some, were insufficient. "I doubt whether a course of lectures on moral conduct will revolutionize the morality of the Bar," declared a member of a Texas law faculty. "The evil . . . is not so much a professional as an American fault. It has its source in our inordinate love for the almighty dollar."[21] But law teachers might still contribute to professional uplift. In a warmly received address in 1906, William Draper Lewis of the University of Pennsylvania Law School criticized the "total absence of any idea that there exists any obligation on the part of the Bar towards the community." The need was for lawyers who looked beyond their practice to "the administration of justice in its broadest sense." But, he suggested, existing patterns of legal education failed to produce such lawyers. Apprenticeship rested on the assumption that mastery of the minutiae of private practice was the only requisite skill. Night education, a comforting extension of the principle of equal opportunity, rewarded individual mobility at the expense of professional responsibility. And the curricula of university law schools were heavily weighted toward private rather than public law. Lewis urged these schools to undertake the reforms necessary for the training of lawyers attuned to modern needs. "If, as a profession, we are awake to our failure to perform our public duties, it is the small class of men who are devoting their lives to legal teaching who must point the way."[22]

Lewis' address, stressing the role of law faculties as the counter-weight to professional sloth and as inculcators of social idealism, struck a responsive chord.[23] But the issue transcended what law teachers might do to redeem the legal profession; as a Chicago professor asked, "What could be more useful or more fitting than that in our law schools should be found a body of men, competent, trained, impartial and honorable, ready and willing to give their aid and counsel in the formation and settlement of public questions having a legal aspect?"[24] For many law teachers, the commitment to reforming the legal profession was merely part of a larger obligation to reform the nation. Louis D. Brandeis captured this spirit when he described the value of a law school professorship "as a fulcrum in efforts to improve the law and through it—society." (His appreciation of these possibilities was held against him in 1916, when one of his detractors criticized him for his affinity to law teachers.[25]) It was articulated by Roscoe Pound, who taught at Nebraska, Northwestern, and Chicago law schools before joining the Harvard faculty in 1910. Pound's call for a sociological jurisprudence rested upon explicit notions regarding the purposes of legal education, which carried him beyond the narrow boundaries drawn earlier by Langdell, Ames, and their contemporaries. It made no sense to berate practitioners for the sorry state of jurisprudence, as Pound did, unless one was convinced—as Pound was—that a new generation of lawyers might be trained to assume their social obligations. "So long as the one object is to train practitioners who can make money at the Bar, and so long as schools are judged chiefly by their success in affording such training," Pound warned, law would lag behind life. Yet law teachers must not become "legal monks," who retreated to cloisters "from which every worldly and human element is excluded." The need was for teachers trained in economics, sociology, and politics, who were thereby equipped to "fit new generations of lawyers to lead the people."[26] Pound's professional colleagues concurred. William R. Vance of Yale insisted that not only must teachers prepare students for a career at the bar, they must also adapt law to social needs. It was unreasonable to expect the practitioner to slight the interests of an individual client for broad social interests. But precisely this task was the "ultimate function" of "the new class of lawyers, just emerging into group consciousness, the law teachers in our great universities, [who] will ultimately rise to the accomplishment of this work of adaptation." Through their direct influence on students, through research and publication, and through advice to government agencies and commissions, teachers might assume the public responsibility that practitioners had abdicated.[27]

Professorial reform rhetoric had its obvious self-serving qualities. Teachers—even the university elite—were not born social activists, nor did many of them ever assume that role. But there is no denying the emerging reform posture of the professoriat. In his presidential address to the Association of American Law Schools, Henry M. Bates of Michigan spoke of the "strategic position" occupied by teachers in an era of social change; they were, he declared, "fortunate, perhaps beyond any who have preceded them, in the possession of opportunity for conspicuous service in the cause of social justice." Modernizing the law, Felix Frankfurter insisted, required the efforts of "those in position to give their entire time to the arduous task. Here the growing body of teachers of the law find a natural field." Samuel Williston, Frankfurter's Harvard colleague, urged teachers to prepare students "to practice the law as it is going to be in the future, not simply as it is now." Teachers, Williston insisted, enjoyed more freedom, and therefore more power, than even judges did to shape the law toward broad social ends. Practitioners, like judges, were bound by *stare decisis;* only a teacher could become "the bold or ingenious theorist."[28]

An elitist claim of expertise, far more than any bold commitment to theory, accounted for the teachers' self-assertiveness. The profession of law teaching surfaced at a time when the complexity of government functions created a demand for the expert who would commit his specialized knowledge and skills to public life. Universities became the bellwether institutions in this process. First in Wisconsin, and then elsewhere, they served the state by training administrators, gathering information, and advising public officials. The Wisconsin idea appealed to the new generation of law teachers, who knew that practitioners were inclined to dismiss them as impractical theorists. "The 'cult of incompetence' associated with early stages of democracy is giving way to a belief in expertness," declared Henry Bates, who concluded that "the long-sneered-at scholar and theorizer is coming into his own." Felix Frankfurter envisioned an academic career as a means to an end, as an opportunity to modernize jurisprudence to serve the nation's industrial and economic needs. He told Learned Hand: "I have long thought that, juristically, the Wisconsin idea should be nationalized, and that it was up to the Law School to do it." When Roscoe Pound counseled Senator Robert M. La Follette, the Senator's secretary expressed his delight that law teachers were involving themselves in public life. "The great hope is with the new lawyers," he wrote in grateful reply. "Would that more of our law schools had men of your type."[29]

There were enough men like Pound to reshape the identity of American law schools. Men might still pursue academic careers to escape the strain and pressure of practice, but generational differences within the teaching profession meant that in the twentieth century the law school could no longer be considered an idyllic retreat.[30] An experience of Pound's underscores this point. In 1915 he attended a dinner in honor of a retired Harvard Law School professor, who at the end of the nineteenth century had taught Massachusetts law and practice. There Pound heard "an outpouring from the old guard" who berated the modern law school for replacing "the human element" with logic, and democracy with elitism. Older teachers knew that they had been superseded. With some disdain Pound observed that modern classroom methods made demands "which the old type is quite unable as well as unwilling to meet. His mind is not trained for class-room logical acrobatics. Naturally he claims to compensate by the 'human' element that consists better with indolence. The wail of the unfit is very apt to be made in the name of Demos."[31]

Pound's elitism was another identifying trait of law teachers who, in common with other progressives, viewed reform by experts as a vehicle for the reestablishment of elite ascendancy in public life. (Indeed, their commitment to elitism would long survive their commitment to reform.) As social engineers they would exert mastery over the complex and disintegrative forces of social change. Law as a science could not help but elevate the expertise, and therefore the power, of those who were trained to master its secrets (and those who did the training). The professional expert was a key figure in Progressivism, as the conspicuous role of economists, engineers, social workers, and especially lawyers clearly demonstrates. The expanding administrative apparatus of government provided lawyers with the opportunity to apply their expertise and to shape policy under the guise of detached scientific objectivity. The notion of expertise provided similar camouflage for law teachers who could hide their elitist claims in their mysterious science which, by definition, remained unfathomable to the uninitiated. As so much of their rhetoric suggested, university teachers (like corporate attorneys) viewed themselves as the "best men," whose service would assure orderly progress—and (not incidentally) elite hegemony. But the methods and values of "objective science" were considerably less neutral than its proponents claimed; they were designed to provide conservative solutions by stressing stability, control, and order. Progressive practitioners like Brandeis concurred. Worried about waning respect for law, he expected law schools to train society's managers and to instill lawyers with the skills required to preserve existing institutions by making them responsive to contemporary needs.[32] Viewed from this perspective the cry for reform, trumpeted by law teachers, enhanced the prerogatives of those who fully accepted the basic contours of the social system and trained young men for success within it. University law schools, by certifying elite aspirants, sanctified the solutions offered by their graduates for the amelioration of social conflict.[33] For this reason the mistrust between practitioners and teachers remained within clear, and rather narrow, limits.

Even those who held traditional notions of the teacher's function had to concede the primacy of university law schools in legal education. The modern law office had no place for the untrained aspirant. Its stenographers and secretaries handled the paperwork; its busy partners had neither the time nor the inclination to tutor neophytes. Upon the law school, declared Dean Harlan Stone of Columbia Law School, fell the responsibility to transmit "the ideals, sentiments, and traditions, the professional spirit, in short, of those who have gone before."[34] Clearly, Stone's notion that the school should convey tradition departed from Frankfurter's insistence upon its innovative role. But their agreement upon the primacy of school and teacher in legal education enabled them to share a spirit of professionalism that separated both of them from their teaching predecessors and from the practitioners with whom teachers were increasingly in conflict.

After the turn of the century a widening no-man's-land separated the two groups, even though the sense of schism always exceeded its extent. In 1901, when Ames described the vocation of law teachers, three quarters of American law professors, by his own estimate, retained their ties to bar or bench. At Harvard, where Eliot had hailed Ames as the prototype of a new breed, Langdell, Thayer, and Gray all had practiced for at least a decade before joining the faculty; the young men appointed during Eliot's regime almost invariably had practical experience. At Columbia, no faculty member appointed between 1891 and 1919 lacked some experience in practice. But the momentum had clearly shifted toward full-time teaching as a separate career. William Howard Taft, appointed with four other men in mid-career to comprise the Law Department at the University of Cincinnati, admitted to "a fundamental error in our organization in that we are all of us busy with other matters and cannot devote our entire time to the work of educating ourselves to educate others." In time, Taft hoped, the faculty would consist entirely of teachers. Concurrently, John H. Wigmore gleefully reported that his president at Northwestern had agreed to replace two part-time local practitioners with one full-time teacher.[35] Unquestionably, a separate profession was emerging, encouraged by the decreasing time lag between graduation and appointment to a law faculty, and the expectation that teachers would devote substantially all of their time to the work of a school. By 1915 Dean Harry S. Richards of the Wisconsin Law School referred to the "radical change" in the character of law faculties, which comprised "mostly young men . . . [who] have had a very limited, if any, experience in practice."[36]

Nothing did more to confirm, and to reinforce, the sense of separateness than the birth, in 1900, of the Association of American Law Schools. Just as teachers branched off from practitioners, so their professional association was the child of the American Bar Association. In a marginal way the ABA had been concerned with legal education since its founding in 1878. One of its earliest committees, on legal education and admission to the bar, issued a lengthy report the following year but then, perhaps exhausted by the effort, it lapsed into silence until 1890. At its 1892 meeting the Association established a Section of Legal Education to provide a forum for members wishing to discuss educational matters. Law school men formed the nucleus of the Section; each year one of them was elected chairman. Although other ABA members were free to participate, they seldom did so. In effect, the Section became the special preserve of the teachers.[37]

But teachers soon grew restless under the ABA rubric. The Association's indifference to legal education was constricting; one teacher complained that his colleagues "found it impossible to secure . . . any adequate expression of their legitimate aspirations." In 1899 the Section appointed a committee to investigate whether "reputable" law schools should draw closer to each other; the committee answered affirmatively and invited schools to send delegates to the next meeting. In 1900 representatives of thirty-five institutions organized the Association of American Law Schools, whose object was "the improvement of legal education in America, especially in the Law Schools." With the founding of this association of "reputable" schools, "the era of the professional law teacher" formally opened.[38]

It opened on a discordant note. The law school community was a motley group, with three- and two-year schools and day and evening institutions jostling for position. From the outset, the law school world (like the profession at large) was caught between divergent models: while university schools perceived their function in elitist terms, night schools served the aspiring urban Lincolns. Many night schools remained outside the Association altogether. In 1901 barely half of all American law students attended Association schools; during the next fifteen years the proportion actually declined by more than 10 percent. Disagreement over membership requirements exposed these divisions. Everyone agreed that member schools must demand of their students a high-school education. But the established institutions insisted upon a three-year law course, while the aspiring two-year schools resisted, forcing a compromise which delayed the three-year requirement until 1905.[39]

The issue had an iceberg quality: during the next two decades law teachers were battered by the underlying social policy questions. Their professionalization, and the multiplication of law schools, occurred during the peak years of immigration to the United States, between 1905 and 1914. "New" immigrants from southern and eastern Europe poured by the millions into American cities. For many of them, and especially for their children, access to the legal profession became an index of American democracy. The nexus between law and politics made a career as attorney personally attractive and politically important. In a twentieth-century replay of the struggle for easy professional access that characterized the Jacksonian era, law schools were pressured by those who demanded a freely swinging door to the profession. Increasingly the question of higher educational standards and requirements for admission to the bar became a critical professional issue. University law teachers found themselves caught between their own notions of educational elitism and the clamor for easy access; and they could never be certain whether practitioners would stand with them as allies or fight against them as adversaries.

These stresses were apparent during the early years of the Association of American Law Schools. In its infancy it seemed to enjoy a filial relationship with the American Bar Association. A law school dean described it as "the creature and mouthpiece" of the ABA. An English visitor to America observed a "close intimacy between the school and the profession."[40] Between 1900 and 1913 the associations met jointly, giving further credibility to the image of amicability. As in any family, however, tension existed. Teachers complained that the ABA shunted them into the background by not allotting sufficient time for their meetings. They assured each other that they no longer required the paternal protection of the parent association.[41] In 1914, after the ABA shifted its annual meeting from August to October, an inconvenient time for teachers, the law school association declared its independence by scheduling separate meetings.

Formal separation encouraged the teachers to voice their complaints. Henry Bates blamed the ABA for the lack of cooperation. "Their idea," he told Roscoe Pound, "is a silk-hatted, frock-coated gathering of functionaries and stuffed prophets who shall attend dinners and other functions. They are rather deliberately and cavalierly cutting out the Law School men." Walter Wheeler Cook noted that under ABA rules many prominent teachers who did not practice were ineligible for membership. A Wisconsin professor spoke of "a feeling . . . of the futility of accomplishing any definite and constructive action through the American Bar Association." Many teachers, he said, believed that the ABA was indifferent to legal education. "We lived with [the ABA] for fifteen years," recalled Harry S. Richards. "What did they do for us? They gave us a poor place on the program, and then often only after a fight, and paid no attention to us, so we took our things and moved out."[42]

Richards and his colleagues demanded explicit recognition of the distinctiveness of law teachers and their expertise in matters affecting legal education. But the growing incompatibility of the

ABA and the AALS, a symptom of the cleavage between teacher and practitioner, cut far deeper than this. In its inception an expression primarily of functional differentiation within the bar, the rivalry soon touched sensitive professional nerves. One of these was the purpose of legal education: should it train practitioners or should it make law into a responsive social institution? The same teacher might indeed attempt both—at the risk of "intellectual schizophrenia" as he tried to be "an authentic academic and a trainer of Hessians."[43] But practitioners suspected teachers for their commitment to reform; teachers distrusted practitioners for their hostility to it. For two decades they shouted at each other across a widening gulf.

The dialogue of disagreement was set when teachers first began to stir professionally. As an Illinois attorney declared, "Now, the lawyer is one thing, and the legal scholar is another. All lawyers know the difference between the scholar and the man of affairs." John F. Dillon, the prominent railroad lawyer, was more explicit. Law reform, he told the American Bar Association, was a constant necessity; indeed, "doctrinaires, jurists and legal scholars" often were the first to perceive this. But, Dillon warned, "no mere doctrinaire or closet student of our technical system of law is capable of wise and well-directed efforts to amend it. This must be the work of practical lawyers." At stake, clearly, was the allocation of professional power. Dillon's preference for the guiding (i.e., restraining) hand of practical lawyers followed from his assumption that they would be more cautious than mere "closet students" on law faculties. As the self-consciousness of teachers intensified, the resistance of practitioners stiffened. A Chicago attorney admitted that "we all . . . feel a sort of contempt . . . for our academic brother out at the university,—we can't help it." Contempt and suspicion—even fear—had their common source in the perception that teachers, rather than practitioners, now held a monopoly on the power to mold future generations of lawyers. When the practitioner sneered at the professor for being impractical and theoretical, for being a man "whose learning is derived solely from books," he voiced his unease lest legal education turn young men away from the virtues and values of private practice.[44]

On the other side, the bold claims of law teachers exacerbated the division. "The practicing lawyer unquestionably tends to become a conservative, and we need a few radicals," declared Samuel Williston of Harvard, anything but a radical himself.[45] The willingness of teachers to adapt law to social needs, argued William Vance of Yale, shocked practitioners, "for they fear the college professor as a reformer." Their image was of "a person possessing a brilliant, but erratic, mind, full of useless and obsolete legal lore, but without knowledge of the practical affairs of life."[46] Vance alluded to an interesting paradox. The practitioner, presumably enlightened by experience, was untroubled by social problems, while the teacher, presumably buried amid his books, perceived the need for reform and worked to achieve it. Although practitioners dismissed teachers as impractical theorists, they clearly feared the influence of teachers in public affairs.

Hostility between teacher and practitioner, as members of both groups knew, was the result of divergent conceptions of law. Thomas Reed Powell, the brilliant constitutional scholar who taught at Columbia and then at Harvard, referred to law as "a cultural study." But practitioners, he noted, were least equipped to grasp this truth. "The able lawyer is a busy lawyer," Powell observed. "The compelling interest of the work which comes to his desk leaves little time or energy for the contemplation of law as a social institution." The tasks of law teachers, however, "permit or require them to regard wider interests than are committed to the care of the practitioner." Powell wished that lawyers would "cease to be content with promoting or protecting the particular individual interests that chance to offer them retainers."[47] Precisely this approach to law offended the established practitioner. In 1915 William D. Guthrie, a leader of the corporate bar, complained to Harlan Stone, Powell's dean at Columbia. Guthrie reported conversations with lawyers resentful of professors at the school who criticized courts and disparaged judges. These practitioners, Guthrie added, claimed that students received "a decidedly wrong slant and prejudice at Columbia"; consequently, they were urging prelaw students to go elsewhere. A year later, referring explicitly to Powell's biting critiques of recent judicial decisions, Guthrie bluntly drew the issue: "It would be most regrettable for the young men who are being trained at the Columbia Law School to go out into the world prejudiced against our judicial system, which is the very *keystone* of the whole arch of American constitutional government. There is much talk among lawyers about the radical attitude of Columbia professors and their leaning toward pure democracy and socialism."[48] Similarly, William Howard Taft, who joined the Yale Law School faculty after leaving the White House in 1913, tried to prevent his students from "sliding into this easy but erroneous path toward change for the sake of change and socialism, instead of individualism." Troubled by his first exposure to the work of Roscoe Pound, Taft expressed the uncertain hope that the Harvard professor "remains still in favor of the rights of personal liberty and of private property." Sociological jurisprudence, he confessed, "excite[s] my indignation, and . . . shake[s] the foundations of law as I have been trained to know them. . . ."[49]

At a time when the judicial recall movement reached its peak, and charges of judicial nullification of the majority will resounded through the nation, such jittery responses were understandable. But their implications transcended Guthrie's complaints about Powell, or Taft's about Pound. Inherent in the professionalization of the law teacher, as Ames and his disciples reiterated, was the notion of the detached expert who scientifically weighed the utility of existing laws against social needs. Whenever law was deficient it was the self-assumed responsibility of the expert to restore the sociolegal equilibrium. Through his research, writing, teaching, and advising—all under the guise of scientific objectivity —the teacher presented a model that disquieted the practitioner.

Yet as sharp as the schism seemed, and in fact was at times, the rivals behaved more like siblings than like Hatfields and McCoys. "Radical theorists" did, after all, train and certify young lawyers for business practice, which suggests that teachers and practitioners shared important values. Often separated by their attitudes toward public regulation of corporate enterprise, they were always bound by the strong commitments of their respective institutions (the university law school and the corporate law firm) to the social system, to governance by a legal elite, and to the exclusion of social and ethnic outsiders who lacked the necessary qualifications for elite membership. Furthermore, the proportion of academics with experience in practice remained high; prominent teachers, like Harlan Stone, not only moved easily between the two worlds but argued the necessity of practical experience as a precondition of successful teaching; members of both camps urged reconciliation at the same time that they hurled imprecations; and both sides knew that "the living law," as Frankfurter wrote, "cannot be spun wholly out of a closet."[50] It was one of the ironies of the professionalization of law teaching that in its first phase, before World War I, *separation* from practitioners moved teachers closer to public life; but in a subsequent phase, legal realists would demand *reconciliation* between education and practice as the opposite means to the identical end. For a time, however, the divergent goals of the Guthries and Powells kept the professions apart. The struggle between them for primacy and power raised the stakes. (As Judge Cuthbert Pound would say of the teachers: "They have definite ideas and a class consciousness of their own and a legitimate ambition to impress their influence on the Bench and Bar."[51]) The desire for control over social policy could transform an internecine flurry into a war to save the Republic. The Guthries of the profession, suspicious of teachers, found it easy to accuse a distinguished professor of constitutional law (Thomas Reed Powell) of undermining constitutional government.

After two decades of sniping, the prospects for reconciliation seemed dim. In 1916, when Elihu Root returned from fifteen years in public life to private practice and to the presidency of the American Bar Association, a dinner was given in his honor. Root described two startling changes in the professional culture since he last had practiced. One of these was the presence in law schools of "half-baked and conceited theorists" who, Root observed, "think they know better what law ought to be . . . than the people of England and America, working out their laws through centuries of life." The second disturbing development was that 15 percent of New York lawyers were foreign-born; an additional third of the metropolitan bar had foreign-born parents. Alien influences, Root warned, must be "expelled by the spirit of American institutions."[52]

These changes were not as unrelated as they might appear. Both pointed toward legal education and admission to the bar as issues of transcendant professional importance. The more important they became, the closer teachers and practitioners edged to a common crusade in which higher educational standards would serve as an instrument of professionalization and, simultaneously, professional purification. Since the 1890's the legal profession had engaged in a vigorous internal debate over qualifications for admission to the bar. The ostensible issues—the sufficiency of high school, college, or even law school training for entry to the profession—were of negligible importance compared with questions of mobility, stratification, and structure that underlay the dispute. At stake was nothing less than the identity of the profession. The debate was triggered by the proliferation of law schools and the corresponding increase in the number of lawyers. From 28 schools with 1600 students in 1870, the number jumped to 54 schools with 6000 students by 1890, and to 100 schools with 13,000 students by the turn of the century. These figures were momentous. They pointed to a major shift to law-school (as opposed to law-office) training. There was an increase of 196 percent in the number of lawyers educated in law schools during a period when the total number of lawyers increased by less than half that figure. In 1870 one-quarter of those admitted to the bar were law school graduates; by 1910 two-thirds would be.[53]

Notwithstanding the growing popularity of formal training, and what one academician hailed as "the changed sentiment of the profession towards the schools," the educational picture had its gloomier side.[54] At Harvard Law School, in the mid-nineties, three-quarters of the students were college graduates, but at Columbia fewer than half were, and the figures for Northwestern (39 percent), Yale (31 percent), and Michigan (17 percent), which ranked next, revealed that an academically educated bar still was more a hope than a reality. In New York State, where the percentage of lawyers with college training was high, 35 percent of bar applicants in 1899 held a college degree; nationally, a decade later, only 8 percent of lawyers admitted to the bar were college graduates. Furthermore, the impressive increase in full-time law schools was overshadowed by the vastly accelerated growth of part-time institutions. Between 1890 and 1910 the number of day schools increased from 51 to 79, or 60 percent, while the number of night schools soared from 10 to 45, or 350 percent.[55]

Buried among these figures were vital questions of social policy. In a society where the professions presumably provided avenues of social mobility, any effort to raise the barriers against easy access would incur the wrath of those who equated democracy with accessibility. This was especially true for the legal profession, which provided direct access to careers in public life. The links between law and politics, and the ease with which lawyers entered political careers and dominated public life, had group and individual implications. When an ambitious Italian, Jew, or black vaulted the bar into the legislature he often carried his group identity with him and found himself advantageously situated to serve the group's needs while advancing his career. Any movement to limit access to the bar might easily become (or indeed originate as) a device to deny political power to specific ethnic or religious groups.

Such a movement, in fact, correlated precisely with the era of mass immigration and the resulting influx of the foreign-born and their children into the legal profession. Census returns after 1900 revealed a profession transformed. In city after city the percentage of new lawyers born abroad or with foreign-born parents greatly exceeded the general rate of professional growth. In Boston, between 1900 and 1910, the number of lawyers increased by 35 percent but the number of foreign-born lawyers increased by

77 percent and the number with foreign-born parents by 75 percent. In New York City, during the same period, the number of lawyers showed an identical percentage increase, but the number of foreign-born lawyers jumped by 66 percent and those with foreign-born parents by 84 percent. In the following decade the number of lawyers in Chicago with foreign-born parents almost doubled the rate of professional growth; in Boston there was a 20 percent increase; in Philadelphia, 12 percent; and in St. Louis, where the increase was 10 percent, there was also a 27 percent spurt in the number of foreign-born lawyers.[56]

Clearly, educational standards and policies of bar admission had explosive potential. They raised, in another guise, precisely the issue of elitism vs. democracy that had plagued the profession before the Civil War and had returned to torment it after the United States entered the urban industrial age. With the metropolitan bar growing rapidly, efforts to restrict access often expressed anti-urban impulses. Since immigrants and first-generation Americans, in large measure, accounted for this growth, such efforts also expressed the hostility of a besieged Protestant culture toward the changing ethnic and social contours of American society. Elitist opposed democrat; country lawyer mistrusted metropolitan attorney; established Protestant fought aspiring Jew.

The newly conspicuous role of university law schools made it impossible for law teachers to escape these professional and cultural collisions. As early as 1900, Dean William Draper Lewis of the University of Pennsylvania Law School proposed a minimum of two years of college as a prerequisite to law school admission. (At the time, member institutions in the newly organized Association of American Law Schools required only a high-school diploma.) Lewis' suggestion sparked angry responses, destined to recur whenever higher standards were proposed. A member from New York warned that an "artificial standard" would eliminate those "who are particularly endowed by nature" for careers as lawyers. "What would this country have lost," he asked, "if Abraham Lincoln had been kept away from the bar, and he certainly would have been kept away if any artificial standard had been enforced in his case."[57] The answer was obvious: the bar would lose the talents of men from "the humblest surroundings of life . . . without means, hereditary or otherwise. . . ." As an opponent of the Lewis proposal asked at another meeting: "Shall we stop those men as they climb single-handed by the force of native will?"[58]

At stake were memories of individual success and professional progress that depended upon human qualities rather than formal training. "What our profession needs," declared an Iowa lawyer, "is moral stamina, sterling integrity and recognized noble manhood." Higher educational standards, he warned, "will shut out from us the sons of farmers and mechanics, occupying that position in society from which come the moral principles and sentiments which preserve our profession." Here was the voice of an older, displaced small-town and rural culture, restating the shibboleth of equal opportunity (especially for those who worked with their hands) and praising natural man at the expense of his artificially trained offspring. The country lawyer, overshadowed professionally by his urban brethren, interpreted the campaign for higher standards as still another wicked attempt to subvert the profession in which he once had held an honored place. He was, after all, "a responsible man of high position in his community," but the urban practitioner "is too often an ambulance chaser, a pettifogger, a contriver of evil and a menace to public decency and order."[59]

Although the country lawyer fired the first shot against higher standards, spokesmen for night law schools eagerly joined in the fray. In 1905 approximately one law student in three attended a night school. Two-thirds of these schools had been established within the previous decade to provide inexpensive education to urban young men who held full-time jobs (and, not incidentally in many cases, to provide income for their founders). Fees at night schools were considerably lower than those charged by the prestigious private university schools. A substantial proportion of their students did not intend to practice; they studied law for a career in business. By the standards of university schools, the education was deficient: part-time instructors, the older lecture method, inadequate library facilities, and students who were often too weary to concentrate on their studies. The difference, appropriately, was as between night and day.

Yet the night schools offered some compelling arguments in their struggle for acceptance. First, by making legal education accessible and inexpensive, they kept the profession from becoming either an aristocratic or a plutocratic enclave. Traditional virtues, it seemed, still had their place—even in modern America. Night students, declared a professor at John Marshall Law School in Chicago, compensated for their deficiencies with "pluck, energy, perseverance and enthusiasm." According to the dean of another night school, the students were "men of heroic mould." Victorious over adversity, they comprised "a goodly race, full of enthusiasm, industry, perseverance, and that fine courage which scoffs at difficulty and welcomes the fray, confident in the ability to win success and reach the goal of achievement." Second, night schools kept the fires burning under the American melting pot. The ethnic heterogeneity of their students demonstrated that American institutions "are capable of making all men of one blood, if not originally created so, and that . . . America is still another name for opportunity." Finally, the night school served as an important control mechanism, which socialized men of diverse backgrounds into the ways of American legal and political institutions. Each graduate might become "a factor for law and order in his immediate neighborhood. His example and influence help to mold the different elements of a cosmopolitan city into one composite mass of law-abiding citizens."[60]

For dissimilar reasons teachers at university law schools and established practitioners aligned themselves in defense of higher standards and against the night schools. A better educated bar was, of course, eminently desirable. Self-conscious professionalism also spurred the teachers, whose own self-esteem was bolstered by more exacting standards for admission to their institutions. Furthermore, by imposing higher standards as a qualification for membership in the Association of American Law Schools, they could consign the night schools to the nether world of academic unrespectability. Successful practitioners, fearful of thrusts from below within the profession, wielded higher standards as a weapon in defense of the elitism that enhanced their own stature; they used them in their war on commercialization and unethical behavior and also, with the assistance of some teachers, to beat back the flow of newcomers to the profession from ethnic minority groups. Since neither the American Bar Association nor the Association of American Law Schools fairly represented all practicing lawyers or all law schools, each association gained strength from cooperation with the other on this issue.

Defenders of higher standards presented the superficially convincing argument that men of intellect and energy would not be excluded by the requirement of a high-school diploma, or even a college degree. They properly accused their opponents of an anachronistic reading of nineteenth-century experiences, when titans of bench and bar made their way unencumbered by formal education. Aspiring Lincolns, they insisted, would gain entry whether the obstacle was Blackstone or a college curriculum. Only "the idle, the lazy, and the unprepared" would fall by the wayside. Law teachers, in their enchantment with scientific expertise, dismissed "the absurd belief" that high educational standards were undemocratic or un-American. Inadequate training would produce nothing but "incompetence and slovenliness."[61]

Night schools were a tempting scapegoat. A faculty member at the University of Pennsylvania Law School dismissed evening education as "a grotesque perversion" of the notion that any man might enter his chosen profession. Although lawyers spoke the language of professionalism, their vocabulary often masked hostility toward those who threatened the hegemony of Anglo-Saxon Protestant culture. Professionalism and xenophobia were mutually reinforcing. In a rehearsal for postwar efforts, teachers and practitioners began to play variations on the themes of anti-urbanism, anti-Semitism, and nativism. In 1911 a New Yorker described the "lamentable" conditions at the metropolitan bar, where a "competitive struggle for existence" lowered professional ethics. Because New York had become "the dumping ground of the world," it needed more stringent standards than did "more favored states" which avoided such a high proportion of immigrants. A senior partner in the prestigious New York firm of Evarts Choate & Sherman was deeply incensed that so many Polish and Russian immigrants—whose first names were Abraham, David, Hyman, Israel, Isidore, Isaac, and Morris—had legally changed their last names to "Sherman." He predicted that unless the legislature restricted the freedom to change names, there would be "a great deal of trouble and inconvenience in the future, to say nothing of other disagreeable features." Night law schools, as the Dean of Wisconsin Law School reminded his professional colleagues, enrolled "a very large proportion of foreign names. Emigrants . . . covet the title [of attorney] as a badge of distinction. The result is a host of shrewd young men, imperfectly educated . . . all deeply impressed with the philosophy of getting on, but viewing the Code of Ethics with uncomprehending eyes." A newcomer to the Columbia faculty expressed his relief at finding more bright students, and fewer Jews, than he had anticipated. And a Pennsylvania attorney observed: "We have in the Eastern cities representatives of the most ancient race of which we have knowledge coming up to be admitted to the practice of law." Although intellectually gifted and persevering, they came to the bar "without the incalculable advantage of having been brought up in the American family life," and, therefore, they "can hardly be taught the ethics of the profession as adequately as we desire."[62]

Not only might higher educational standards immunize the bar against immigrants and Jews; such standards would also provide antibodies against the virus of radicalism. In a plea for higher standards Dean George W. Kirchwey of Columbia Law School advocated a "new and enlarged conception of legal education" as the prerequisite for a socially responsible profession. Kirchwey meant that in an era so beset by social and industrial problems law schools must offer education that would "conserve the law of the land, [and] establish its principles on unshakeable foundations." William Howard Taft pleaded for higher standards "for the good of society." More stringent requirements would deter "radical and impractical changes in law and government." Only lawyers highly trained, especially in sociology and history, could comprehend "the present thinking of the people who are being led in foolish paths," and, presumably, offer them less foolish alternatives.[63]

It was impossible for teachers and practitioners to implement their xenophobic fears or their educational hopes in the prewar years. Their mutual suspicions militated against concerted action for higher educational standards. Corporate lawyers still were too insecure about their place in American society to launch an aggressive campaign for a restricted bar. More important, the social climate of the country had not yet made nativism, anti-Semitism, and anti-radicalism entirely respectable. World War I and postwar hysteria swept away the remaining impediments to professional stratification.

Four

Cleansing the Bar

World War I dissolved tolerance, legitimized the suppression of dissent, aroused ethnic tension and hostility, and stimulated feelings of super-Americanism. The Bolshevik Revolution strengthened and focused these impulses, which erupted in the Red Scare of 1919 and lingered for years thereafter. Lawyers were not uniquely susceptible to anti-radicalism and xenophobia. Once they succumbed, however, their role as custodians of legalism and constitutionalism enabled them to manipulate the symbols of patriotism and law for professional and political advantage. From the beginning of the war until the end of the postwar decade, professional leaders waged a crusade against radicals, aliens, foreign-born citizens, and native-born members of ethnic minority groups. Shielded by their wartime patriotism and armed with the weapons of higher educational standards and stringent bar admission requirements, they attempted to purify their profession and their country of alien influences.

Although one prominent attorney complained that he "belonged to a class which had no function in war," and another lamented that "law, lawyers and law practice seem so banal and trivial in these vital and terrible days," some lawyers perceived the war as a professional godsend. After the buffeting of the prewar years, when the profession had been rent by internal discord and criticized as an obstacle to change, the national unity demanded by the war effort provided lawyers with an opportunity to regain lost status and cohesiveness. Lawyers sensed that they might return to grace by becoming advocates for the state. Since "just now the business of America is war," the lawyer's obligation was clear: to explain and defend American policy. The dialogue over American intervention—especially the rhetoric of saving the world for democracy—provided lawyers with their opportunity. Leaders of the bar described the war effort as a crusade to defend the sanctity of liberty under law. The war, declared Solicitor General John W. Davis, "is primarily a battle for law." According to Elihu Root, the protagonists were "divine right" and "individual liberty." Lawyers, observed Charles Evans Hughes, were especially interested in the outcome because they were "fighting the battle of the law, the battle for the rule of reason against the rule of force."[1]

Yet the battle for reason, liberty, and law was curiously waged by the legal profession. Lawyers responded to the declaration of war more as patriots than as professionals. Commendably, the American Bar Association quickly established a special committee for war service which functioned as a clearing house for government departments in need of legal talent, placing more than four thousand attorneys in government positions. In one bar association after another lawyers organized preparedness committees to care for the practices of attorneys in military service, to provide legal assistance to dependants of military personnel, and, not incidentally, to defend American war policy. But these activities had their darker side. A Boston committee of lawyers reported that it had distributed thousands of copies of the proclamation of war, taking pains to place them in neighborhoods of the foreign-born. In a declaration of loyalty to the government, the American Bar Association (at Elihu Root's urging) unanimously adopted a resolution condemning "all attempts . . . to hinder and embarrass the government. Under whatever cover of pacifism or technicality such attempts are made, we deem them to be in spirit pro-German and in effect giving aid and comfort to the enemy."[2] Such an expansive notion of treason was endemic throughout the profession.

An Oklahoma attorney announced that the lawyer's role as government advocate required him "to condemn every man or woman who, by word or deed, assails the purity and integrity of of country's motives," because the opponent—even verbal—of the government was "a traitor." A North Dakota judge confessed at the annual dinner of the American Bar Association that when he read about war critics he felt "that there are men in this country that I would like to hang." And Attorney General Thomas Gregory, citing an Illinois Bar Association resolution that condemned a lawyer's defense of a draft evader as unpatriotic and unprofessional, told the ABA: ". . . speaking not as a lawyer, but as an American citizen, I wish to express my admiration of the action taken."[3]

Lawyers who criticized the war effort or defended war critics incurred punitive professional sanctions. A radical Pennsylvania attorney was disbarred for associating with groups whose views conflicted with those of the American government and for organizing opposition to the Selective Service Act. In Washington a lawyer was disbarred for speaking publicly under the auspices of the Industrial Workers of the World. Another lawyer, who disregarded Selective Service guidelines and charged fees for counseling registrants, was disbarred and instructed by the court that it was "the duty of the lawyer, above all others, to serve his country. . . ." In Idaho, disbarment was ordered and a jail term imposed upon a lawyer who advised eligible young men not to register for the draft. And in Texas, where a German-born lawyer was disbarred for saying "Germany is going to win this war and . . . I hope she will," the state supreme court, reversing on the ground that his statement was outside the scope of his professional activities, praised the residents of San Antonio for resisting vigilantism and noted approvingly that the recently enacted Sedition Act had remedied the prior defect in the law which had permitted such "dangerous expressions."[4]

Lawyers permitted the dictates of patriotism to determine the contours of professionalism. To define an act as unpatriotic was *ipso facto* to render it unprofessional. One attorney, for example, distinguished between a court-appointed lawyer and the lawyer who voluntarily defended someone accused of aiding the enemy. The latter, he maintained, "has forgotten his oath and should surrender his license." Henry Taft, brother of the former President, secured unanimous approval from the New York Bar Association for a resolution that denounced opposition to the war effort "springing from political motives, indifferent patriotism, or downright disloyalty." The nation, read the Taft resolution, "has no time to listen to the subtle distinctions of men whose speech gives aid and comfort to the enemy." When a Chicago attorney, acting on behalf of his client, challenged the constitutionality of the draft law, the *Central Law Journal* accused him of subversion, contending that although an individual enjoyed the right to his own opinions, he could not delegate that right to his attorney—to whom, presumably, the client's views were imputed. When Samuel Untermeyer received a court assignment as defense attorney in a treason case, he eloquently expressed the lawyer's professional obligations, stated his personal duty to accept, and cited the need to satisfy Americans of the impartiality of law—before enumerating "personal considerations" that he hoped would disqualify him. Among these were his long support of Woodrow Wilson's policies, which the defendant had maligned with the allegation that the President's declaration of war was a crime greater than the sinking of the Lusitania. Untermeyer, who regarded the war as "the most sacred and unselfish conflict ever waged," insisted that if he were assigned as counsel he must retain the freedom to express his own opinions in court, lest he be assumed to acquiesce in "obnoxious" views. Members of the profession seemed unaware that their patriotism intruded upon their professional obligations. Lawyers prejudged guilt, equated dissent with treason, and accepted the notion that by taking a case they committed themselves to the defense of a cause. Professional stature took precedence over professional responsibility. As the president of the California Bar Association declared in 1917: "The air is now pregnant with loyalty and devotion to, and sacrifice for country. Professionally we must absorb this atmosphere, or we will slip backward as a profession, and others will take our places."[5]

The Armistice prompted lawyers to proclaim proudly their wartime achievements. "Who dominated the war—and controlled it—made possible the winning of it?" asked Lucien Hugh Alexander, chairman of the ABA membership committee. "LAWYERS! And what a proud record it is for our profession in America!" Alexander used this record to launch a postwar membership drive. It was, he said, the "supreme duty" of every lawyer with pride in his country and in his profession to join the American Bar Association. Lawyers hoped that their wartime service would erase public criticism of the bar.[6] By equating military intervention with the defense of liberty and law, and by associating themselves so closely with the war effort, elite lawyers wriggled out of their embarrassing prewar image as defenders of privilege and opponents of reform and wrapped themselves in the comforting folds of the American flag.

Their exuberant nationalism, reinforced by xenophobia and anti-radicalism, remained potent after the Armistice. But the target changed: the Hun yielded to the radical; the enemy abroad became the enemy within. The Bolshevik Revolution, labor-management conflict (including the menacing strike by Boston policemen), race riots, and a series of bombings contributed to the Red Scare of 1919. The nation's chief law enforcement officer, Attorney General A. Mitchell Palmer, led the anti-radical onslaught, while the professional elite arrayed itself in defense of Anglo-Saxon patriotism. Within professional circles, the primary instrument of purification was higher educational standards for admission to the bar. After two decades of dislocation and intense criticism, elite lawyers, strengthened by their patriotic war record and encouraged by postwar nativism, moved in concert to set their professional house in order.

For years the tradition of virtually free access to the bar had troubled lawyers who watched uneasily while immigration and urbanization transformed the nation and their profession. Cries of overcrowding swelled the chorus of complaints about commercialization and declining ethics. The culprits were the "unlearned, unlettered and utterly untrained young lawyers with no *esprit de corps* and little regard for the traditions of the profession." In translation, this meant immigrants and Jews. A veritable flood of lawyers with foreign names, concentrated in cities, who often entered the professional portals through night law schools or even correspondence courses, threatened the image of the legal profession as an aristocratic enclave. Even before the war Theron Strong complained sourly about "the influx of foreigners." Strong was especially troubled by the rising proportion of Jewish lawyers, which was "extraordinary, and almost overwhelming—so much so as to make it appear that their numbers were likely to predominate, while the introduction of their characteristics and methods has made a deep impression upon the bar. . . ." In an apocalyptic memorandum another lawyer warned of "the great flood of foreign blood . . . sweeping into the bar." Eastern European immigrants, "with little inherited sense of fairness, justice and honor as we understand them," were committed only to their own "selfish advancement." How, the author inquired, "are we to preserve our Anglo-Saxon law of the land under such conditions?" Other observers were less explicit, but no less distressed by the social origins of the newcomers. Harlan Stone referred to "the influx to the bar of greater numbers of the unfit," who "exhibit racial tendencies toward study by memorization" and display "a mind almost Oriental in its fidelity to the minutiae of the subject without regard to any controlling rule or reason."[7] An ABA committee proposed closing "the easy doors through which many an ill-prepared, unfit and improperly educated man has passed into the profession." The New York *Tribune* cited "too indiscriminate an influx" into the profession, while the president of the Carnegie Foundation decried low admission standards which opened careers in law to "the poorly-educated, the ill-prepared, and the morally weak candidates." When the American Bar Association approved a rule excluding aliens from the bar, one member bluntly stated: "It is a matter of patriotism, and a national and political question."[8]

With the Armistice and the Red Scare came the opportunity for proscription.

The changing social structure of the bar posed a painful dilemma for the profession. The Tocqueville and Lincoln models were compatible, and indeed mutually reinforcing, only as long as the profession recruited primarily from a homogeneous Anglo-Saxon Protestant society; notions of the profession as an accessible democratizing institution which fostered social mobility became suspect once the origins of its newest members changed. Tenement dwellers from Manhattan's Lower East Side must not usurp the places once reserved for Kentuckians who studied Blackstone by candlelight. The "privileged body" described by Tocqueville, which occupied a "separate station" in society, hardly could survive inundation by the masses. The legal profession would no longer be "the most powerful, if not the only counterpoise to the democratic element" once the demos poured into its ranks.[9] The threat was, in a word, democratization.

Once democratization came to mean the admission of foreigners and Jews, higher educational standards surfaced as a remedy that could reassure the public about professional responsibility. Higher standards might also deflect attacks against the judiciary by offering an alternative target—an urban bar "filled with untrained or ill-trained men"—for public discontent with the legal profession.[10] Educational reform was an effective vehicle for the exclusion of ethnic minority-group members whose access to the profession was eased by minimal educational requirements. University law-school teachers were understandably troubled when their own students failed bar examinations which placed a premium upon memorization of black-letter law, while night-school graduates, products of institutions indifferent to inductive reasoning, easily passed. Such examinations sacrificed intelligence to accessibility. To redress this imbalance members of the professional elite struggled to secure more stringent educational requirements and bar admission standards. It was not coincidental that their effort climaxed during the period that encompassed the Red Scare, virulent nativism, and the imposition of draconian restrictions on immigration.[11]

The problem for law teachers was to win the support of practitioners, whose influence would be decisive in bar associations which drafted recommendations, and with state legislatures which enacted them. A postwar alliance was formed between these uneasy allies; within three years they won major victories. Teachers, especially at university law schools, appreciated the need for rapport with the once-scorned American Bar Association. As one dean reminded his colleagues: "We know and freely admit that we are the leaders of legal education in this country, and that all of the advances in legal education are due to us." Yet to secure higher standards, "we have got to connect up, and connect up closely, with the American Bar Association." For these men, perhaps the strongest attraction of an alliance was the boost it would give their own efforts to beat back the night law schools, whose enrollments continued to climb. By raising educational standards for admission to the bar, and by diluting the value of a degree earned in a night school, the university schools might win their competition for student enrollments and their struggle to control the formulation of standards.[12]

The ABA's willingness to cooperate was fed by different motives. For nearly a decade lawyers had looked enviously at the medical profession, which succeeded in eliminating schools and restricting professional access at the very moment when the number of law schools and new lawyers was increasing. Reform in medical education had been stimulated by publication in 1910 of the Flexner Report for the Carnegie Foundation. The report endorsed medical education as a university, not a proprietary school, function. In pursuit of educational excellence it proposed severe restrictions upon the number of schools and graduating students. But the most conspicuous victims of the rapid implementation of Flexner's recommendations were blacks (by 1923 six of the eight black medical schools in the country were closed) and the poor, who confronted rising educational costs and the disappearance of night schools. In 1913 the American Bar Association, doubtlessly wishing for lightning to strike twice, requested a similar study of legal education.[13] Alfred Z. Reed's *Training for the Public Profession of the Law* appeared late in 1920, a most opportune time for restrictionist educational reform in the legal profession. The Reed report hit the profession with an explosive impact which spread shock waves through the American Bar Association and the Association of American Law Schools. When the storm had subsided, the associations stood arm-in-arm against their common enemy: night law schools and the immigrants who crowded into them.

Reed, a Carnegie Foundation staff member, described the practice of law in the United States as a public profession. Lawyers were "part of the governing mechanism of the state. Their functions are in a broad sense political." Yet these political functions were crippled by the existence of a "differentiated profession," a profession "so disunited within itself as seriously to impair its capacity to formulate—let alone to realize—professional ideals."[14] Legal education in the United States should reflect this public function; instead it perpetuated social stratification. Pulled in one direction by educational considerations that pointed toward rigorous training, it was pushed in another by political factors (flowing from the public nature of the profession) requiring that the administration of law not become a class privilege. In the early nineteenth century the political had been much the stronger force, but in the twentieth class stratification had come to predominate in professional life. Law schools, like practitioners, were split into rival camps. Full-time schools sought to "make American law what American law ought to be." They served the community by producing well-educated lawyers. But night schools also served, Reed insisted, by recognizing "that the interests not only of the individual but of the community demand that participating in the making and administration of the law shall be kept accessible to Lincoln's plain people." They kept the privilege of practicing law from becoming a class monopoly. According to Reed, no law school could satisfy both objectives. Functional divisions in legal education produced "the genuine differentiation of lawyers which, in defiance of legal theory, now exists, and makes of American legal practitioners, not a united bar but a heterogeneous body."[15]

Reed was sympathetic to the objectives of night law schools but critical of their achievements. They were little more than "cheapened copies" of university schools and did "more harm than good to legal education." Yet it was neither possible nor desirable to abolish them. Many aspiring lawyers could not attend full-time schools, not because they were intellectually inferior to those who did, but because "they spring from an economically less favored class in the community." Here was the nub of the problem: a *public* profession must guard against class exclusiveness but the structure of legal education perpetuated it; at the same time the theory of a unitary bar ignored the fact that law schools produced "radically different types of practitioners." Reed saw two avenues of escape: legal education might be restructured to produce a unitary bar; or the theory of unity might yield to the reality of diversity. Reed chose the latter: a unitary bar, he concluded, "not only cannot be made to work satisfactorily, but cannot even be made to exist. . . ." Consequently he proposed that the bar formally divide along functional lines according to services rendered. Night schools would train lawyers for probate work, criminal law, and trial practice; day schools would train lawyers for more prestigious and lucrative work in business practice. Reed conceded that lawyers might then be explicitly differentiated "by the economic status of the client rather than by the nature of the professional service rendered." This left him untroubled, since "the general principle of a differentiated profession is something that we already have, and could not abolish if we would."[16]

Although Reed proposed nothing more startling than formalization of the status quo, his book deeply unsettled academics and practitioners alike. Chicago attorney Albert M. Kales, reviewing it for the *Harvard Law Review*, minced no words: "It is superficial. It is false. It is impolitic." Arthur L. Corbin of Yale, in his presidential address to the Association of American Law Schools, declared that Reed's proposals were "so out of harmony with the history of the bar and of legal education in this country that they cannot hope to make any substantial progress." Harlan Stone wrote scathing reviews for two legal periodicals, condemning it with equal passion in both.[17] Teachers and practitioners rejected the notion that a stratified bar existed, because its implications for the provision of legal services and for the impartiality of the adversary process were too painful to confront. Yet by committing themselves so energetically to higher educational standards (whether to eliminate night schools or to deny their graduates access to the bar), the elite tacitly acknowledged the very underclass whose existence they publicly refuted.

Publication of the Reed study came at a propitious time. Just four months earlier, in August 1920, the Association of American Law Schools and the American Bar Association had held their first joint meeting in seven years. Both groups were troubled. Two-thirds of American schools remained outside the law school association, many of them with standards far below the AALS minimum. Teachers could not hope to win their fight for higher standards without assistance from the ABA. The bar association was now willing to provide the "cordial and close cooperation" requested by the teachers. A year earlier its president had reminded members of the presence in the United States of millions of immigrants from southern and eastern Europe, and had urged them to "start the fires under our melting pot and keep them burning, until every man . . . has either become thoroughly and wholly American, or . . . is driven back to the country from which he came." With social unrest linked to the presence of aliens, and with bar membership swollen by the newcomers, the ABA was ready to act. At the joint meeting in 1920 it authorized a special committee, chaired by Elihu Root, to make recommendations within a year to "strengthen the character and improve the efficiency of persons to be admitted to the practice of the law."[18]

The Root committee tried to utilize Reed's facts to undercut Reed's recommendations. Committee members had received privileged access to the Carnegie study; Reed himself had appeared before them to answer questions. The committee implicitly accepted the notion of professional differentiation—it referred to "a class of incompetent practitioners"—but it urged abolition rather than formalization. The bar must "purify the stream at its source" by raising educational requirements. Assuming that "no man who wants a college education need go without," the committee proposed that every candidate for admission to the bar must graduate from a law school that required two years of college for admission and three years of full-time law study. Furthermore, graduation from law school should confer no right of admission to the bar; a fitness examination by public authorities was endorsed.[19]

The committee's report sparked an acrimonious debate in 1921 in the Section of Legal Education, with ABA leaders aligned against spokesmen for the night schools. For William Howard Taft, who seconded the Root proposals, the question involved nothing less than "saving society from the incompetent, the uneducated, and the careless, ignorant members of the bar. . . ." But the night-school representatives rejected that formulation, labeling the Root report "dangerous, uncalled for, unnecessary and un-American." Edward T. Lee, dean of the John Marshall Law School in Chicago and spokesman for the night schools, attacked the AALS which, he charged, was attempting to eradicate non-member schools "and to put an anathema upon them." Adoption of the Root recommendations would place control over legal education "in the hands of the deans of a few large day law schools." Their adoption would also close the legal profession "to all save the leisure class of youth . . . and would bar hundreds of naturally well-endowed, zealous, and industrious youths from attaining an honorable ambition." Furthermore, city-dwellers from ethnic minority groups would be denied legal assistance "for want of lawyers familiar with their language and distinctive customs." Lee, noting the patriotic war record of the foreign-born and their sons, tried to cut the link between ethnic origins and radicalism. Indeed, he warned, if minority group members were prevented from attending law school or from entering the profession they would cease to be "a safe and conservative influence in our midst." Lee understood that ethnicity was the real problem underlying the Root proposals. Yet, he predicted, this problem soon would be resolved under the new stringent immigration laws. Those immigrants already in the country would be assimilated, thereby rendering superfluous the barrier of higher education standards.[20]

Defenders of the report met their opponents on the common ground of nativism. A West Virginian on the Root committee defended the college requirement as an instrument of Americanization. Men from diverse backgrounds and regions should gather together in colleges, "where proper principles are inculcated, and where the spirit of the American government is taught." The "influx of foreigners" in the cities comprised "an uneducated mass of men who have no conception of our constitutional government." Surely, he concluded, the American Bar Association did not wish to lower standards "simply to let in uneducated foreigners." Root, defending his committee's report on the floor of the Association, was more oblique. The honor and dignity of the bar were at stake, he warned. The bar could not secure the administration of justice "in this disturbed country" nor "meet the new conditions that confront us" until it raised admission standards. The Root forces were successful, defeating an attempt by the night-school group to postpone action and securing Association approval for the report.[21]

The ABA meeting was but a rehearsal for the full-dress performance two months later, at a conference of bar association delegates in Washington attended by representatives of more than one hundred and fifty state and local associations and numerous law schools. The Washington Conference on Legal Education was called to rally support for the ABA proposals from those professional groups with maximum influence on state legislatures, which bore ultimate responsibility for bar admission requirements. Root urged association presidents to select as delegates men of "standing and influence," preferably practitioners. The impetus for the conference, he assured them, came not from the law schools but from "a recognition by the members of the American Bar Association of the bar's obligation to protect the public and the profession from incompetent and unethical lawyers." For Root the two-year college requirement assured a process of cultural osmosis, in which aspirants "will be taking in through the pores of [their] skin American life and American thought and feeling." This was necessary because so many newcomers to the bar, especially in the cities, were eastern European immigrants with "all those predilections and fundamental ideas which differentiate the Continental systems of jurisprudence from the Anglo-American system." As Root declared: "I do not want anybody to come to the bar which I honor and revere . . . who has not any conception of the moral qualities that underlie our free American institutions; and they are coming, today, by the tens of thousands."[22]

Root's allies came primarily from the professional elite in metropolitan centers and in the American Bar Association. Silas Strawn of Illinois, a future president of the ABA, stressed the wisdom of college training if lawyers were "to meet the requirements of the modern captain of industry" as clearly they must. To a New York delegate, it was "absolutely necessary" that the bar comprise men "able to read, write and talk the English language—not Bohemian, not Gaelic, not Yiddish, but English." Charles Boston, representing the ABA, expressed the "common knowledge that a poor man can get as good an education as a rich man. . . . It is but a question of whether the applicant is willing to undergo the requisite discipline." George W. Wickersham, attorney general in the Taft administration and senior partner in a prestigious New York firm, conjectured that an overwhelming proportion of urban immigrants who sought access to the profession were motivated solely by the desire for social mobility. Wickersham found it "appalling" that such men, lacking the "full realization of the meaning of our law historically," should become judges and legislators. "To think that those men, with their imperfect conception of our institutions, should have an influence upon the development of our constitution, and upon the growth of American institutions, is something that I shudder when I think of."[23]

The converging shadows of Reginald Heber Smith and Alfred Z. Reed fell across the Washington conference, but the implications of their findings were quickly dismissed. Just as Smith's evidence of the failure of legal aid was submerged by Americanism, so the professional elite dismissed Reed for disguising incompetence as democratic dogma. The overriding urge was to make the bar safe for Americans, not to provide equal justice or equal access. The inferior quality of the bar was blamed upon easy access; the denial of justice to the poor, in turn, was blamed upon an inferior bar. To elitists, in practice or in the professoriat, the remedy was obvious: the quality of the bar and the quality of legal services would improve only if professional access was restricted. Here was the panacea for every professional problem in the era of mass immigration. Legal institutions were blameless; alien adulteration had subverted them.[24]

Opponents of the ABA recommendations—especially the college requirement—were a more motley group. At one extreme, delegates from primarily rural associations in Tennessee, Maryland, and Indiana (a state where "good moral character" was the sole test of fitness) resisted their elitist and "un-American" bias. They would, it was argued, discriminate against "the self-made man . . . who has made this country what it is." At the other extreme, however, a representative of the Pennsylvania Bar Association and the dean of Fordham Law School opposed the proposals because "radicalism and socialism were very widespread" in American colleges and universities. More education, therefore, meant more virulent radicalism.[25]

Only the president of Yale, who labeled "fatuous" the assumption that a college education was accessible to all regardless of economic circumstances, scratched the surface of prevailing assumptions that will and desire were sufficient to unlock university doors. The full resolution before the conference warned that the legal profession must not become a class monopoly, but concluded that the college requirement would not contribute to such a development because young men with "energy and perseverance" could meet the ABA standards. In fact, however, ethnicity, class, and status determined educational opportunity far more than willpower. Barely 5 percent of American children born between 1895 and 1904 were finishing high school and college. Between 1915 and 1925, those high-schoolers whose fathers had less than eight years of schooling had 8 chances in 100 of entering college; those with fathers who had entered college had 47 chances in 100.[26] Unless one assumes, first, that 95 percent of American youngsters lacked "energy and perseverance," or, second, that the correlation suggested by these figures for college attendance was accidental, the only tenable explanation is that factors other than pluck accounted for a student's presence in or absence from college. Conference members were assured that at Harvard funds were sufficient to enable "the poor deserving boy" to spend at least two years there. But if the poor deserving boy was Jewish his chances after World War I were considerably diminished.[27] If he was Italian, Polish, or Greek—or a black American—his chances were virtually non-existent.

When opponents of the ABA recommendations proposed amendments that would eliminate the college requirement and impose merely good character and literacy as the standard, Root regained the floor, announced his imminent departure from the conference, and delivered a dramatic—and successful—plea for adoption of the ABA resolutions. Root's oratory was consistent with the outlook of a man who had felt most comfortable in the McKinley era, and who endorsed immigration restriction and the popular racist theories expounded in Madison Grant's *The Passing of the Great Race* in his attempt to return to the bygone age of Anglo-Saxon Protestant hegemony. But there was more to the struggle for higher standards than the bias of Elihu Root. Other established practitioners, who shared his eagerness to turn the clock back, also grasped higher standards as the best available means to that end. Significantly, five of the seven members of the Root committee came from New York, Philadelphia, and Chicago —the cities with the highest proportion of lawyers from foreign-born families. The other two were from West Virginia and Nevada, states with stable bars untroubled by urban or ethnic stress. Those least accustomed to change, and those most troubled by their exposure to it, joined hands in an attempt to halt it.[28]

Measured solely by state statutes, the fight for higher standards was mostly sound and fury. In 1922 fifteen states had no general education requirements for admission to the bar, while twenty-three states required nothing more than graduation from high school. Four years later, not a single state had subscribed to the ABA recommendations for two years of college and a law degree.[29] Once again, as during the Brandeis confirmation fight, the limits of professional power were not infinitely expansive. State legislatures represented broader constituencies than did the American Bar Association. Yet the fight for higher standards had tangible significance for the organized bar and for law teachers. For lawyers organized in bar associations, educational reform elevated their own professional respectability and status. Elite lawyers did not succeed in excluding undesirables—state legislatures were an insurmountable obstacle—but they did maintain professional stratification to correspond to their own ethnic and social preferences.

If practitioners used higher standards to preserve an Anglo-Saxon professional elite, university law teachers committed themselves to higher standards to create an educational elite. This represented the culmination of their efforts, dating from the founding of the Association of American Law Schools, to achieve professional recognition and power. For two decades law schools were tormented by the persistent dilemma of universities. "On the one hand, they have served to channel the children of the elite into elite positions, and thus to harden the social structure. . . . On the other hand, they have provided avenues, broad or narrow, by which talented children of relatively humble origins . . . may enter the ranks of the elite. . . ."[30] It is important to acknowledge this dualism, lest either the university's innovative or its stabilizing role be improperly exaggerated. Yet it is imperative to recognize that during the critical postwar years the legal profession emphatically committed itself to restrictive stabilization—to shoring up the Anglo-Saxon foundations of the professional structure to resist the incursions of immigrant newcomers. With some college education required for admission to a university law school, and with a law school degree required for admission to the bar, the desired social barriers would be set in place. The two-year college requirement, in conjunction with the exclusion of night law schools from the AALS, meant that university schools would control both the formulation of educational standards and access to the professional elite. By 1923 Roscoe Pound could safely conclude that night schools were "on the run." For a generation, Pound reminded Harlan Stone, Harvard and Columbia had fought "a difficult but, on the whole, effective battle against this sort of institution. . . . We [have] only to stand by our guns a little longer to see everything we have been struggling for . . . realized."[31]

Filial reconciliation between teachers and practitioners was not too high a price for such an achievement. The AALS, obedient offspring from 1900 until 1914, had tested its adolescent yearning for independence during the following six years. With the confidence born of maturity, it entered into an equal postwar partnership before striking out once again on its own. As equals, parent and child received their due share: the bar associations could not resist social change without the assistance of teachers who, in turn, could not police their own domain without help from the associations. Higher educational standards was the bridge between them. But teachers and practitioners crossed it moving in opposite directions. Pausing temporarily as they met, they created an illusion that they might travel together when in fact, before long, they would be further apart than ever before.

Stimulated by the efforts of the American Bar Association, various state bars wrestled during the 1920's with the political and professional implications of the changing social structure of the bar. Two statistical trends lent a sense of urgency to their debates. The first was a sharp spurt, exceeding 80 percent, in law-school enrollments between 1919 and 1927. The second, confined to a few Eastern cities but known and feared elsewhere, was a substantial proportional increase in the number of foreign-born lawyers. In New York, where the number of lawyers increased by 57 percent between 1920 and 1930, the number of foreign-born lawyers increased by 76 percent. In Philadelphia, during the same decade, the bar grew by 21 percent while the number of foreign-born lawyers increased by 72 percent. "Overcrowding," the euphemism for these trends, vividly expressed the cultural claustrophobia of those who used it.[32] In the search for a cure state bar associations sampled various purgatives. Higher educational standards was one. Another, in Pennsylvania, was a rigorous registration and preceptorship plan. A third, bar integration or incorporation, promised strengthened professional disciplinary power by compelling every lawyer to belong to his state bar association. The debates on each of these restrictive devices exposed a deeply fractured profession. In some states the primary division was between rural and urban lawyers; in others, between an urban elite and the mass of unorganized practitioners; in still others, between attorneys of Anglo-Saxon and Eastern European backgrounds. Everywhere, however, the protagonists confronted the identical structural issue: stratification versus democratization.

Bar integration—compulsory bar association membership—had originated during the prewar years as a progressive reform. Its leading advocate, Herbert Harley, was a lawyer turned newspaper editor who envisioned compulsory bar membership as a restrictive device to limit professional competition (thereby mitigating unethical behavior), and as a means to increase the power and influence of the bar as a lobby for judicial and legal reform. In 1913 Harley organized the American Judicature Society, which was his public voice on behalf of bar integration.[33] Each of the earliest states to integrate—North Dakota (1921), Alabama (1923), Idaho (1923), and New Mexico (1925)—had a small homogeneous bar unaffected by severe rural-urban conflict. But in New York, where integration failed, and in other urbanized, industrialized states with a heterogeneous bar, the integration movement became a pawn in the struggle for power within the legal profession.

In New York, higher admission standards and bar integration were debated simultaneously, with nativism the dominant ingredient in both. George Wickersham led the fight for higher standards, warning that "a pestiferous horde" would be set loose upon the profession if standards were not raised. The horde included men whose knowledge of the English language "is of the most imperfect character" and aspiring lawyers not from Anglo-Saxon stock who sought admission to the bar "without having the faintest comprehension of the nature of our institutions, or their history and development." Wickersham's fears were shared by Austen Fox, Brandeis' antagonist during the confirmation hearings, whose experience on the bar's character committee elicited from him expressions of dismay over "the ignorance and low general calibre" of applicants without college training, a group of young men who "as a class, acquire very rapidly but do not assimilate—quick to learn and quick to forget." There were, in addition, "the many immigrant boys . . . [who] can hardly speak English intelligibly and show little understanding of or feeling for American institutions and government. . . ."[34]

Identical arguments were offered in the debate over bar integration. An address by Chicago judge Clarence N. Goodwin in 1922 to the New York State Bar Association, in which the problem of unassimilated foreigners figured prominently, led to the appointment of a special committee to consider integration. The following year a model bill, drafted by Goodwin, was circulated for discussion. Opponents of the plan, led by William D. Guthrie, made an impassioned defense of professional elitism. Guthrie, the association's waspish president, condemned compulsory incorporation as an attempt to democratize the bar at the expense of "the elite of the Bar, the best part of the Bar," who dominated the association. Incorporation might be appropriate for rural, homogeneous states, but, Guthrie predicted, it could not succeed among New York's "teeming population of recent immigrants and their progeny." Indeed their progeny bore responsibility for precisely "the difficult and grave problem and menace . . . arising from the admission to our bar in recent years of large numbers of undesirable members." With incorporation, a state association of four thousand members would increase sixfold and power would pass to "this mass." The result would be "a public calamity."[35]

Guthrie's rhetoric offended some of his own supporters, who shared his elitism but repudiated his nativism. Among them was Louis Marshall, the nationally prominent Jewish lawyer who had married into the aristocracy of German-American Jewry and developed a remunerative corporate practice while litigating important minority-rights cases before the Supreme Court. Marshall, the product of an earlier generation of Jewish immigrants by now securely established in American business and professional life, was no stranger to anti-Semitism at the New York bar. A member of the state bar association since 1885, when he practiced law in Syracuse, he had been startled to learn, after his arrival in New York City a decade later, that "few Jews were admitted to membership, that men of the highest character and ideals had been ostracized, that others who would have been very glad to have become members shrank from having their names proposed because of similar indignities . . . [and] that it was only in exceptional cases that men of my faith were appointed on committees of the organization." Although Marshall reassured Guthrie that he made no allegations of prejudice, he urged the association president not to weaken his sound position by giving needless offense.[36] Yet Marshall's towering reputation and his successful practice as a partner in the nation's most prominent Jewish law firm (Guggenheimer, Untermeyer & Marshall) gave him a strong stake in the preservation of the professional elite to which he belonged by virtue of every attribute but religion. Bar integration, Marshall warned, would shift the balance of power in New York from metropolitan to upstate lawyers.[37] Fearful of the consequences of integration for city lawyers, he cooperated with Guthrie, William Nelson Cromwell, and other pillars of the metropolitan elite to defeat the measure. Marshall would defend his co-religionist lawyers against anti-Semitism, but he was hardly prepared to assert that stratification had no place in his profession.

Although the New York bar association twice declared its

support of incorporation in principle, it could not overcome vigorous opposition from those metropolitan members who wanted to maintain their distance from new-immigrant lawyers and from upstate lawyers whose numbers would dominate an integrated state bar. The legislature considered and defeated the incorporation bill in 1925; enthusiasm for it subsided thereafter. But social stratification as a professional issue continued to plague the New York bar. Members like Henry W. Taft found it inconceivable that the bar could retain its influence, to say nothing of its special identity, if its doors remained open to uneducated, uncultured, and unethical aspirants, whose "gestures are unwholesome and overcommercialized" and whose "historical derivation is such . . . that it is impossible and always will be impossible that they should appreciate what we understand as professional spirit." Similarly, Harlan Stone concluded from his service on the bar admissions committee of the Association of the Bar of the City of New York that the profession was attracting "the undesirables and unfit" when half the metropolitan bar comprised lawyers who were born abroad or to foreign-born parents.[38]

In Illinois, Chicago was the nativist analogue to New York City. Before World War I the committee on legal education of the state bar association had recommended law-school training as a requirement for bar admission. But no connection was then drawn between education or status and ethnic origins. By the early twenties, however, social and ethnic factors predominated. In 1922 the state board of law examiners, reporting that a significantly higher proportion of successful applicants to the bar was foreign born, urged higher standards as a restrictive weapon against members of ethnic minority groups. References to the "morally unfit" began to punctuate debates. Judge Goodwin, who had spoken so effectively before the New York association, told his Illinois colleagues about ethnic enclaves in Chicago which were populated by generations of foreign-born inhabitants "who are almost as divorced from American life and American traditions as though they and their parents had never departed from their native lands."[39] The chairman of the legal education committee observed that recently naturalized citizens comprised an overwhelming majority of the bar's newest members. "It would not be so serious if . . . [they] were all from English speaking countries or countries with institutions similar to ours." But, he noted, more than 80 percent were not.

As strong as feelings were against the foreign-born, some association members were equally resentful of the fact that adoption of ABA standards would discriminate against every law school in the state except those at the University of Chicago and the University of Illinois. Why, asked the representative of a downstate institution, "should we be deprived of giving legal education because . . . in Chicago the Czecho-Slovaks, the Russians, and so forth, come into the law schools?" Indignantly, he branded higher standards as "the easy way to get rid of that element, and to do that they ask the rest of the state must be crucified." Night-school delegates, still incensed with their treatment at the Washington conference, tried to overcome their handicap as educators of minority group members by playing on the resentment of downstate lawyers toward large metropolitan firms. These firms, Dean Edward T. Lee declared, represented "Mammon introduced in the Temple of Justice. . . . Where in such an office is the directing will, the guiding conscience, the personal responsibility, the professional pride of the old time practitioner?" Graduates from prestigious law schools were eager to "enter the employ of a legal octopus"; only the night schools, he implied, still trained lawyers who were, like their nineteenth-century forebears, "free, independent and unsubsidized." Similar expressions of animosity punctuated the subsequent debate over bar integration. Chicago lawyers in the Illinois Bar Association chastized newcomers to their profession for lacking "the American viewpoint." They expressed concern lest they be compelled to enter into "a close, intimate social relationship" with lawyers whose integrity and honor they questioned. In particular, one lawyer declared, he was not prepared "to welcome to social intimacy, to social intercourse, to social association," members of the all-black Cook County Bar Association.[40]

The ethnic hostility so blatantly voiced in New York and Chicago seemed almost subdued by contrast with the Pennsylvania experience. The configuration was identical: a state with at least one large city densely populated with foreign-born inhabitants; a state bar torn between metropolitan and outland factions; and a metropolitan bar, in turn, split between a dominant Protestant elite and a vast majority of unorganized practitioners from ethnic minority groups. Conditions peculiar to Philadelphia, however, exacerbated these divisions. An especially rigid metropolitan social and professional structure, and a particularly high influx of foreign-born, especially Jewish, lawyers into the legal profession (more than three times the rate of professional growth generally) produced a unique instrument of control and a shamelessly crude defense of it.[41]

The objective in Pennsylvania, as one lawyer expressed it, was to keep the bar "clear and clean." The primary source of difficulty, explained another attorney, was with those who entered the profession "with a background of the ideas of continental Europe." As James Beck, former solicitor general of the United States, wrote: "if the old American stock can be organized, we can still avert the threatened decay of constitutionalism in this country." In 1925 a special committee was appointed to recommend appropriate changes in Pennsylvania admission requirements that would maintain the "honor and dignity" of the bar. To identify the problem, names were read from a *New York Times* list of graduates from New York Law School; by the reader's estimate 250 of 279 were men "who perhaps by their disposition, their character and their training, would be naturally disposed to move along the line of a commercial or industrial life." They were, in a word, Jewish. "What concerns us," said another member forthrightly, "is not keeping straight those who are already members of the Bar, but keeping out of the profession those whom we do not want." Merely to raise educational requirements was risky, because "if we do that we keep our own possibly out." An alternative solution (a variant of the grandfather clause once utilized in the South to prevent blacks of slave ancestry from voting) was to require that every applicant to the bar be the son, and preferably also the grandson, of native-born Americans. It was proposed but not seriously considered—perhaps because it, too, might disqualify some elite aspirants.[42]

The final solution was an elaborate, covertly discriminatory preceptorship and registration system. It required every prospective law student to secure a preceptor with five years' experience in practice who would agree to provide him with a six-month clerkship upon graduation from law school. He was also compelled to register with the State Board of Law Examiners, supporting his application with three sponsors, two of whom were members of the bar. Before admission to law school his county board of examiners would interview him to determine his fitness and qualifications for a profession he was three years from entering. At each stage in the new admissions process a prospective Jewish lawyer—the association's primary target—confronted a high wall of resistance. The preceptorship-registration system placed a premium upon social standing, family connections, and upon those personality traits defined as desirable by examiners who were unrestrained in their power to screen candidates through their own prejudices. The student's first problem was to find an experienced lawyer who would make the commitment, in advance of any tangible evidence of his legal abilities, to employ him as a clerk. Then he had to convince his county board of examiners that he was familiar with the ethics of a profession he had yet to enter—indeed, he had yet to enter law school. In fact, the examination tested neither character nor ability, but background. Since the critical decisions were made prior to a candidate's admission to law school, no record of excellence there was permitted to overcome the handicap of inferior social origins.[43]

The justifications offered on behalf of the Pennsylvania plan stressed professional ethics and integrity, good character, and the public interest. The new regulations were described as a necessary and proper response to changing conditions. Within a year of their adoption in 1927, a Pennsylvania attorney declared, "many have had difficulty in finding preceptors"—the beneficial result, he conjectured, of the refusal of local examiners to accept as preceptors "ambulance chasers" or "criminal lawyers of a certain type." Henry S. Drinker, an authority on legal ethics who chaired the Philadelphia Law Association grievance committee, described the malefactions of those who, having come "up out of the gutter . . . were merely following the methods their fathers had been using in selling shoe-strings and other merchandise. . . ." Many "Russian Jew boys," Drinker noted, had been the subjects of complaints before his committee prior to adoption of the preceptorship plan. And Robert T. McCracken, who subsequently became chairman of the ABA Committee on Professional Ethics and president of the Pennsylvania Bar Association, explained to New York lawyers how Pennsylvania had handled "the question of the social origins of the men." Before the character examination was instituted, 76 percent of the applicants were immigrants, predominantly Russian Jews. "That is a pretty high percent," McCracken observed. Within a few years, however, with the character exam as a filter, that proportion had slipped to 60 percent. Even established Jewish lawyers in Philadelphia were swept along by the xenophobic spirit of the times. Several hundred of them pledged their services as "Big Brothers" to advise "worthy young men," promising that in the interests of Jewish lawyers, the profession, and the public, they would discourage the ambitions of "unworthy" aspirants. They would, in other words, assist in the elimination of those prospective Jewish lawyers who, in their judgment, would cast discredit upon them.[44]

Although the Pennsylvania solution was extreme in the thoroughness of its controls, it represented for the organized bar a culmination rather than an innovation. At least since the drafting of the Canons of Ethics the professional elite had hitched ethics and stature to ethnic and social origins. A self-appointed elite devised criteria for admission which gerrymandered the boundaries of legitimate professional behavior to exclude, or to anathematize, members of ethnic minority groups. Throughout the postwar decade the legal profession was haunted by the relationship between the foreign-born and American legal institutions. Immigrants traditionally had served as surrogates for anxieties about urbanization, industrialization, and accelerated social change—anxieties that lawyers did not monopolize. But immigration, which directly affected legal ethics, education, admissions, and ultimately the social structure of the bar, presented urgent problems for the profession. Whether the organized bar looked at society, or at itself, its fearful gaze settled upon the foreign-born, whose presence shaped its perceptions of law and order, criminal justice, legal aid, civil liberties, and, ultimately, its own professional identity.

With total exclusion impossible except in rare instances (under the Pennsylvania plan, for example, not a single black was admitted to the bar between 1933 and 1943), stratification was an acceptable alternative. The distribution of wealth, status, clients, and power within the profession, with Wall Street practice at one level and solo practice at another, correlated closely with ethnic stratification that separated white Protestants from Russian Jews, Polish Catholics, and blacks. Professional barriers were high, but not insurmountable, for those young men who could afford to attend college and who excelled at Harvard, Yale, or Columbia Law School. But, by design, the exclusiveness of these schools increased in direct proportion to diminishing financial resources and prevailing definitions of ethnic inferiority. The legal profession did serve as a channel for upward social mobility, but in a restricted way. Between 1920 and 1930, in cities like Chicago, New York, Philadelphia, and St. Louis, the profession grew at a greatly accelerated rate, with the absorption of the foreign-born and their children largely responsible for this growth. But gross numbers are deceptive, since professional preferment patterns determined that, although every member of the bar was entitled to the name "attorney," ethnic and social origins dictated the nature of his practice, his clients, and the standards by which his conduct would be measured. Occasionally, aspirants from disfavored groups might find alternative mobility opportunities within the profession—for example, in the scattered Jewish corporate law firms. But as long as merit was defined by the professional elite to correlate with social class and ethnic origins, mobility and fluidity were tightly controlled, and the availability of legal resources and the possibility of equal justice were sharply curtailed.

Professional stratification defined social policy. Any symbol of social change—the immigrant, the city, the Jewish reform lawyer nominated to the Supreme Court—was opposed as a threat to those interests and institutions whose defense the organized bar deemed essential. The implications, however predictable, were troublesome. How would a profession so uniquely situated vis-à-vis American public life, yet structured to resist change, cope with the wrenching transformations endemic to modern America? How would a stratified profession—whose public voice was the voice of its restrictive associations and whose reform function had passed by default to law teachers—justify its traditional claim to leadership? What were the consequences of its failure to cope, or to lead?

Lawyers subscribed rhetorically to the notion of law as a public profession; but invariably they guarded it as a private preserve, practicing caste politics under cover of patriotism and professionalism. They deflected criticism from deficiencies in the legal system, for which they bore a major share of responsibility, to the inferiority of minority-group lawyers. Why, they asked repeatedly, did a public profession require lower standards? The point, however, was not that standards should be lowered because law was a public profession, but that because it was public, only those requirements that did not discriminate against particular classes or ethnic groups should be imposed. It was disingenuous at best for lawyers to dismiss stratification as a specter nurtured in the recesses of the Carnegie Foundation when they had themselves designed their profession to perpetuate it. Within the legal profession in the postwar decade Tocqueville's aristocrats still sniffed haughtily at Lincoln's plain people.

Five

Babbitry at the Bar

In the twenties corporate lawyers enjoyed unchallenged professional hegemony and unsurpassed opportunity to articulate their wishes as professional values. They spoke for the profession, asserted their clients' interests as professional and national interests, and served as role models for those who aspired to the most rewarding professional careers. A wartime legacy of virulent patriotism, lingering after the Armistice, provided an important boost to their efforts: corporation merged with country as the object of their allegiance. Legal periodicals waged a vigorous postwar law-and-order crusade, warning of the imminence of revolution unless lawyers educated Americans to the dangers of radicalism and the values of Americanism and corporate capitalism. One law journal, applauding the anti-radical raids led by Department of Justice agents, concluded: "There is only one way to deal with anarchy, and *that is to crush it.*" It urged punishment that would be *"not a slap on the wrist, but a broad-axe on the neck."* Another professional periodical, warning of "foreign intermeddlars," casually dismissed the Bill of Rights by urging the bar to condemn the heresy "that a man can drive a dagger at the very heart of his government and then expect it to be able and willing to furnish him with that protection to his rights which he so vigorously demands." Early in 1920 *Bench and Bar* asserted that "the need for repression is great, and the time for repression is now."[1]

The American Bar Association was addicted to xenophobic patriotism. Its presidential spokesmen, especially during the first half of the postwar decade, were tormented by visions of national subversion by alien radicals. In his presidential address of 1922 Cordenio A. Severance, a railroad and corporate lawyer from Minnesota, voiced in uneasy detail the anxieties that beset so many of his colleagues. In a world shaken by revolution, he warned of "large organizations of men extending to every industrial center of America, who are at work carrying on an active propaganda directed to the eventual destruction of our Constitution and the substitution therefor of a government such as brought chaos to the great Russian people. . . ." These "mere theorists" disregarded the incentive of individual success, "which history has shown to be absolutely essential in the development of the world. . . ." They flooded American cities with literature calculated "to stir up hatred, produce discontent, and in many cases . . . incite to violence." The country had become a haven for such men, "whose chief mission has been to plot and agitate against the free institutions under which they had enjoyed liberty and opportunity such as were undreamed of in the lands of their birth." Severance asked the bar to assert its influence against their "challenge to civilization."[2]

Severance's successor, Robert E. Lee Saner, rose to the presidency from his chairmanship of a special ABA committee on American citizenship, established in 1922 "to stem the tide of radical, and often treasonable, attack upon our Constitution, our laws, our courts, our law-making bodies, our executives and our flag, to arouse to action our dormant citizenship, to abolish ignorance, and crush falsehood, and to bring truth into the hearts of our citizenship."[3] Saner attacked the "immense and unassimilated foreign element" and those who would "tamper with the Constitution so as to throw power at once and completely into the hands of a mere majority, at a time when that majority's composition itself involves a most serious problem. . . ."[4] His committee described a struggle between a "government under a written constitution . . . and a government by the mob—or, if you please, the proletariat. . . ." In this confrontation, it concluded, the conservative influence of the bar was essential.[5] Committee activities included advocacy of laws requiring the teaching of the Constitution; the celebration of Patriot's Day, Memorial Day, and Independence Day with lawyer-spokesmen serving as "Minute Men of the Constitution"; and preparation of a "citizenship creed" to remind Americans that theirs was the best and freest government ever created.[6] Saner urged a vigorous membership drive that would strengthen American institutions (especially the bar) against radical attack. The lawyer "who steers the course Commerce, and keeps Business on an even keel," he insisted in his presidential address, must simultaneously assume the burden of defending the nation against its enemies.[7]

Lawyers eagerly proclaimed their stabilizing role. "In times of excitement," members of the Alabama Bar Association were told, "the lawyer's voice ought to be almost controlling. He becomes a breakwater between a reckless or a lawless multitude and the forms and rules of the law." A Minnesota judge, decrying socialism as a "social disease," called upon the members of his state bar association to "stand for law and order and against class rule and class hatred." A federal district judge, speaking to the Iowa Bar Association, summoned American lawyers to "the standard of conservatism." In Missouri, lawyers were urged to "stand firm and unshaken and be true to the eternal and unchanging verities." In Ohio, the president of the state bar association suggested that his colleagues draft laws so that "no Trotsky, Haywood or their like can find even a soapbox on which to stand and vomit forth their treason." And in Virginia the Vice President of the United States, a member of the Indiana bar, exonerated lawyers from responsibility for the disturbing fact "that we have permitted to drag their green trunks across and along the planks at Ellis Island thousands and hundreds of thousands of anarchists, revolutionists, . . . fellows who propose to take charge of this republic of ours."[8] At the end of the decade the American Bar Association issued a "citizenship map" of the United States: agrarian regions, inhabited by native Americans, were white; industrial areas, populated by the foreign-born, were black.[9] The organized bar, certain that good and evil correlated precisely with geography and ethnicity, renounced the darker regions of American life. In doing so it tried to repudiate much of what the United States had become in the twentieth century: an urban industrial society inhabited by a heterogeneous population.

The shrill super-patriotism of war lost its urgency after the Red Scare subsided, but in the Dollar Decade after 1919 professional cohesion was sustained by the symbols and values of the American business system. Loyalty to business was the new standard of patriotism, to which lawyers pledged their allegiance as enthusiastically as they had to the American flag. The business corporation was one symbol of modern America that elite lawyers could eagerly embrace. The defensive tone of the prewar debate over commercialization yielded to a confident affirmation of business principles and practices. The corporate lawyer, once the target of professional criticism, now wore his practice as a badge of honor and patriotism. "Is it still a disgrace or something contemptible to be a man of business?" asked a venerable Boston lawyer in 1923. No, he answered with assurance. Successful modern lawyers might be less oratorical or philosophical than their predecessors but they possessed the ability to solve practical problems of banking, commerce, and manufacture. In a materialistic age, he asserted, it would be an act of folly for lawyers to remain aloof from business values.[10]

The legal profession resolved its prewar identity crisis by merging capitalism, boosterism, and Americanism with legalism. In an era when Jesus was praised as the greatest businessman of all time, for selecting twelve men from the bottom ranks of business and forging an organization that conquered the world, it is hardly surprising that Babbitry was rampant at the bar. Salesmanship and service became professional bywords. "A lawyer *sells* his services, just as the merchant sells his goods. The good lawyer is a good *salesman* without knowing it. If he gives *service* the people will come to him, just as they will flock to the merchant who gives service." To meet the needs of businessmen—the paramount obligation—lawyers must be practical, prompt, efficient, and systematic. "Busy businessmen don't want excuses," warned the president of the Illinois Bar Association in 1924. "They don't want delays. They don't want alibis. They want results. . . ." His association's committee on office management urged lawyers to adopt the business point of view and to "deliver the goods." It contrasted the old-school lawyer, who worked at a cluttered roll-top desk in a dingy room, answered a wall telephone, greeted the mailman, and entered his fees in his ledger book, with the new breed of attorney, who belonged to a firm housed in a bank building with clean offices, an efficient receptionist, and elaborate files. "Everything is clean; no dust is visible; no papers are exposed to view." There the client received service—"the thing he wanted above everything else—and is willing to pay the price." After all, the committee report concluded, "a successful lawyer is merely a good salesman."[11]

For decades lawyers had told each other to become more practical and efficient. At the turn of the century this meant little more than neatness, promptness, and sobriety. After World War I, however, the message assumed a special urgency. In part a reflexive application of principles of scientific management which swept through American industry, it was also a strident call to progress as businessmen defined that term. "In this business world," declared Dwight G. McCarty, an Iowa attorney who wrote one of the early law office management manuals, "the lawyer must be a businessman to keep pace with all this progress." As businessmen maximized profits by maintaining constant overhead while increasing production, so the lawyer "by eliminating waste and lost effort, and increasing his day's output, substantially increases his income." By adopting modern office methods—time records, efficient filing and bookkeeping, delegation of routine tasks—the lawyer prepared himself to advise businessmen "and help in the solution of the many problems of present day business life." No lawyer could realize "the full measure of success" if he failed to "back up his professional attainments by a thoroughgoing business system." Proper management, McCarty concluded, "is the new freedom for the lawyers."[12]

Whether lawyers practiced in small towns or in large cities, they proclaimed their concert of interest and identity of outlook with the businessmen whose techniques they borrowed. They referred to their work as their business; they noted without embarrassment that they did substantially what bankers and merchants did; their stated objective was "to run a law office like a business office"; and they warned professional colleagues to heed "organization, system, equipment, efficiency and other like terms formerly applied to business exclusively." They told graduating law students to systematize their offices as though they were banks and to keep the same hours that businessmen kept. "We know how helpful it is to have the businessman's point of view," declared one prominent lawyer who advocated business systems in law offices. Special words of praise were reserved for lawyers who launched business careers or who entered larger metropolitan law firms, which were described as "important business institutions." Lawyers were even told that they might wriggle out of their traditionalist skins by committing their energies to business objectives. Businessmen were depicted as innovators who were "always growing, changing, progressing," and as men with little use for those lawyers whose fondness for precedent made them "intellectual slaves." Because businessmen worshipped efficiency they would not tolerate "the dilatory methods of the old time practitioner." Instead they needed lawyers who could "move forward with business, or at least . . . help business with its forward movement." It was the lawyer's responsibility to accelerate "the onward march of business" by removing legal technicalities and delays.[13]

Occasionally a lawyer expressed unease at the organized efficiency of corporate practice. A country practitioner from Georgia confessed: "I know nothing of large law offices, with their commanding generals, their regiments of salaried assistants, their companies of stenographers, their squads of lay employees, their drum majors of office boys, their liaison officers . . . , their batteries of typewriters. . . ." Clearly, however, he knew that he wanted to avoid an army that seemed to him as devastating as Sherman's. But in the twenties his fright was atypical. Lawyers were eager to shuck their parasitic image and to replace it with something more contemporary, more positive, and more progressive. Thus one attorney-turned-businessman boasted that the modern lawyer was "the self starter of business, the carburetor that mixes the fuel and the air which makes the gas that drives the engine. The world should certainly concede that the lawyer furnishes the gas." At the end of the decade, in a retrospective view of postwar developments, New York attorney William L. Ransom declared: "The business lawyer seems to be an outstanding factor in many of the most hopeful developments of this generation."[14] The business ethos enabled lawyers to defend stability yet welcome progress. In postwar America the corporate lawyer was as flexible in the service of his business clients as he was rigidly resistant to changes in American society that threatened business values.

The presidential candidacy of John W. Davis, a prominent corporate lawyer, provided appropriate symbolic expression of postwar values. Davis, a West Virginian by birth but a New Yorker by choice, had enjoyed a distinguished ten-year public career, first as congressman, then as President Wilson's solicitor general, and finally as ambassador to England. After the war he returned to private practice as senior partner in what quickly became one of New York's largest and most prestigious corporate firms. In 1922 he was chosen president of the American Bar Association. Two years later the Democratic party, after a tortuous convention struggle, nominated him for the presidency. Davis was not the first corporation lawyer to be selected by a major party; his twentieth-century predecessors included Alton B. Parker, William Howard Taft, and Charles Evans Hughes. But each of these men received the nomination while holding public office; only Davis was plucked directly from private practice.

Davis' career as a lawyer had a Horatio Alger quality about it;

the distance between Clarksburg, West Virginia, and Wall Street was not measurable only in miles. Yet Davis, even at the pinnacle, could not accept his own repudiation of his past. The more financial success he achieved the more he craved—and the more he craved the more enthusiastically he defended those country lawyer attributes that he had left far behind in Clarksburg. Notwithstanding his lucrative practice in a Wall Street firm, Davis never shed his rhetorical commitment to the virtues of Clarksburg practice as his father's partner at the turn of the century. "A lawyer who fails to engage in a country practice loses something," he explained. "In a country practice, or even practice in a smaller center, he gets more variety and better opportunity for an all around development." The more renown that Davis won as an urban specialist in corporate law, the more diligently he reiterated his pride that he was "born and raised a country lawyer, and a country lawyer, I suppose, I shall remain. . . ." The wealthier he became, the greater was his fondness for quoting the aphorism that a lawyer "works hard, lives well, and dies poor." He conceded that New York City was "unquestionably the greatest market for legal brains in the world," and that the rewards at the top were substantial. But, he invariably added, "the road up is long and the discomforts of living in New York are great."[15]

Davis' professional education had been thoroughly conventional. He recalled of his professors at Washington and Lee Law School, where he crammed a two-year course into one during the 1890's: "They were more concerned that you should learn what the law was, than that you should be invited to speculate on what the law *ought* to be." Davis learned the lesson well. Speaking of his railroad accident litigation under the archaic master-servant rule, he observed: "I never conceived at that time that it was my duty to reform the law. It was my duty to find out what the law was, and tell my client what rule of life he had to follow. That was my job. If the rules changed, well and good."[16]

There was nothing conventional about Davis' legal career during the 1920's. Davis expressed retrospective surprise that the law firm he reorganized with his friend Frank Polk grew as it did.

"Nobody in the world anticipated so large a firm. It came about simply because clients demanded service." That simple explanation slighted not only Davis' ample talents but his relentless pursuit of clients. Soon after he returned from England he disavowed political ambitions until, as he expressed it, "I can rebuild my shattered fortunes." Not even a place on the United States Supreme Court could tempt him from private practice. When Justice Van Devanter inquired about Davis' receptivity to such an appointment, Davis replied: "I established my present connection in New York . . . with the fixed intention not to be tempted into official life again until I had accumulated . . . some economic independence. The latter seems to be measurably within my grasp." Davis described his surroundings as pleasant, his partners as able and loyal men, and his clientele (which included the J. P. Morgan Company and Standard Oil) as of "a very satisfactory character." He refused, therefore, to consider a Supreme Court appointment. "I am a hard-working lawyer," he subsequently wrote, "with no ambition in the world except to have good clients and plenty of them." Even Chief Justice Taft was provoked to complain that "If you people in New York were not so eager for money . . . you might have some representatives on our bench."[17]

Yet early in the spring of 1921 friends of Davis pondered his presidential prospects. Within the American Bar Association some of them worked diligently in his behalf in the hope that one presidency would lead to another. But, as a member of the Yale Law School faculty told Davis, he risked the danger of being "eternally damned" as a corporate lawyer. Unless he exercised considerable discretion, he was warned, he would suffer from "the curse of his evil associations."[18] In 1924, when Davis became a serious candidate, his vulnerability showed.

As Davis edged closer to the presidential nomination, the Morgan retainer, his major asset, became a political liability. Davis took refuge in the duty "to serve those who call on him" without regard for the implications of such service for personal popularity or political reward. "Any lawyer who surrenders this independence or shades this duty . . . disparages and degrades the great profession to which he should be proud to belong."[19] Critics did not question his right to choose corporate practice for the ample financial and professional rewards it assured; they asked only that Davis be held accountable for his choices. But Davis (like Charles Evans Hughes several years later in similar circumstances) evaded accountability by claiming professional duty, which was sufficiently resilient to accommodate any demand upon it. Still, a lawyer who was obligated to serve well those who called was hardly compelled to engage in practice which virtually eliminated non-corporate callers. It was sophistry for Davis to claim that he was asked to betray his professional independence when, in fact, he was being urged to demonstrate it by broadening his clientele.

No one attacked him, and all that he represented, more relentlessly than Felix Frankfurter, who declared war on Wall Street from his academic redoubt in Cambridge. Davis was a convenient foil for Frankfurter, who was disgusted with the moral tone of the postwar bar. Its leaders, he complained, had opposed the Brandeis nomination but not the Palmer raids. They had disregarded violations of constitutional rights and professional standards by powerful lawyers, "but they have been regular sleuth hounds against petty corruption and the delinquencies of the small fry." The postwar ambiance troubled Frankfurter; the complicity of lawyers pained him. He asked his friend Charles C. Burlingham: "Don't you think it would be like a breath of fresh air in our dank national atmosphere if a few lawyers who did matter would say we don't like all this degradation and enveloping commercialism and general corrupting atmosphere?"[20]

Davis' candidacy released a torrent of criticism from Frankfurter. In unsigned articles for the *New Republic*, published in the spring and summer of 1924, he told "Why Mr. Davis Shouldn't Run." Davis' corporate connections, especially his retainer from the J. P. Morgan Company, left him a prisoner of his associations. "Latterly clients have had lawyers and not lawyers clients," Frankfurter complained. "The 'great lawyers' as a rule are inextricably implicated in and with big business." He dismissed Davis as "the employe of Big Business" who identified his clients' interests with the national interest. Worse yet, in Frankfurter's eyes, Davis had declined to commit his prestige and position as ABA president to "sanity and reason" amid the postwar "lawlessness and intolerance" which the bar association had abetted. "From the time that public office gave him a prestige to be capitalized, he capitalized it in terms of big money-making for himself and of service to the powerful." His rejection of a Court appointment especially dismayed the Harvard professor, who found in the rejection an example of the "pernicious sense of values" that characterized "an opulent New York lawyer" who had become "a 'big business' man." But Frankfurter, who believed that Davis' career was an unsavory model for aspiring lawyers, was most distressed by what he described as "the far-reaching psychological and political consequences" of Davis' professional posture. "It is good neither for these lads that I see passing through this School from year to year, nor for this country," Frankfurter wrote, "that we should reward with the Presidency one to whom big money was the big thing." Decades later, reminiscing about the 1920's, he returned to the same point. It still pained him that "the whole drive, the whole propulsion, or compulsion almost" for the best Harvard Law students of that generation was Wall Street practice.[21] The strength of the Wisconsin idea—commitment to disinterested state service in the public interest—had ebbed; until the New Deal years, Wall Street practice would define the limits of a lawyer's dream.

Other observers, from varied angles of vision, corroborated Frankfurter's conclusions. Chief Justice Taft conceded that "able lawyers have yielded to the inducement of large salaries and embraced exclusively the cause of large corporations." Large urban firms, not only in New York but throughout the country, recruited the best available talent. Even in Middletown, the Lynds reported, corporate lawyers enjoyed higher prestige than judges. The war and postwar economic booms stimulated the reorganization and combination of law firms to handle the expanded volume of business. The Davis, Polk reorganization of the Stetson firm in New York was one example; another was the merger of

Shearman and Sterling with Cary and Carroll. In 1918, before the merger, the old Shearman firm had two partners; the new one had twelve and a heightened appeal to ambitious young lawyers who, according to the firm's historian, desired immersion in "the operations of Big Business, with its excitement, stimulation and training. . . ." When President Harding was asked whether he would risk nominating a corporation lawyer as attorney general, he replied that he would not appoint anyone else, because the only lawyers capable of filling the position were corporation lawyers.[22]

For Harvard law students, whom Frankfurter knew best, the anticipated reward of achievement was employment on Wall Street. Frankfurter capitalized upon his friendship with Henry Stimson, Emory Buckner, Charles C. Burlingham, and other Wall Street partners to send many students there. Most Harvard students, David E. Lilienthal noted in the journal of his student days, "invariably refer . . . to their hope or plan to enter the employ of some corporation, as one of their counsel, or join some firm of lawyers doing corporation work." According to Henry J. Friendly, a member of the Class of 1927, "practically everyone thought there was only one career that was worth pursuing, namely, private practice." A scattered few went into law teaching; occasionally men with mediocre records might work for the government "because they didn't have quite the grades to get the jobs they wanted in private firms. . . ." Friendly's own career decisions fit the pattern precisely. A law review editor, he clerked for Brandeis and then, like Brandeis before him, spurned an offer to join the Harvard Law School faculty. He considered a position with the Interstate Commerce Commission where, he conceded, lawyers had the power to decide important questions. But the salary in government was low, although profitable employment with a railroad was a possibility once government service paled.[23] With as wide a choice of career options as any of his contemporaries, Friendly joined the Root, Clark firm in New York.

As easily and inexorably as so many Harvard graduates joined a New York firm, some recoiled at the implications. Dean Acheson told Frankfurter, "New York raises the most serious questions—

financial, physical and philosophical." Acheson was pulled in opposite directions. The financial advantage of New York practice
was a strong inducement; but so, too, was the opportunity "to go
to a place in which one can live— . . . find an office which is not
sold out, learn to use legal tools, and then use them toward ends in
which I believe." Lilienthal, who went to work with Chicago labor lawyer Donald Richberg, was similarly torn between law "as
a medium whereby I may do a work in the labor movement" and
"striking out frankly for enough money to subdue my worries."
When Herbert B. Ehrmann asked Frankfurter's advice he was
told: "If you like the illusion of greatness that comes from great
excitement, if your view of the law of chances is that you're
destined to be a Guthrie or a Cravath . . . why then try N.Y."
Otherwise, Frankfurter suggested, Ehrmann should remain in
Boston where his network of professional and business associations
would provide alternative satisfactions.[24]

The financial rewards of law-firm practice were undeniably
substantial. In 1920, for example, the average annual income of a
member of the Class of 1905 at the Harvard Law School who
practiced on his own was slightly more than seven thousand dollars. His classmate who held a partnership earned approximately
twice as much. Five years later, in 1925, the disparity was considerably greater: the average Harvard solo practitioner earned ten
thousand dollars, and the income of his classmate in a firm was
nearly three times as high. The lure of larger firms cut across
geography and age. A Texas lawyer advised his son, a third-year
law student who years later would become president of the
American Bar Association, that the large firm "has much advantage over a little business"—specifically "the advantage of being
able to get business and attend to business anywhere over the
country." In New York Elihu Root, pondering the resumption of
his public career, told Henry Stimson that "to have a flourishing
law practice in New York was a great *point d'appui* for anything
that might come up." The private elevator that connected the
Davis, Polk firm with the offices of J. P. Morgan and Company,
"like an umbilical artery" according to *Fortune*, was a tangible

symbol of the rewards of New York corporate practice. So, too, was Davis' annual income of $400,000.[25]

Throughout the twenties the career choices of law students reflected and reinforced the business values of the decade. Law review editors from the elite schools, who constituted the most talented stratum of the legal profession (as the profession itself defined talent) and who enjoyed the greatest freedom of choice in managing the directions of their careers, consistently and overwhelmingly entered private corporate practice upon graduation from law school. Between 1918 and 1929, 81 percent of a sample of nearly three hundred law review graduates from Harvard, Yale, and Columbia chose employment in private practice immediately upon graduation. If anything, this figure errs on the *low* side, since an additional 8 percent, who became clerks to judges, may only have remained outside the private sector for one year after they graduated. Between World War I and the Crash of 1929, only nineteen editors from these schools took positions with the federal government. Eleven entered in 1925, nearly all of whom went to work as assistant United States attorneys in positions in the Southern District of New York filled by Emory Buckner, an attorney on leave from the Root, Clark firm who possessed a singular talent for recruiting able lawyers, whether he was in the private or public sector.[26]

Career patterns during the 1920's prompted Felix Frankfurter's dispirited observation that professional leadership meant "being a key figure in a New York law office. . . ." Decisions to enter private practice reflected its obvious appeal, the paucity of alternatives, and the low status these alternatives bestowed. In the federal government, for example, the demand for lawyers was relatively slight. Although Harlan F. Stone, who left the Columbia deanship for Sullivan & Cromwell and shortly thereafter became attorney general, tried energetically to recruit graduating law students of high calibre to the Department of Justice, his efforts were in vain. The department's reputation under Stone's two predecessors—A. Mitchell Palmer, who exploited the postwar Red Scare, and Harry Daugherty, who was implicated in the scandals of the

Harding administration—may have undercut Stone's efforts. But more likely, lawyers found nothing in Washington as enticing as the opportunities in Wall Street firms for financial gain, professional distinction, and legal craftsmanship. Lawyers who, in Frankfurter's sardonic phrase, searched for the "holy grail" only in New York corporate firms accurately reflected the values of the age in which they reached professional maturity.[27]

These values were seldom challenged during the twenties. The patriotism of business was the dominant professional and national creed. Only the Sacco-Vanzetti case, the *cause célèbre* of the postwar decade, rippled professional tranquility. The case might plausibly be dismissed as an aberration—hence its notoriety. But its implications were too entangled in the threads of professional life to be so easily discounted. The defendants, doubly vulnerable as aliens and as radicals, confronted social and legal institutions in which nativism, super-patriotism, and business values were deeply entrenched. Within the legal profession—as in John Dos Passos' *U.S.A.*—the Sacco-Vanzetti case exposed two nations: one, comprising law teachers, warned of its implications for the legal process and the administration of justice; the other, the organized bar, preserved its silent decorum except to criticize the critics. Teachers, the professional "radicals," defended the sanctity of legal institutions while practitioners, the traditionalists, weakened them by their silence.

Felix Frankfurter's role in the case illuminates this paradox. By 1927 Frankfurter enjoyed some notoriety as a committed defender of unpopular causes—to the extent that even his old mentor, Henry Stimson, criticized him for his postwar radicalism. Yet Frankfurter's "radicalism," as the Sacco-Vanzetti case so clearly demonstrated, was nothing more than the insistence that legal officers and institutions honor their professed commitment to due process and equal justice under law. This was the most truly conservative of positions, made to appear radical only by the ease with which so many lawyers departed from these standards in the name of saving American institutions. "Just because I believe in law and in our Constitution," he told Stimson, "I believe a peculiar *noblesse oblige* rests upon our officials, particularly our law officers and the courts, to observe the laws and the Constitution, however strong the winds of passion may blow."[28] Frankfurter's praise was reserved for those who were committed to "the permanence of our essential institutions"—quite the antithesis of radicalism.[29]

The winds of passion blew strongly in 1927. Listening to them howl, Frankfurter described the case as a test of law as an instrument of justice. With the forces of intolerance, suppression, and materialism (as Frankfurter described them) arrayed against the defendants, it was American political and legal institutions, not two anarchists found guilty of murder, that were on trial. The real adversaries were justice and injustice. Revealingly, Frankfurter rarely dwelled on the defendants; indeed he dismissed Sacco and Vanzetti and proclaimed his concern for law, not for them. Similarly, Herbert Ehrmann, who struggled relentlessly on behalf of the defendants during their final year of life, found "the absurd and pathetically impracticable views" of his clients far less disturbing than the visible cracks in "a strong and well-ordered society . . . at a time of passing public prejudice."[30] Frankfurter and Ehrmann were galvanized by a discomforting sense of the fragility of law at a time of social stress. Their mission was to strengthen legal institutions, not to further undermine them.

Frankfurter was especially concerned with the implications of the case for the legal profession. It cut "to the root of the moral foundations of our profession," exposing an "abdication of mind and conscience" bordering on professional immorality. In the journal of the American Bar Association, for example, the solitary sustained reference to the case came in an article solicited from a Canadian appellate judge who defended the fairness of the trial and attributed its notoriety to the illogical efforts of Socialists to secure executive clemency for the defendants. Association president Charles S. Whitman, in his annual address (delivered several days after the executions), included an inaccurate one-sentence reference to the case, in which he declared that it was ABA policy to remain aloof from "controversial matters of a political nature" and to avoid "purely local" questions. The silence of the organized bar appalled Frankfurter, although it hardly surprised him. ("A bar which has Charles S. Whitman as its president is, of course, morally bankrupt," he told Roscoe Pound.) He was disgusted by the willingness of lawyers to permit executions rather than to confront the possibility of judicial fallibility. During the spring preceding the executions Frankfurter made the fullest statement of his position: "It's because I care so deeply about law and the capacity of the judicial system to operate according to law, that I so fervently hope that a demonstrable professional immorality will be authoritatively rectified. I only wish that my profession were more earnestly devoted in practice to the vindication of its pretensions about law and the legal system."[31]

Frankfurter's sense of professional malfeasance was heightened by his belief that members of professional elite groups—"lawyers with standing" and with "education and culture and the freedom from fear that position and power should engender"—must assert themselves in times of social frenzy. That responsibility, he insisted, was ample justification for his own activities, which included publication of an article about the case in the *Atlantic Monthly* (bitterly resented by lawyers who criticized his intrusion while a decision was pending) and an acrimonious public exchange of letters with Dean John Wigmore of Northwestern Law School, the evidentiary expert who took issue with Frankfurter's conclusions. As Frankfurter explained to a partner in the Boston firm of Ropes and Gray: "I conceived it to be my duty to speak out when what I believe to be a demonstrable miscarriage of justice involving the honor of the Commonwealth is impending." But it was more than a matter of Massachusetts honor. Frankfurter's interest in the case, he reiterated, was "the interest of loyalty to our legal institutions. . . . It is of the essence of law to give it to those whom we do not like." His zeal in holding these institutions to the letter of their obligations set him apart. It did not shock him to find error in a criminal trial. But the failure to rectify error "works a lasting damage to confidence in our whole system of law." To demand loyalty to legal institutions, regardless of their performance, surely would undermine trust in them. No "perversion of our established legal procedure," Frankfurter maintained, should become "the test of loyalty to our legal system."[32]

Frankfurter's public indignation and the public silence of the practicing bar suggest the divergent professional roles of teachers and practitioners. The lines of division were not precise: Frankfurter fought Wigmore as vigorously as he criticized any practitioner; William G. Thompson and Arthur D. Hill, respected members of the Boston bar, formed the legal backbone of the Sacco-Vanzetti defense; and the senior partners in two prominent Boston firms divided over the issue of a review commission. Generally, however, the existing boundaries of a divided profession held. The silence of established practitioners suggested that judgments of the courts must be supported at all costs—especially when radicals were on trial and the verdict was guilty. To support the defense cause, a former New York associate of Frankfurter's told him in a representative complaint, was tantamount to "aiding the attempt to disorganize society." Similarly, Chief Justice Taft wrote that law schools "might be about better business" than scrutinizing trial conduct in Massachusetts. The teaching profession, he complained, "does not always exercise the best judgment in keeping out of fields in which their members are apt to make egregious mistakes." Predictably, the reaction against Frankfurter was most intense in Boston legal circles, where criticism of the conduct of the trial was equated with sedition. Harvard alumni communicated their intention to withhold financial contributions from the Law School unless Frankfurter was silenced. They accused him of sins ranging from "bad professional taste" to siding with "the Reds and Radicals." But colleagues like Pound and Zechariah Chafee, Jr., and alumni like Thompson, stood firmly by him. It was a choice, Thompson wrote, "between tolerating injustice for the sake of placating possible contributors to the School, and standing for what we believe to be right regardless of the consequences to the School or to any of us as individuals." It turned out to be a costly choice for Thompson's firm, which lost clients and considerable income as a consequence of his defense work.[33]

Law teachers provided the most insistent criticism of the trial. At Columbia Karl Llewellyn, like Frankfurter upset by the failure of legal institutions to honor their principles, launched a law-school petition campaign for complete review of the case, which majorities on the faculties of six law schools supported. From Yale Charles C. Clark reported that the Connecticut bench and bar were becoming increasingly aroused over statements by some of Clark's colleagues. But, Clark added, "I do not see how we can claim to be critics of the law and back down where something more than commerce and money is involved." "Critics of the law" —Clark's phrase captured the identifying trait of law teachers which so upset practitioners. Yet the criticism, as Frankfurter invariably noted, originated in the deepest loyalty to the legal process. Frankfurter, Llewellyn, and Clark insisted upon nothing more—or less—than the rigorous application of venerable standards of justice. They could hardly have promulgated a more Burkean conservative position. Their plea for institutional continuity, their fear lest tradition be abandoned, and their reliance upon the *noblesse oblige* of a professional elite sounded radical only because so few professional leaders lifted their gaze beyond the business values of the day. Professors shared the conviction of ABA presidents that nothing must be permitted to harm the legal system, but they disagreed over the primary source of danger. For the bar it was criticism that questioned the fairness of the judicial process; for the professoriat it was silent acquiescence in the failings of that process. The professor, critic of the law, was an elitist with commitments to a traditional ideal. The practitioner, defender of The Law, was too deeply immersed in the politics of client caretaking to respect other traditional values. He seldom grasped the point, raised by one of Frankfurter's colleagues, that "it is by correcting abuses in a legal system rather than by ignoring them that the conservative elements in a community can most effectively prevent the very radicalism which in this instance seems to blind them to these abuses."[34]

As the case moved through its final stages, Frankfurter pondered the role of law and lawyers in American society. Knowing that "the law is what the lawyers are," he yearned for lawyers who understood social and economic forces. Yet Frankfurter also knew that "the law and lawyers are what the law schools make them."[35] Here was the recurring role ambiguity which university law teachers could never escape. Their institutions made lawyers, but their curricula equipped them to make business lawyers better than any other kind. The corrosive skepticism and constricted focus of the case method impeded the development of that broader social consciousness which might relate law to an entire society, not only to its most privileged sector. Consequently the best graduates—or at least those among the best who were not disqualified by their ethnic origins—moved inexorably into Wall Street firms and into the value system these firms embodied, while other graduates distributed themselves in the pyramid of job opportunities which had Wall Street at the apex.

Other signs of a divided profession appeared toward the end of the postwar decade, as the cement between legal and business values in American society lost some of its adhesive force. At periodic intervals since Oliver Wendell Holmes, Jr., had called for law nurtured on experience rather than on logic, legal educators had urged that law be viewed as a *social* institution. As late as 1909, however, more than a quarter-century after Holmes wrote *The Common Law*, Roscoe Pound observed that "the movement for the adjustment of principles and doctrines to the human conditions they are to govern rather than to assumed first principles . . . has scarcely shown itself yet in America."[36] Pound did more than any of his contemporaries to emphasize the social effects of law and to relate legal thinking to the social sciences. In the twenties members of a younger generation of teachers, clustered in Columbia Law School, tried to carry the standard to new intellectual and institutional terrain. Failing at Columbia they moved on to Yale Law School and to the short-lived Johns Hopkins Institute for the Study of Law; from there they directed a withering attack against mechanical jurisprudence and traditional analytical scholarship. These scholars, subsequently labeled Realists, shared little more than skepticism toward law as a "symmetrical structure of logical propositions, all neatly dovetailed."[37] But their skepticism was sufficient to redirect energy from law as a pillar of the status quo to law as an instrument of social reform.

Disenchantment with received dogma had increased after World War I. As Pound noted: "Professor Gray's famous dictum, 'We must obey the Gods, Lord Coke says so' has lost its point." Under pressure from a younger generation of teachers, law-school curricula expanded to absorb courses with social science content. Emphasis upon social policy considerations made it apparent that legal education might be shaped as much by what professors taught as by what judges decided. With some exasperation Harlan Stone referred to the "constant pressure" from his faculty at Columbia to experiment and innovate. Until all too recently, a colleague complained, Columbia students had graduated "with only a dim picture of society as it actually exists." The most striking development in legal thought in a generation, observed another Columbia faculty member, was the growing realization that law was "made by man to serve human interests, and can and should be changed as those interests changed." From Yale Robert M. Hutchins, the precocious dean, reported that his school was beginning to train students "to see the rules of law in contact with life as it is being lived in the United States today."[38]

These changes were welcomed—and often instigated—by younger lawyers whose legal education had been stultifying and parochial. Herman Oliphant of Columbia had left comparative philology for law study, "hoping thereby to get in touch with something more closely allied to life." But he graduated from law school "with a pretty keen feeling as to the law's detachment from life." Thomas Reed Powell recalled that when he graduated from Harvard he still believed that "the eternal principles of the common law were encased in the Year Books and that all we needed was right reason to find from them the law on any point today. . . . I thought that the past had determined all." Law and theology, Powell concluded, shared an "emphasis on received doctrine." William O. Douglas, who came to Columbia in 1922 as a student who believed "that perhaps one trained in law could be an effective voice for human rights," discovered that law practice required "predatory qualities" and that law faculties "were encrusted with heavy-footed traditionalists." And Jerome Frank, whose *Law and the Modern Mind* formally ushered in the Realist decade in 1930, confessed: "It was not until I had been practicing law for several years that I began to see that I was practicing ethics, political science, and economics. And much time elapsed before I became aware that judicial decisions are inherently value judgments."[39] The skepticism of Oliphant, Powell, Douglas, and Frank—and others like Thurman Arnold, Karl Llewellyn, and Underhill Moore—did not have its full impact on legal education or public policy until the 1930's; it was easy enough to disregard revealed truth once the social structure that sustained it tottered precariously. But the seeds of doubt already were being scattered during the 1920's when, to all but a few, The Law still resembled an indestructible monolith.

It took the Great Crash of 1929 to weaken the dogma that had sustained business values within the legal profession during the postwar decade. Nothing was more indicative of the swiftness of change than the unexpected flurry of protest that erupted over the nomination of Charles Evans Hughes as chief justice just months after the stock market collapse. Upon Chief Justice Taft's resignation, President Hoover nominated Hughes, who had left the Court in 1916 to accept the Republican presidential nomination. After his defeat he enjoyed a lucrative corporate practice in New York, which he interrupted to become secretary of state in the Harding-Coolidge administration and resumed upon leaving public office in 1925. Hughes' nomination triggered an acrimonious debate in the Senate, reminiscent of the Brandeis confirmation struggle fourteen years earlier. This time, however, the symbolism was reversed: Brandeis, the "people's attorney," had been pilloried as an outsider by the professional elite; Hughes, a prominent member of that elite, was taken to task by progressive senators for his corporate counseling.

Hughes was doubly vulnerable. Not only had he left the bench for politics, but his roster of wealthy corporate clients suggested political and professional preferences suddenly rendered unfashionable by the Crash. "No man in public life so exemplifies the influence of powerful combinations in the political and financial world," declared George W. Norris, who led the Senate opposition. During his years in practice, Norris charged, Hughes had "associated with men of immense wealth and lived in an atmosphere of luxury which can only come from immense fortunes and great combinations. . . . These influences have become a part of the man. His viewpoint is clouded." William Borah thundered his concurrence. Citing Hughes' advocacy for oil companies, radio corporations, meat-packers, and companies accused of violating the Sherman Act, he asked when the nominee had not appeared "for organized wealth and against the public?" Hughes' economic views, charged Burton Wheeler, "are those of the great combinations and trusts. . . ." It was a striking fact, another senator added, that "if a man is to be thought a notable lawyer he must be in the employ of great corporations. . . ."[40]

Hughes' defenders, confronting an unexpected assault, barely denied the charges and, in their efforts to deflect them, only substantiated them. A Massachusetts Republican summarized the professional lures to which Hughes had responded: "The call is inevitable and irresistible for every lawyer of extraordinary ability to go from the country to the city where the great professional prizes are, and if he succeeds in the city he is bound to get as clients what every lawyer is seeking for, those who control the most important interests. So Mr. Hughes attracted as clients the great business interests of the country. They are the ones that naturally demand the highest talent; they can pay for the highest talent. . . ."[41]

Critics of the profession could not have made the point more cogently. The inexorable movement of legal talent to the financial and commercial centers of the nation; the inevitable rewards of money and prestige; the correlation between the wealth of a client and the availability and quality of legal services—for these developments Hughes, like John W. Davis, was an appropriate postwar symbol whose luster during prosperous times quickly tarnished after 1929. Hughes was deeply pained by the unexpected criticism. Like Davis during the 1924 campaign, he casually dismissed any inference that a lawyer might be judged by the clients he kept. In his *Autobiographical Notes* he cited Professor Zechariah Chafee, Jr., approvingly for declaring that Hughes had merely fulfilled his "duty to represent loyally the client for whom he happened to be working."[42] But the question was not whether loyal representation was provided. No one doubted that. It was whether the recipients of loyal representation constituted a restricted, identifiable clientele whose interests shaped a lawyer's practice, values, and politics.

Nothing required a lawyer to balance his ledger by counseling all comers, or by ferreting out needy clients. Indeed both Hughes and Davis, on special occasions (doubtlessly embedded in professional lore because they were so rare), defended politically unpopular defendants. Within broad professional limits and narrower structural ones, every lawyer exercised the freedom to choose clients from the pool that his type of practice made available. But the Depression suddenly compelled the lawyer whose public identity and professional esteem rested upon service to a restricted corporate clientele to confront the implications of his choices. Such was the lesson of the Hughes contretemps; it was repeated three months after Hughes was confirmed when allegations that President Hoover's subsequent nominee, John J. Parker, held anti-labor and anti-Negro views doomed him to defeat. Elite lawyers might protest against attempts to identify attorney with client, but they honored separation more in theory than in practice—as their behavior during the Brandeis confirmation struggle had so clearly revealed. The opposition to Hughes was a harbinger for the professional elite. In the Depression decade professional values would receive their sternest test since corporate lawyers had upset the nineteenth-century equilibrium thirty years earlier.

Hughes, like Davis, was momentarily embarrassed by his professional identity but hardly impeded by it. Davis remained the acknowledged leader of the appellate bar; Hughes was confirmed as chief justice. But their similar responses to identical challenges demonstrated how the professional elite was prepared to evade the political implications of corporate practice. The notion of "craft" had already become the *sine qua non* of elite professionalism. It offered a definition of the lawyer's role that disguised volition and values under the cloak of technical proficiency. To Davis the lawyer was merely a technician ("He does not create. All he does is lubricate the wheels of society.") wearing a surgeon's mask ("The lawyer must steel himself to think only of the subject before him & not of the pain his knife may cause"). Such professional tunnel vision was designed to obliterate those disturbing substantive issues that Davis preferred to ignore: his lubricant was selectively sold and applied; the clients whose retainers he avidly pursued were not the victims of the pain he inflicted. Exaltation of craft eliminated the political and social implications of a lawyer's work from consideration, sustaining the illusion that law was science, not politics, and detaching process from purpose. The point is not that Davis was a hired gun, but that he consistently sold his craft to the highest bidder while claiming that the practice of law was "an avenue for service and not a means for private gain."[43] Hughes and Davis committed themselves to the style of practice dictated by their hunger for success and money. As they ascended to elite status they responded to the demands of a society that rewarded achievement and wealth and was oblivious to their cost.

The twenties represented both flood tide and ebb tide for the professional elite. A pyramidal structure was securely built, in which the hegemony of Protestant corporation lawyers was unquestioned, and business values were the predominant professional values. Infidels might crowd the professional portals, but structural, ethical, educational, and admissions barriers deflected all but a minuscule few from elite enclaves. Although university law teachers chanted their litany of criticism, their *modus vivendi* with practitioners permitted them to speak their conscience on public issues while they prepared their best students for corporate practice. Yet pressure for change welled up during the late twenties—socially from the expanding recruitment base of the profession and ideologically from the Realists. But not until the crash and the Depression temporarily weakened professional allegiance to corporate capitalism and compelled re-examination of the economics of the profession was change possible. And not until a reform administration provided alternative employment opportunities for the best young lawyers from ethnic minority groups was change inevitable.

On the eve of the Depression American society seemed acutely torn between the irreconcilable values of two quite different eras. It was difficult to reconcile modern materialism with arcadian innocence, but Americans could not stop trying. Jay Gatsby still believed that the past was worth recapturing and that money could buy it. Middletowners, the Lynds observed, planted "one foot on the relatively solid ground of established institutional habits and the other fast to an escalator erratically moving in several directions at a bewildering variety of speeds." Charles Lindbergh flew a new and complex piece of machinery across the Atlantic, yet he was endlessly praised for his self-sufficient individualism. Henry Ford's assembly line revolutionized American life, but Ford built Greenfield Village as a monument to those nineteenth-century small towns that his Model T had consigned to oblivion.[44] The legal profession was similarly tormented. John W. Davis counseled the J. P. Morgan Company as senior partner in a reorganized, modernized Wall Street firm as he spoke reverentially of the virtues of country practice. Felix Frankfurter was damned as a radical for defending traditional professional ideals. Aspiring young lawyers, desperately eager for access to the legal system, were stigmatized by its elite guardians as an alien menace to

Anglo-Saxon liberty and law. Bar associations laboriously constructed a reactionary professional structure, while their members became salesmen for contemporary materialism.

These tortured responses expressed the ambivalent values of a new professional culture that had grown from infancy to maturity since the turn of the century. Lawyers clung to nineteenth-century moral homilies on professional virtue while they scrambled for twentieth-century business opportunities. They replicated the community cohesiveness of the horse-and-buggy era only by reorganizing law practice to provide specialized services to a restricted corporate clientele. Eager for the professional opportunities available in an urban industrial age, they nonetheless mourned the disappearance of a homogeneous society and sought to recreate it within their profession. It was as though horses could pull the Model T, or Daniel Boone could fly "The Spirit of St. Louis."

The Wall Street firm institutionalized these contradictions in its upper echelons. John W. Davis and John Foster Dulles (senior partner, by the mid-1920's, in Sullivan & Cromwell) were cut from the same cloth of late Victorian Protestant culture. Both were small-town Protestants who linked success to individual effort and hard work, yet who entered professional life with a powerful family boost. Davis' father was an established, well-known West Virginia lawyer; he told his son that no young attorney in the state enjoyed more opportunities at the outset of practice. The father's clientele provided initial security; the surging prosperity of local industry offered subsequent opportunity. Dulles, who searched in vain for a Wall Street position, finally procured one by capitalizing upon his grandfather's much earlier association with Sullivan. Neither man rose emotionally above his origins. Even at the apex of his Wall Street career, Davis described himself as a simple country lawyer; Dulles, according to his sister, was always a man of upstate Watertown, not New York City. Both men were relentlessly, even obsessively, diligent workers whose professional lives exacted identical costs from their families: Dulles' absorption with his career left him with little emotional energy for his children; Davis' commitment to work elicited a poignant lament from his daughter: "What I wanted from him was his time, and he had little to spare." Both men emulated the style of living set by their wealthy clients: club memberships, expensive clothes, and Manhattan apartments and Long Island estates amply staffed with servants.

Davis and Dulles were, additionally, intellectual twins. "The Dulles mind," wrote one of his biographers, "was fundamentally shrewd and practical, but quite narrow in range, seeking always an immediate and a tangible result."[45] The same might have been said of Davis; it was that kind of mind that successful corporate practice demanded and rewarded. Yet these supposedly shrewd and practical men were also dogmatic and intellectually rigid. They learned early that law embodied right and prohibited wrong; they mastered the ability to reason from fixed principles whose validity they rarely questioned; they (like Hughes) defended corporate capitalism yet convinced themselves and their fellow professionals that they were merely proficient craftsmen. The Wall Street law firm was their monument.

But as it soared during the prosperity decade, so it trembled during the Depression. Like the Empire State building, another monument of prosperity that became its mausoleum when the highest floors went unrented after the crash, the professional edifice displayed signs of hollowness once prosperity vanished. Its elite tenants were badgered, even displaced. A new upstart organization challenged its national association. Many young lawyers spurned Wall Street for Washington, repudiating private practice for public service. Critics decried the restricted availability of legal services, a deficient adversary process, and a partisan judiciary. Romantic memories were interrupted by a contemporary nightmare.

Six

A Great State Service

During the Depression decade the legal profession, structurally and ideologically committed to stability, confronted wrenching change. Economic catastrophe prompted a reconsideration of the nexus between corporate capitalism and professional values. It disrupted and redirected career patterns. Enmity between teachers and practitioners, stirred by the currents of legal realism, increased with the migration of law teachers and their students to Washington. In mid-decade the birth of the National Lawyers Guild marked the first serious challenge to the hegemony of the American Bar Association. Neighborhood law offices were designed to provide lawyers with clients and clients with low-cost legal services. The thrust of these changes was to undercut temporarily the deference afforded corporation lawyers, to reshuffle professional elites, to weaken the correlation of ethnicity with professional opportunities, and, briefly, to direct professional energies into unexplored channels of social activism.

The misshapen social structure of the profession, divided into its "blue-stocking . . . respectable bar" and its "catch-as-catch-can bar," distributed economic hardship unevenly among lawyers.[1] Those in the lower professional strata faced economic annihilation. From one end of the country to the other the vulnerability of young lawyers, solo lawyers, and lawyers from ethnic minority groups was reported. In California, among lawyers admitted to the bar between 1929-1931, 51 percent did not earn enough during their first year in practice to support their families; 37 percent did not do so in their second year of practice; and 33 percent still did not in their third year. Three new California lawyers in ten earned less than half their income from law work. In Missouri lawyers suffered in direct proportion to their economic status. Those who earned less than $5,000 annually saw their income sliced by more than half between 1929 and 1933, while the incomes of those who earned in excess of $20,000 declined by one-tenth. Nearly half the country lawyers in the state lived at or near the subsistence level.

In New York City more than one-third of those with incomes below $2,000 were solo practitioners; only 17 percent of firm lawyers earned that little. In 1933 the median income of lawyers in Manhattan was below $3,000; nearly half the members of the metropolitan bar earned less than the minimum subsistence level for American families. One year later 1,500 lawyers in New York City were prepared to take a pauper's oath to qualify for work relief. Jewish lawyers (approximately one-half the metropolitan bar) discovered that their practice had become "a dignified road to starvation." Regardless of the number of years they had spent in practice, their income was "strikingly less" than that of their Christian colleagues. At every income level below $5,000 the proportion of Jewish lawyers exceeded the proportion of lawyers generally; above that figure, at every level, the proportions were reversed. Jewish lawyers were disproportionately concentrated among solo practitioners and excluded from law firms; they were confined to the least lucrative fields of practice and to a preponderantly Jewish clientele; and they dependend upon other sources of income to supplement their meager rewards from practice. But their economic plight soon spread to the entire profession. Nearly half of American lawyers, in the mid-thirties, earned less than $2,000 annually. Young attorneys, the American Bar Association concluded after a study of the economic condition of the bar, comprised "a severely handicapped group," while "the proportion of lawyers in large cities who earn so little as to constitute a serious professional problem is very great."[2]

At elite levels partners flourished but aspirants languished. Some firms doubtlessly lost lucrative corporate retainers after 1929, but bankruptcies, receiverships, and corporate reorganizations took up the slack until New Deal regulatory legislation stimulated litigation. As the historian of the Cravath firm observed, "depression-induced bankruptcies and New Deal agencies engulfed business and created such demands on the profession that competent legal assistance was at a premium." A prominent Seattle corporation lawyer recalled that his firm was "very little affected" by the Depression; in fact, its business increased. And Harrison Tweed of Milbank, Tweed in New York concluded that law firms always earn more "when times are very bad or very good. . . ." On Wall Street those lawyers most likely to be adversely affected by the Depression were young men seeking access to corporate firms. The Cravath firm and Sullivan & Cromwell, to cite two examples, sharply curtailed recruitment between 1931 and 1933. Exceptional applicants, observed a partner in Sullivan & Cromwell, would be hired, "but some of the others may be up against it."[3]

Young professionals were not the only ones to view the future with foreboding. With economic collapse as the prod, the role of the legal profession in American society was re-examined. Frustration and disenchantment bred sharp criticism. In a climate of hostility toward "economic royalists," lawyers who had enjoyed immunity from public criticism during a decade of prosperity suddenly found themselves vulnerable. Any reassessment of the American business system was bound to generate a parallel reconsideration of the profession that served it so conspicuously.

The tone of this criticism was set by Supreme Court Justice Harlan F. Stone, who delivered a widely publicized address in 1934 decrying the diminished public influence of the bar. Stone, nearing the end of his first decade on the Supreme Court, could view his profession with uncommon perspective. His experiences at Sullivan & Cromwell, as dean of Columbia Law School, and as attorney general in the Coolidge administration had exposed him to the major sources of professional opportunity: private practice, legal education, and public service. From these, however, Stone drew scant comfort. Even before World War I he had begun to mourn "a deterioration of our bar both in its personnel, its corporate morale, and, consequently, in the public influence wielded by it. . . ." Professional leadership had already passed into the hands of the business lawyer who, at his best, was a "skillful, resourceful solicitor," but at his worst was "the mere hired man of corporations." Postwar developments heightened Stone's unease. Soon after his appointment to the Supreme Court he declined an invitation to contribute an article on the bar to *Harper's*, explaining that a respectable article would require sharp criticism. After sketching the outlines of such an essay Stone drew back, telling editor Frederick Lewis Allen that if the author "is influential, you are not likely to get very much of an article; if he is critical, he is not likely to be influential."[4]

By 1934, however, Stone could no longer contain himself. The subject of his address, he confessed, "had been festering in my insides long enough so I had to get it out." In a blistering attack, delivered during dedicatory exercises at the Law Quadrangle of the University of Michigan and subsequently published in the *Harvard Law Review*, Stone deplored professional defects that inhibited the ability of the bar to resolve social problems. In a stratified, specialized profession the lawyer no longer served as the "representative and intepreter of his community." Instead, with success measured by income, he managed "a new type of factory, whose legal product is increasing by the result of mass production methods." His primary allegiance went to business clients, not to the professional ideal of disinterested service that once had elevated lawyers to a position of public influence and leadership. Stone sadly concluded that these changes had transformed "the learned profession of an earlier day [into] the obsequious servant of business, and tainted it with the morals and manners of the market place in its most anti-social manifestations."[5]

Stone's nostalgic *cri de coeur* for the golden age of American lawyers, before the country had swallowed the forbidden fruits of industrialization, urbanization, and commercialization, had a familiar ring. Yet Stone's lament touched a raw professional nerve during the thirties. Throughout the decade lawyers were raked by criticism—and by searing self-criticism (which reached its apogee when Yale law professor Fred Rodell advocated the abolition of the legal profession by making it a crime to practice law for money). The most frequent complaint, predictably, was complicity with big business. Harvard law teacher Calvert Magruder told Maryland lawyers that the profession "must cease to take its ethics, its economics, and its political ideals from the banker." Adolf A. Berle of Columbia accused the profession of having become "virtually an intellectual jobber and contractor in business matters." Commercialization had "stripped it of any social functions it might have performed for individuals without wealth." Dean Charles Clark of Yale, cognizant of the preparatory training that law schools provided for metropolitan firms, warned of a profession that might "sink to the role of servitor of modern business." Little wonder that a New York corporation lawyer, stung by criticism, nostalgically recalled his own law-student days when "legitimate money-making was not . . . thought of as a crime, and proper service to business was regarded as service to society."[6]

Although there were many critics, few diagnosed corporate counseling (actually a negligible dimension of the problem) as merely a symptom of structural and organizational defects within the legal profession as a whole. One scholar who did offer such an analysis was Karl Llewellyn, an acute observer of his professional culture. He described the bar as "an almost meaningless conglomeration" containing thousands of lawyers "without unity of tradition, character, background, or objective. . . ." Professional differentiation was a product of legal specialization, reinforced by social and ethnic stratification. It encouraged the channeling of the best professional talent into corporate counseling. This meant, in Llewellyn's words, "that the fitting of law to new conditions has been concentrated on *only one phase* of new conditions: to wit, the furtherance of the business and financing side, *from the angle of the enterpriser and the financier.* It has been focused on organizing their control of others, and on blocking off control of them by others." Bar organization, Llewellyn insisted, was the critical issue. "Our bar is organized, theoretically, in terms of an Adam Smith economy: individual initiative, small enterprisers, individual skill, work, and reputation. Its ethics are so organized. Its theory is so organized." But in fact the large law firm, the most creative response to modern organizational imperatives, had transformed "bar-leaders and aspirants to leadership *ever more* into specialized adherents of the Haves. . . ."[7]

Llewellyn's indictment of the bar for its service to wealth was echoed throughout the thirties, when the legal profession presented an inviting target because of its "surrender" to business, and law "factories" were as vulnerable to attack as their industrial counterparts. Sophisticated critics developed anthropological analogies to explain professional defects. Fred Rodell likened lawyers to medicine-men and priests, "who blend technical competence with plain and fancy hocus-pocus to make themselves masters of their fellow men." Thurman Arnold, a student of folklore and symbols, airily compared the activities of the American Bar Association with the "quaint customs" of primitive Philippine tribes. Journalist Ferdinand Lundberg described the American lawyer as a member of "a privileged priesthood" who interprets "tribal customs and superstitions of the dim past with sacerdotal solemnity. . . ." But Lundberg, like others, missed the point when he asserted that the legal profession was "independent of society" and "psychologically, at least, quite outside of the social system."[8] The legal profession all too accurately mirrored American society; and it enjoyed anything but an independent existence. Every essential feature of professional organization and structure reflected prevailing national values: stratification along ethnic lines; recruitment patterns that rewarded corporate counseling with the highest income and status; availability of legal services according to income rather than need; and a skewed adversary process that distributed its benefits in proportion to the money and power of those who utilized it.

Law teachers led the attack, complaining that professional leadership was reserved for corporation lawyers with constricted social vision. Harvard law professor Thomas Reed Powell, who heard one ABA president advise his students to go to church and join Rotary clubs, concluded that the presidency was reserved for men of "no distinction." His colleague Zechariah Chafee, Jr., told the members of a local bar association that professional leaders devoted their attention to "matters just as appropriate to plumbers as to lawyers." Teachers, convinced that practitioners were crippled by self-interest and social myopia, insisted that they, rather than practicing lawyers, possessed the necessary critical detachment and enlightened awareness to cope with social ills.[9]

Teachers also re-examined their own role and function. Dean Clark, in his presidential address to the Association of American Law Schools in 1933, reminded his colleagues that financiers and businessmen might bear the brunt of blame for the Depression, but "at their right hands as counselors and advisers stand the ablest of the men we have instructed and we ourselves are not too far away." Law teachers, he suggested, no longer could comfort themselves with the assurance that they had taught professional proficiency. "Have we," Clark asked, "taught civic responsibility?" It was difficult to answer affirmatively. Schools had progressively raised their admissions requirements and academic standards, but legal educators rarely questioned the uses to which legal craftsmanship should be put. This might have provided reassurance as an acceptable standard of professional neutrality, but law schools, by avoiding such questions, tacitly acquiesced in the answers provided by others. Their ablest graduates, equipped with the finely honed skills of their craft, had moved inexorably into corporate law firms and the service of business. With the swift discrediting of business institutions that accompanied the early Depression years, this service function was challenged. Law schools, a critic asserted in the *Columbia Law Review*, "must assume a responsibility that they have hitherto neglected in the education of lawyers for decent citizenship."[10]

Law schools found it difficult to assume this burden. "Civic responsibility" and "decent citizenship" were not easily defined by legal technicians, nor was educational inertia easily overcome. Indeed, the author of a study of legal education prepared for the Russell Sage Foundation concluded that even in the 1930's any "wrench from traditionalism" was quite painful. Most difficult was the effort to view law "not as *The Law*, but as a flexible, ever-changing instrumentality created to aid in the resolution of social and economic problems."[11] Schools continued to train eminently practical lawyers, to certify their best students for positions with corporate firms, and to offer courses that emphasized the resolution of corporate problems.

The most unnerving critique of professional values came from a cluster of legal Realists at Yale and Columbia. Their revolt against formal jurisprudence antedated the Depression, but Realists attracted their most enthusiastic following and drew their sharpest criticism after 1933, when the success of New Deal reforms depended upon constitutional flexibility, judicial tolerance for legislative experiments, and an instrumental approach to law. The Realist persuasion rested upon three basic propositions: that law and society were constantly in flux; that law must serve social ends; and that judges made, rather than discovered, law. Realists—Jerome Frank most conspicuously—hammered at illusions of legal certainty held by traditional lawyers. ("*I'm nauseated,*" Frank confessed shortly after publication of *Law and the Modern Mind,* "*by the excessive use of wishful descriptions of the judicial process.*") Before the Depression decade ended, Realism would become vulnerable for its relativism, for the affinity of its most prominent spokesmen for the New Deal, and for its excessive focus on judicial lawmaking. But in the early thirties, skepticism toward tradition was a liberating antidote to what Thomas Reed Powell called "the usual cant" about revealed law in bar association addresses. Realists demystified judicial lawmaking and professional pontificating, which had thwarted a generation of social reformers.[12] Although the Realist critique did not sweep traditionalists from the battlefield of jurisprudence, Realists enjoyed sanctuaries in university law schools and in New Deal agencies to which they retreated in safety after their sallies against bench and bar.

Once the Roosevelt administration had begun to recruit from law faculties and among recent graduates, rather than from the established bar, elite lawyers knew that their public influence and control had waned. Legal Realism rubbed salt in their wounds. At a time when courts and Constitution were vital bulwarks against change, Realists, ensconced in law schools, doubtlessly sounded like the bar's bolsheviks when they insisted that judicial decision-making did not rest solely upon syllogistic reasoning from rules and precedents. John W. Davis put it bluntly when he referred to "wild men" at Yale and Harvard law schools "whose social, economic and legal principles I distrust." James Beck, confessing high regard for law professors for their "philosophic detachment" and for their renunciation of high fees, found them prone to "visionary ideas" which "are not helpful in the development of sound public opinion. . . ." An ABA member, reporting extensive correspondence with practitioners regarding their evaluation of recent law-school graduates, described "a schism between the thought of the law schools and the thought of the Bar." The notions held by professors might be "scintillating in their brilliance and evince profound learning, but at the same time display an utter lack of touch with the realities of the law." One venerable lawyer, reminiscing nostalgically about "the bar of other days," was convinced that law schools were becoming more political than professional; he regretted the "arraignment of distinguished members of our Bar" by law teachers. Yale's decision, under Charles Clark's deanship, to permit law students to take courses in the social sciences so upset one lawyer that his firm refused to employ Yale graduates for years thereafter.[13] It was one thing to instill reverence for law; it was quite another to train students who might become critics rather than defenders of the old order.

The discord between teachers and practitioners was acute, but its implications were susceptible to exaggeration amid the turmoil of the thirties. No basic redirection of legal education occurred; with the exception of some experiments in clinical education, no alternative model of law teaching emerged. If young lawyers began to display "decent citizenship" and "civic responsibility," social forces released by the Depression and the New Deal, not a conspiracy of law teachers, made the difference. Yet university law schools did re-emerge as critical public institutions. Once the Depression prompted a reappraisal of the practicing bar, some teachers rediscovered their public calling. The service function of some university law schools swung perceptibly away from private corporations, toward the state. And law students displayed a sudden and striking degree of generational consciousness.

Felix Frankfurter stood at the intersection of these developments. The New Deal climaxed a drama that had begun for him during the Progressive era, when he had turned to law teaching as the career *par excellence* for involvement in public life. Frankfurter had envisioned a role for law teachers and law schools as participants "in a great state service." During the postwar years, when he often expressed his disappointment at the model provided by bar leaders, he waited in vain for lawyers to protest against the commercialism and corruption of the postwar decade. It pained him that "the attractions of New York" lured the best Harvard law students of that generation, although Frankfurter himself had been responsible for placing many of them. He served virtually as an employment broker for his old friend Emory Buckner, partner in the Root, Clark firm in New York, and he performed similar services for the Cravath firm, for his former mentor Henry Stimson, and for others who came to rely upon his appraisal of legal talent.[14]

The conjunction of the Depression, with its attendant dislocations in the private sector of the legal profession, and the Roosevelt administration, with its special dependence upon legal skills, propelled Frankfurter close to center stage. The personal and political affinity between the professor and the President provided Frankfurter with a unique opportunity to focus his energies and experience as social critic, as teacher, and as job broker. "One would be a complacent optimist," he had written, "who would take pride in the influence exerted by the Bar upon our public affairs in recent times." He often referred to the antediluvian attitudes of bar leaders, citing the "obtuseness" of the bar and "the inaccessibility of its mind to the needs of a rather rapidly changing world." Disdainfully, he described bar leaders as "about the least educable portion of the community."[15]

One year before Roosevelt's inauguration Frankfurter expressed his conviction that "never has there been greater need in this country for the quality and the talents that the best in the law can give to society." Yet, he conceded, "I am more and more compelled to the conclusion that I am spending my time the better to fashion minds whose chief concern is the making of money." Frankfurter's model of state service, inspired by the Wisconsin idea, was nurtured by his admiration for Louis Brandeis and Henry Stimson, strengthened by his conviction that law school graduates were especially competent to assume the responsibilities of governance, and reinforced by his envy of the respect afforded to British civil servants. Expertise, Frankfurter insisted, was indispensable to efficient government in a modern society. But, he had written midway in the Hoover administration, "the whole tide of opinion is against public administration as a career for talent." When one of his superior students rejected a position with the Reconstruction Finance Corporation for private practice, Frankfurter chastised "those worldly wise men in New York" who advised young lawyers to spurn careers in teaching or government. When another student, who chose the favored path, reported about his work in utilities regulation, Frankfurter took obvious delight in "what 'the Wisconsin idea' meant in action—the continuous and systematic utilization of the best available intelligence on the complicated social problems of the day."[16] It was this opportunity and challenge that the New Deal provided on a national scale.

As Frankfurter surveyed available legal personnel during the early weeks of the Roosevelt administration he expressed dismay at "the lack of capable, free men among lawyers" between the ages of thirty and fifty. Freedom meant "the resiliency of mind and judgment . . . that is ready to take in new facts and new forces and to realize that new accommodations have to be made, though made in the organic unfolding of valid past traditions and techniques." Lawyers, Frankfurter believed, should be "experts in relevance"; yet all too often they were practitioners of obstructionism. In an unusually sharp letter, he reminded Henry Stimson that prominent law firms "were in some cases the architects and in others the agents of practices which you would be the very first to regard as indefensible and anti-social." Accusing them of efforts to "chloroform" New Deal legislation, he concluded with a denunciation of "exploiting business men and their leading lawyers. . . ." When James Landis, who served on the Federal Trade Commission and the Securities and Exchange Commission to enforce the new securities laws, described "the genius of the New York bar for devious misconstruction," Frankfurter repeated his comment approvingly and referred to their belief that "they serve God when their client is Mammon." Inevitably, Frankfurter concluded, financial lawyers would retard or defeat New Deal legislation—"unless those lawyers are matched at their own game, in advance, by the use of lawyers equally astute in the public interest and as ready to devote their time to the public cause as are Wall Street attorneys to the cause of Wall Street interests."[17]

By 1933 this no longer was an idle hope. Frankfurter was in contact with "some of the ablest younger lawyers in the big New York offices," who derived little satisfaction from their work. "Literally by the score they are sick of it all," he told Walter Lippmann. "That's one of the heartening things about the times— that the Government can avail itself of an abler lot of younger lawyers for key junior positions than has been true of any time since I left the Law School." In addition, there was the annual crop of fresh recruits, who were drawn to Washington by the special attractions of the New Deal and by restricted opportunities in private practice. Frankfurter quickly found "impressive evidence that the more recent generations of law school graduates care about the law as a great social process and as opportunities for having a share in the effort to solve some of the most complicated riddles of modern society." Few individuals deserved as much credit for this development as Frankfurter himself. To a generation of law students, wrote one among them, Frankfurter conveyed the conviction "that there was no more challenging and exciting business in the world than the responsible, craftsmanlike handling of the power of the state."[18]

Frankfurter derived immense pride from the migration of lawyers to Washington. He knew that the success of the administration, to which he was deeply committed, depended upon its ability to compete with private interests "for the command of brains." Its outstanding achievement, in Frankfurter's judgment, was the extent to which Roosevelt "stirred the imagination of younger people to the adventure of, and the durable satisfactions to be derived from, public service. . . ." As Frankfurter told one young protegé whose presence in Washington delighted him: "You know very well that I regard the building up of the equivalent of the British Civil Service in our country as second in importance to nothing affecting the public life of our nation." Aboard the *Britannic*, sailing to England for a year at Oxford, he expressed his satisfaction "that there are more intelligent and more purposeful and more disinterested men in the service of the country than there has been for at least half a century."[19]

The precise nature of Frankfurter's role in the New Deal was difficult to calculate amid the political hyperbole of the 1930's. Roosevelt's opponents depicted Frankfurter as a sinister, diabolical schemer. *"The most influential single individual in the United States,"* whose " 'boys' have been insinuated into obscure but key positions in every vital department," Hugh Johnson thundered after he left the National Recovery Administration. According to journalist John Franklin Carter, Frankfurter dominated "the infant industry of legal liberalism," supplying lawyers who were "sufficiently ingenious to justify the New Deal to the Courts and sufficiently radical to sympathize heartily with its purposes." Critics on the left, seizing upon the continuity of Frankfurter's job brokerage function, bolstered their contention that New Dealers were providing only mild medicine for economic and social ills that could be cured only by drastic surgery. Thus Fred Rodell of Yale drew attention in an article entitled "Felix Frankfurter, Conservative" to the number of Frankfurter protegés who had served the Harding, Coolidge, and Hoover administrations or profited from Wall Street practice.[20]

Frankfurter's self-assessment varied according to circumstance and audience. He persistently claimed that he initiated no recommendations regarding personnel or policy "unless asked." When public exposés of his role appeared, he backed away even further. In the wake of Hugh Johnson's denunciation of him as the Iago of the administration, Frankfurter observed that many more of his former students held positions in Wall Street firms than in New Deal agencies. He took vigorous exception to similar criticism from editor Raymond Moley, a displaced member of the Roosevelt "brain trust"—prompting Moley to remind him that "whether or not you regard yourself as the leader of 'the boys,' they certainly regard themselves as your disciples. . . ." The presence of his "boys" in such profusion—Dean Acheson, Benjamin Cohen, Thomas Corcoran, Paul Freund, Alger and Donald Hiss, James Landis, David Lilienthal, Nathan Margold, and Charles E. Wyzanski, Jr.—enabled Frankfurter to play a role and deny it too. Corcoran, who functioned as his personal (and personnel) ambassador to New Deal administrators, acknowledged his mentor's role and cheerfully carried out Frankfurter's instructions. Others performed similar services at Frankfurter's instigation. He could thereby honor his "fixed rule . . . not to make any requests of any of the officials." But he relied upon Corcoran, Freund, Wyzanski, and others to "insulate [his] intervention."[21] By intervening covertly through intermediaries Frankfurter maintained both his private influence and his public distance.

Several considerations dictated his need for camouflage. He anticipated a diminishing willingness for public service if "this silly, uncritical, wholesale gibing at the 'brain-trusters'" persisted. He often cited the respect and esteem directed toward British civil servants, in contrast to the carping criticism that characterized the Washington scene. "One of the worst of American traditions," Frankfurter complained, was "that anybody could do everything, and that the government is no place except for the drone and the politician." Eager to transfer the British model to the United States, and anxious lest young lawyers be deflected from government service, Frankfurter sought to minimize his own involvement in order to remove a potential irritant upon public opinion. In his judgment he could serve the President and his own policy objectives best by working covertly through others.

Although Frankfurter gained notoriety as "the Jew, the 'red,' the 'alien'" within New Deal councils, the allegations of radicalism once again were wildly inappropriate. Frankfurter was the prototypical immigrant boy who thirsted for recognition as an American. He unequivocally embraced the traditions and symbols of his adopted land, yet he could never forget that he was an outsider. His continuing reiteration of patriotic loyalty, most evident in his unabashed worship of Franklin D. Roosevelt and the institutions of American government, demonstrated a yearning that never was fully assuaged. Frankfurter revered tradition, not innovation; he was more inspired by the British civil service than by revolutionary soviets. Scornful of egalitarian government, he supported rule by an elite—preferably a legal elite. His ideas had not changed since the Progressive era, but during the Depression they won their most sympathetic hearing. Frankfurter's strategic location within the profession and his unique access to the administration provided him with the opportunity to cultivate a counter-elite that would mitigate the influence of the dominant professional leadership he despised, and to serve an administration he revered, thereby identifying himself, finally and fully, as an American. "Let people see how much I loved Roosevelt, how much I loved my country," he would say to the editor of his letters near the end of his life.[22]

The demographic and social patterns of the 1920's made possible Frankfurter's achievement. Originating beyond the profession, they first affected university law schools, the key institutions of professional access. In the postwar decade there was a sharp spurt, exceeding 80 percent, in law school enrollments. And between 1920 and 1930, in a few Eastern cities with a heavy concentration of immigrants, there was a substantial proportional increase in the number of foreign-born lawyers.[23] The professional elite easily defended law firms and professional associations against intruders. But the Depression, which restricted employment in the private sector, generated pressure for alternative career opportunities. The Roosevelt administration offered the most promising (and, to the old elite, the most threatening) option. New Deal agencies were enemy country. They attracted lawyers with relatively weak commitments to private practice in corporate firms; lawyers trained in administrative law, skilled in legal draftsmanship, who had been exposed to the stimulating currents of legal Realism and were eager to apply their expertise; and younger lawyers, drawn disproportionately from ethnic minority groups, who satisfied personal, professional, and ideological needs in public service for a liberal reform administration.

"Public service" meant different things to lawyers who reached Washington at various stages of their careers. The attraction of the New Deal resided in the multiple possibilities it afforded for personal and professional opportunity. It could simultaneously appeal to a settled attorney who seized the opportunity for new professional experiences; to an independent country lawyer whose model was nineteenth-century practice; to a legal Realist who made bold intellectual leaps beyond most of his twentieth-century professional counterparts; and to a craftsman who saw nothing incongruous in voting for Hoover and then committing his energy to the Roosevelt administration. Above all, it attracted young, upwardly mobile, minority-group lawyers whose professional advancement within traditional channels was thwarted by the Depression and frustrated by the social structure of the bar.[24] Their choices revealed the primary sources of professional frustration and energy during a decade of dislocation, when the social values which had sustained the legal profession were exposed and suspect.

A special fillip of excitement attracted successful, established lawyers who were secure and prosperous in private practice. Francis Biddle left his prestigious Philadelphia family firm, whose clients included the Pennsylvania Railroad, for "the sense of freedom, the feeling of power, and the experience of the enlarging horizons of public work." Exposed to the rewards of government service, Biddle was compelled to concede that private practice no longer satisfied him, because "there was little dedication to ends beyond monetary rewards for the narrower needs of self." Lloyd Landau, president of the *Harvard Law Review* during World War I and subsequently a clerk to Justice Holmes, was prepared to abandon a substantial private practice with a large annual income for the opportunity to serve under Frankfurter if his old mentor became Roosevelt's solicitor general. A New York attorney, who had served in Washington during the war, wanted to return because he sensed that a position with the New Deal "might be even more of a thrill." Established law teachers found novel gratifications. James Landis, who temporarily vacated his Harvard professorship, quickly discovered that he could enjoy "a larger share in the handling of government than I ever had after years in the handling of the Harvard Law School." Landis had dreaded interviews with Dean Pound; he never enjoyed the privilege of an interview with President Lowell; but he anticipated conferences with Roosevelt "with pleasure, knowing that there will be an exchange of views. . . . It is things like this that make life fun." Thomas Corcoran echoed that theme, declaring of his New Deal experience: "We're all having the best time we'll ever have, not because of the power or closeness to great events, although those are exciting, but because we're functioning—using all our muscles to the full."[25] For these established lawyers the New Deal provided satisfactions—notwithstanding their modesty about the exercise of power—which had eluded them in practice or teaching.

Robert H. Jackson was an unlikely New Deal lawyer. The prototypical New Dealer was an upwardly mobile urbanite, a second-generation member of an ethnic minority group with superior academic credentials and, perhaps, some Wall Street experience. Jackson was the obverse: an upstate Protestant New Yorker who never attended college, attended but never graduated from Albany Law School, served an apprenticeship in a Jamestown law office, and incessantly preached the nineteenth-century virtues of the small-town practitioner: "hard work, long hours, and thrift." The son of a Pennsylvania farmer, he was later aptly described as a lawyer whose "bent was to plow old pastures in a new way, not to leap fences and attack virgin soil. . . . It was his job to defend, not to formulate, policies."[26] The consummate advocate, he defended the New Deal as special counsel for the Securities and Exchange Commission, as assistant attorney general in the tax and antitrust divisions of the Justice Department, and as solicitor general and attorney general. Regardless of office, Jackson remained the nineteenth-century liberal in the twentieth century; his anachronistic liberalism was conspicuous, yet as a New Dealer he seemed to march in step with the times. This was less paradoxical than might appear. His critique of the legal profession, a recurring theme in his public addresses, focused on the corporate lawyer as the personification of wrongdoing; for his was the animus of Main Street displaced professionally by Wall Street. Jackson's New Deal colleagues, who voiced similar complaints, fired at the same target for different reasons. Theirs was the cry of contemporary politics; his was the voice of nostalgic betrayal.

The nobility of the legal profession, for Jackson, derived from the work of general practitioners, who were "not submerged in a specialty nor dominated by a single overshadowing interest, who become champions of any just cause. . . ." Such men found no place in metropolitan law firms, whose distinctive attributes were "the wide clientele and the narrow lawyer." Once a leader, the lawyer had become a mouthpiece. "More than any other class," Jackson insisted, "our opinions, as well as our services and talents, are on the auction block." In professional associations "we generally pyramid conservatism until at the top of the structures our bar association officers are as conservative as cemetery trustees." Speaking to law teachers in 1934, Jackson lamented the concentration of legal business and talent in large metropolitan firms and the consequent malaise of the middle-class bar, once the backbone of the profession. Unable to retreat to the nineteenth century, he sought to transplant its virtues to a different era. He spoke of the government's need for lawyers who possessed education and experience, but were not overcome by "mental ossification." The bar, he concluded, "is one of the most stubborn, reactionary and short-sighted groups in our national life. . . ." But, he added with revealing ambivalence, "I should be sadly disappointed if my son should fail to join it."[27] Government service, with the nation as his client, brought Jackson closer to the elusive nineteenth-century spirit of independent practice than anything else he could possibly do. Jackson "embodied a significant part of the American dream," one of his admirers has written—"the storybook American boy who by dint of brains and work and pluck drives himself from an unpromising start to a glorious finish."[28] He won the highest legal prizes the Roosevelt administration could bestow, yet he was an incongruous New Deal lawyer; he seemed most contemporary when he spoke for the bygone liberal professionalism of an earlier era.

If Jackson carefully tilled old soil, Jerome Frank preferred to leap fences. A precocious graduate of the University of Chicago at nineteen, he yearned to write novels, instead became a lawyer at his father's insistence, and developed a lucrative corporate law practice, first in Chicago and then in New York, which he never enjoyed. Psychoanalysis had convinced him that lawyers chose "childish thoughtways in meeting adult problems." He wrote *Law and the Modern Mind*, an exciting venture into the psychology of jurisprudence, out of the desire "to see and have others see and help me see more clearly just what we lawyers are doing daily." Why, he asked Roscoe Pound, was "absolutistic thinking so difficult to surmount in cerebration about law? Why is certainty-hunger peculiarly vigorous in lawyerdom? . . . How make [lawyers] . . . eager to think pragmatically, to use concepts operationally, instrumentally?" Frank, who delighted in tilting against "illusions about legal certainty [that] get the lawyers in bad with the public," rebelled against what he perceived as legal authoritarianism. Most lawyers and judges, he wrote in his book, insisted upon the certainty of law when it was, in fact, "largely vague and variable." They did so because they had "not yet relinquished the childish need for an authoritative father and unconsciously have tried to find in the law a substitute for those attributes of firmness, sureness, certainty and infallibility ascribed in childhood to the father." Frank demanded "a skepticism stimulated by a zeal to reform, in the interest of justice, some court-house ways." He tried, in the words of his close friend Thurman Arnold, "to free the law from its frustrating obsessions. His jurisprudence was the jurisprudence of therapy."[29]

After the crash the financial distress of Frank's clients preoccupied him, and he became frustrated and restless. Although he managed to write his book, maintain a voluminous correspondence, and engage in a busy practice, he complained: "It's hell how practicing law interferes with decent intellection." Shortly before the 1932 election he confessed to being "so fed up with the tawdry aspects of practice" that he would welcome an academic appointment. Roosevelt's victory opened tempting possibilities. Frank offered his services to Adolf Berle; he suggested to Thurman Arnold that Yale, where Frank lectured, organize its own brain-trusters; and he accepted with alacrity an invitation, extended at Frankfurter's behest, to draft farm legislation and then to become general counsel of the Agricultural Adjustment Administration. "Financially, it is a somewhat risky adventure for me," he conceded, "but I couldn't resist the opportunity."[30]

Frank personified the affinity between legal Realism and the New Deal. Realists, he declared in a thinly veiled autobiographical statement, easily became New Dealers because they were "less Procrustean and more flexible in their techniques" and because they judged legal institutions by their human consequences rather than by their Platonic essences. As experimentalists they were skeptical of their own notions but not paralyzed by inaction. The lawyer who believed in "undeviating fixed legal principles,"

Frank's "Mr. Absolute," would be repelled by the New Deal. His adversary, "Mr. Try-it," could run social experiments for sixteen hours each day without strain or fatigue.[31]

Yet Frank's social experiments were designed merely to harness private financial gain to social welfare. This presumably rash, brash experimentalist, freed from authoritarian dogma, only wanted "*the profit system to be tried, for the first time, as a consciously directed means of promoting the general good.*" Here were the limits to his experimentalism—a point never appreciated by those who criticized him as the New Deal's Robespierre. Frank suffered from a reputation exceeded only by Frankfurter's as the radical lawyer-ogre of the administration. Once the Agricultural Adjustment Administration became the battleground for a clash between Southern sharecroppers and their landlords, Frank and his group of talented, socially committed young associates, sympathetic to the plight of the sharecroppers and eager to secure their legal rights, were suspect, vulnerable, and finally expendable. But Frank, who wrestled self-consciously with drawing the boundary between policy preferences and legal judgments, demonstrated considerably greater restraint than his public reputation suggested. He insisted that his own advocacy was usually directed toward inducing his colleagues "to narrow the issues so as to confine the argument as far as possible to controversy on *traditional lines.*" As he told Frankfurter: "I do not believe in trying to vindicate abstract principles and . . . to me the important thing is to win *particular* cases." Frankfurter knew that Frank wanted to win cases; he also saw that Frank was "a damned romantic intellectual." Frank conceded, yet denigrated, his romanticism. He described his work as general counsel as "heartbreaking days and nights spent with almost reckless financial sacrifice in aid of public causes [I] deem desirable. . . ." Yet, he hastened to add, one of his major aims was "to have our job done with legal accuracy—so that it would stand up in court. . . ."[32]

Therein lay the source of the tension that tormented Frank as long as he remained in Washington. His personal commitment to Realism and to experimentalism impelled him toward policy-making; his professional preferences for process and precedent restrained him. An experimentalist as to legal means, he unquestioningly accepted social ends. Standing at the cutting edge of legal thought in the Roosevelt administration, he demonstrated the compatibility of Realism with the New Deal and the strength that each derived from the other. But the conflict between professionalism and capitalism on the one hand, and social change on the other, was painful; by 1935 he felt "functionless." He sensed that his effectiveness in Washington was at an end, but he dreaded returning to private practice, where he anticipated a hostile reception. "I'm badly bewildered," he told Frankfurter, "—and not a little frightened."[33] Frank, like the administration he served with such passionate distinction, was simultaneously liberated by lawyers' skills and inhibited by lawyers' values.

If the New Deal appealed to nostalgic liberals like Robert Jackson and to bold Realists like Jerome Frank (along with Thurman Arnold and William O. Douglas), it also attracted able legal technicians who found matchless opportunities for honing their skills and practicing their craft. Charles E. Wyzanski, Jr., grandson of a successful immigrant peddler and a product of Exeter, Harvard College, and Harvard Law School, was one of these. Drawn to law study after reading Zechariah Chafee's *Freedom of Speech*, he was a law review editor, clerked for both Learned and Augustus Hand, and practiced for three years in Boston's prestigious Ropes, Gray firm. When the Roosevelt administration came to power Wyzanski's career was still in its formative stages, and New Deal opportunities set his course. As solicitor for the Labor Department he enjoyed immeasurable freedom to apply his legal skills to problems of statecraft. Wyzanski, who had voted for Hoover in 1932, hardly went to Washington as a crusading reformer. But he took pride in the fact that "we were a level of employees that Washington hadn't previously seen." And the considerable demands upon his skills were exhilarating. Given twenty-four hours to draft the public works title of the National Industrial Recovery bill, he compared the travail to "plunging into the furnace."[34]

Wyzanski's instinct for craftsmanship made him eager for

"more law work, and less administration." By 1935 he concluded that the solicitorship offered "less play to legal than to political and administrative currents." Moving over to the Justice Department, he compared his earlier government work, which taught him "to analyze quickly, to assume responsibility and to act courageously," with the greater fulfillment provided by "the intellectual satisfaction which comes from a chance to turn problems around so that every angle is displayed. . . ." He subsequently referred to the *process* of drafting the Wagner Act brief as the consummate experience of his Washington service: "We would talk back and forth at each sentence. . . . We were just a crowd hard at work . . . doing our best to understand the kernel of the thought, and then reducing it to the narrowest possible statement. . . ." This was the distillation of his law school training under Thomas Reed Powell who, Wyzanski recalled years later, could "make you think twenty times before you write that sentence quite that way."[35] Wyzanski believed that he was participating in the restoration of the lawyer's role as social mediator, a role weakened by lawyers for corporations who were "too loyal to a *part* of the community to see the new problems in the light of the *whole* community." New Deal social legislation represented an effort to restore the equilibrium between public need and private right. Wyzanski treasured nothing from his Washington experience more "than the feeling that I have been part of a practice (which I hope will become a tradition), under which young men give part of their early manhood to public service."[36]

Relatively established lawyers like Jackson, Frank, Wyzanski—even Thomas Corcoran and Ben Cohen—provided the principal manpower pool for the early New Deal. But it was the law school graduates of the 1930's, those "born to an era of insecurity," who did most for the élan of the Washington New Deal community. It was these younger lawyers on whom the New Deal worked its special magic. They seized the opportunity to shoulder responsibility, to influence policy, to exercise power, and to commit themselves to a reform administration during a national emergency when, they believed, the course of history might turn upon their efforts. Karl Llewellyn observed the diminished appeal of "Wall-Street-flocking" and expressed surprise at the number of young lawyers who "hunger to make law *do* something. The number who prefer a government job. The number who are pestered with the prospect of becoming prostitutes."[37] Eric Temple, lawyer in a Louis Auchincloss novel, epitomizes the spirit of this generation when he says: "I have decided to . . . go to work for Mr. Roosevelt and the future instead of continuing with Messers Arnold and Bovee and the past."[38] The choices of Harvard, Yale, and Columbia law review editors, who comprised the student elite (as the law schools and law firms defined it), illustrated the deflection of talent from Wall Street to Washington. Among editors who had graduated from these schools a decade earlier, 81 percent went directly into private (invariably corporate) practice. Only 6 percent took positions with the federal government. Graduates of the classes of 1930-32, the first Depression generation, entered private practice at a significantly reduced rate (67 percent). Once the Roosevelt administration took office, law review editors began their pilgrimage to Washington. Twelve percent of the editors from the 1933-35 classes took New Deal positions, a figure that tripled the proportion of those entering government service from the 1930-32 classes. The first year of the New Deal marked the flood tide. Almost as many editors took federal government jobs during 1933 as had done so during the preceding fourteen years. The crash pulled lawyers away from private practice; but the turn toward the federal government was distinctly the result of the coming of the New Deal.[39]

The influx to Washington of *Harvard Law Review* editors accounted for the early surge. Their choices underscored the half-jocular contemporary claim (and complaint) that the most direct route to Washington was to go to Harvard Law School and turn left. (Two-thirds of the Harvard editors from the classes of 1930-32 were employed by the federal government at some time during the 1930's. By contrast, slightly more than one-quarter of Yale editors and one-third of Columbia editors from those years went to Washington, although their numbers increased after 1933.)

Something special at Harvard galvanized law review editors: the conjunction of Felix Frankfurter's influence with the Depression and the New Deal. His advocacy of public service as the lawyer's highest calling, which had barely penetrated the dominant private practice aspirations of his pre-Depression students when alternatives to private practice were limited and the motivation to pursue them was lacking, left its mark once the economic structure of the nation collapsed and an energetic reform administration offered careers and a cause. Then the Harvard pattern changed swiftly. Not only did graduates of the Depression generation flock to Washington; men from the twenties generation also went there, despite the fact that they had grown to maturity during a prosperous decade and had experienced the relatively rich rewards of private corporate practice. Frankfurter's conviction that there was nothing more challenging and exciting than public service assumed new meaning for law students after 1932.[40]

Young lawyers who went to Washington were exhilarated and exhausted by the demands upon them, especially in the innovative newer agencies. Abe Fortas, who worked under Jerome Frank in the Agricultural Adjustment Administration, "could see the new world and feel it taking form under our hands." One of his colleagues discovered that for the first time in his life he was working sixteen hours daily under the spur of enjoyment rather than from ambition or compulsion. Another attorney recalled the rare opportunity that he enjoyed to draft a statute, guide it through litigation, and then defend his own handiwork before the Supreme Court. It was, he concluded, "a lot of fun."[41] It was also hard work. Donald Richberg, general counsel for the National Recovery Administration, compared the calm life of a practicing lawyer with the "intolerable burden" endured by a government attorney. Thurman Arnold discovered "a lot of drudgery"; Ben Cohen, tucked away in the Federal Emergency Relief Administration, found administrative work a disappointing contrast to legislative draftsmanship and yearned for his freedom; James Landis learned that even a day that ended at midnight left work undone.[42]

Various motives prompted young lawyers to become New

Dealers. A commitment to government employment, a crude index of "public service," meant different things to different lawyers. Some, doubtlessly, wanted only to earn a living. Others lusted after power. Still others were unabashedly opportunistic: Adlai E. Stevenson, for example, wanted experience in administrative law because it "should prove useful after my return to private practice."[43]

Other attorneys found the New Deal ideologically compatible —or sufficiently fluid to permit them to implement their own political and social commitments. Nathan Witt, galvanized by the Sacco-Vanzetti case, drove a taxi for two years to earn enough money to afford Harvard Law School. His greatest ambition, he told Frankfurter, was to devote his energies to "the public service of the law." This meant work for minority groups, who were "most likely to complain of the failure to be accorded even-handed justice." Witt battled for sharecroppers in AAA and for workers as general counsel for the National Labor Relations Board. Lee Pressman, his classmate and friend, followed a parallel path. Pressman received his decisive push from a course on labor unionism at Cornell. Unable to find work in labor law because no firm specialized in the field, he did corporate receivership and reorganization work with Frank, until he eagerly escaped the "yoke" of private practice after Frank went to Washington.[44] Pressman, like Witt, joined Frank in AAA; like Witt, he moved over to the labor movement; like Witt, he flirted with the Communist Party. These second-generation radical children of immigrant parents found the New Deal congenial to the investment of their legal skills and to the nourishment of their radical political convictions.

But Thomas Emerson, unlike Witt and Pressman, came from a venerable family whose forebears had reached New England in the seventeenth century. Emerson, first in his class at Yale Law School and editor-in-chief of the *Law Journal*, had his pick of offers from Cravath, Sullivan & Cromwell, Root, Clark, and Davis, Polk. Uneasy about the routinized absence of individual responsibility for fledgling attorneys in those firms, he chose instead to work for Walter Pollak, a talented civil liberties lawyer who had argued the *Gitlow* case before the Supreme Court, often acted as counsel to the American Civil Liberties Union, committed his firm to matters of "social significance," and believed in "justice through [the] legal process." Emerson was not disappointed; his first case was the landmark appeal of the Scottsboro boys in *Powell v. Alabama*. But by mid-1933, after two years with Pollak, Emerson responded to the excitement in Washington. Moving along Frankfurter's underground railroad, he reached the National Recovery Administration and journeyed from there to the National Labor Relations Board. In both agencies Emerson delighted in the immediate delegation of responsibility to young lawyers, the tumult and the challenge, and the opportunity to implement his own belief in law as "an instrument by which social change can be effectuated."[45]

Function, freedom, responsibility, power—compensation enough for the endless hours, dizzying pace, and hostile barbs from New Deal critics. But for many young Jewish lawyers who went to Washington there was an additional inducement, less frequently articulated but perhaps more consequential than any other. To them the New Deal offered a unique opportunity for upward professional and social mobility. The newest members of a profession with a legacy (and indeed a future) of anti-Semitism in its elite circles, they were not often permitted to climb the rungs of the Wall Street ladder. But the crash, as Arthur Schlesinger, Jr., has observed, "led to a general discrediting of the older ruling classes . . . and a sudden opening of opportunity for men and ethnic groups on the way up in the competition for position and power."[46] The New Deal did nothing to topple the Wall Street ladder; if anything it strengthened Wall Street firms by providing them with considerable additional business. But it did erect another ladder in Washington and invite Jewish lawyers to scramble for the highest rung. Jewish lawyers were not the sole targets of discrimination, but their ambition and achievement brought the brunt of professional prejudice upon them. Disproportionately concentrated at the top of their law school classes, they were disproportionately clustered at the bottom of the metropolitan bar.

But the New Deal needed legal talent, and Jewish lawyers needed the jobs that the New Deal provided. Lawyers who defended unregulated corporate enterprise were predictably enraged to see young Jewish lawyers in Washington drafting and enforcing regulatory statutes against their clients. The Roosevelt administration challenged the dominant professional culture and its political values and symbols. It enabled a new elite, drawn from different social and ethnic strata, to begin its ascent to professional influence and power.

Professional discrimination and job retrenchment during the depression virtually eliminated the prospects of Jewish, Catholic, and black lawyers for remunerative employment in the more lucrative sectors of the profession—regardless of their qualifications. For Jews, at the time perhaps the most professionally ambitious minority group, the problem was especially acute. Catholics and blacks tended to cluster in ethnic law schools—Fordham, Georgetown, Howard—and follow narrow channels into state and municipal politics, lower-level federal government employment (for example, in the Federal Bureau of Investigation for Catholics), or solo practice. Prospective Jewish lawyers, however, competed successfully with the Protestant elite in the national law schools, only to discover that a law review editorship was insufficient for elite certification. A few German Jews from an earlier generation—Brandeis, Louis Marshall, Julian Mack, Samuel Untermeyer—had securely established themselves. Against high odds, some pre-Depression Jewish graduates—Frank, Wyzanski, Cohen, Pressman—also managed to carve out successful private practices. But the Depression generation of talented Jewish law students was saved from professional extinction, insofar as it was saved at all, only by the New Deal alphabet agencies.[47]

For no group of second-generation Americans did the New Deal serve as a more efficacious vehicle for social mobility and political power than for Jewish lawyers, who in many instances possessed every necessary credential for professional elite status except for the requisite social origins. Teachers and practitioners with experience in job placement received constant confirmation of the national scope of anti-Semitism. In Boston, Frankfurter concluded, "none of the so-called desirable firms . . . will take a Jew." In Chicago, Jewish law review graduates of Northwestern were turned away from elite firms. In New York, Emory Buckner, a partner in the Root, Clark firm, cited "a somewhat restricted area for Jewish boys." Although Buckner referred to his own firm as a "notable exception," he verified the existence of the problem when he described a Jewish lawyer in his firm as "devoid of every known quality which we in New York mean when we call a man 'Jewy.' " When Jerome Frank received a list of Yale graduates seeking employment, those who were Jewish, although highly recommended, were specifically identified. Thurman Arnold, enthusiastically recommending a *Yale Law Journal* editor, emphasized that his Jewishness was his only handicap but added that he was devoid of "Jewish characteristics."[48]

Not only were Jewish and Christian law review editors hired according to religion rather than achievement, but the "social background" of Christian law students who lacked academic distinction received special comment. The hiring partner in one restricted Wall Street firm told Arnold that there were places in his office for those who had not earned high marks—a covert form of reassurance that unexceptional students who were Protestant need not lose out to superior students who were Jewish. Firms that refused to hire Jews, except under "extraordinary circumstances," justified their decision by referring to the prejudices of their clients. Firms with even a single Jewish partner were flooded with applications from the best Jewish students. In the spring of 1936 eight *Harvard Law Review* editors—all Jewish—still were not placed for the following year; Felix Frankfurter subsequently complained bitterly: "I wonder whether this School shouldn't tell Jewish students that they go through . . . at their own risk of ever having opportunity of entering the best law offices." New York attorney Morris Ernst finally urged Supreme Court Justice Harlan Stone to meet with the partners of Wall Street firms in an attempt to ease restrictive hiring practices.[49]

The situation was no better—and perhaps worse—in university

law school faculties. The dean at North Carolina wondered about the wisdom of hiring a Jewish *Harvard Law Review* editor, given the "provincialism" of the community. James Landis, who recommended a Jewish graduate of Harvard for a position on the Illinois faculty, emphasized: "I do not regard him as forward or pushing, and I should have no hesitation in saying that such Jewish characteristics as he possesses are not a handicap." The dean at Northwestern told Thurman Arnold that his colleagues had not agreed to the appointment of a Jewish candidate during his tenure at the school. Although his faculty had appointed a half-Jew as a librarian, any attempt to appoint a Jew to the faculty would be an "idle gesture." (Arnold's unsuccessful candidate was a recent graduate named Abe Fortas.)[50]

The Jewish lawyer from an immigrant family who managed to secure a New Deal position recognized it as his own coming of age as an American. Malcolm A. Hoffmann, a Harvard graduate, described himself at the outset of his government service with the NLRB as "a young neophyte at the bar, a member of a minority religious group, a boy who had never seen the inside of a political club nor had power nor status in our huge egalitarian society. . . ." Government employment provided just that sense of power and status. It legitimized the aspirations of minority group members and assuaged the disappointment that they encountered, or anticipated, in the private sector. Roosevelt, sensitive to the social implications of government service, tried to tap this supply. "Dig me up fifteen or twenty youthful Abraham Lincolns from Manhattan and the Bronx to choose from," he told Charles C. Burlingham. "They must be liberal from belief and not by lip service. They must have an inherent contempt both for the John W. Davises and the Max Steuers. They must know what life in a tenement means. They must have no social ambition."[51]

Except for the absence of social ambition, Roosevelt procured the type of lawyers he sought. Indeed, in Washington the problem was too many Jews, not too few. Nathan Margold, solicitor for the Interior Department, and Jerome Frank in AAA had numerous legal jobs to fill; both, however, were troubled by the overabundance of qualified Jewish lawyers and by the political liabilities inherent in placing too many of them on their staffs. As Adlai Stevenson wrote from the Agricultural Adjustment Administration, "There is a little feeling that the Jews are getting too prominent— . . . many of them are autocratic and the effect on the public—the industries that crowd our rooms all day—is bad." The poignancy of the problem was compounded by the flood of requests from young highly qualified Jewish lawyers who pleaded, usually with Frankfurter, for New Deal employment. Other minority-group members were, if anything, at an even greater disadvantage. An Armenian-born female law review editor from Wisconsin Law School, suffering from the double professional handicap of ethnicity and sex, asked in vain for help. Black lawyers not only were barred from white firms; they also suffered discrimination at the administration's hands. One black attorney, seeking an NRA position, was kept waiting for three hours while every white applicant was interviewed; finally he was told that the position was reserved for whites only. Angrily, he confronted his painful dilemma: "One is driven to hate either his color or his country."[52] Thus while the New Deal did open the door a crack to professional opportunity—especially for Jews, to a lesser extent for Irish Catholic lawyers like Corcoran, Frank Murphy, and Charles Fahy, and, on rare occasions, for a black lawyer like William Hastie—the great wall of ethnic exclusion, made more imposing by Depression conditions, still cut through the legal profession.

The fortunate ones, those young lawyers who capitalized upon New Deal opportunities, shared a special generational experience. Their motives were too tangled for easy resolution. The subsequent careers of some New Deal lawyers might suggest that government service merely offered a novel means for the pursuit of the traditional ends of money and power; that New Deal lawyers were ideological chameleons, opportunists whose affirmations of public service and liberal reform cloaked self-interest. As the demand for lawyers fell in private practice and rose in government service lawyers responded, as they always had, to market factors;

the availability of jobs mattered more than the political identity of the employer. Unquestionably New Deal service paid some lawyers a deferred dividend: ample financial rewards in private practice for the expertise developed in Washington.

Still, though opportunism explains much, it does not explain everything. It does not explain the willingness of established lawyers to leave private practice for government employment when economic self-interest dictated otherwise. It does not explain why law review editors, who (except for Jews) were the least susceptible to the contraction of job opportunities, went to Washington. Only *post hoc* reasoning can support the proposition that because some lawyers subsequently capitalized upon their New Deal experience, they all went to Washington in the first place motivated solely by the knowledge or expectation of later financial or professional rewards. For that generation of lawyers there were few precedents upon which to base such expectations. Similarly, opportunism which construes power rather than money as the dominant motive fails to link power to goals. The desire to exercise power doubtlessly was strong, even dominant in many lawyers, but that desire cannot realistically be considered apart from New Deal politics and values. The special attractiveness of the New Deal to legal Realists, the influence of law teachers (like Frankfurter) who were publicly identified with liberal reform, and the commitments of many young lawyers to the labor movement or to other liberal causes espoused (or tolerated) by the administration suggest that power was sought for particular public purposes.

It is beyond dispute that some lawyers were better paid in Washington than they would have been elsewhere; some enjoyed more power in government than they could have exercised in the private sector; and many found unique opportunities to practice their craft in demanding and rewarding circumstances. It is nonetheless true that a significant number of New Deal lawyers were motivated by a reformist ethic combined with the prospect of opportunities to exercise that ethic in action; that this reform ethic was more prevalent among lawyers in the thirties generation than in the twenties generation; and that lawyers' attitudes were shaped by such common generational experiences as the crash and the coming of the New Deal. Employment in Washington, by itself, is only a crude indicator of social consciousness. But recruitment patterns do point to a shift in lawyers' values from the 1920's to the 1930's, a shift reflecting the phenomenon of generational change. Lawyers, like non-lawyers, are opportunists. But, for better or worse, there are historical moments when people pursue ideologies and values that transcend self-interest or subsume it in the broader currents of public life. The Depression and early New Deal period was one of those special moments. A new generation of lawyers, deeply marked by social changes that coincided with their embarkation upon professional careers, shared experiences and perceptions that defined their special generational identity.

Seven

The New Deal: A Lawyer's Deal

Depression economics and New Deal politics elicited a tenacious defense of the past from the professional elite. For years the American Bar Association had commingled corporate capitalism and patriotism with legalism. It equated reform with revolution and economic regulation with dictatorship, warning lawyers against "rash experiments with American ideals."[1] From its vantage point the New Deal was institutionalized subversion. The sudden proliferation of New Deal laws and agencies exceeded the worst fears of traditionalist lawyers, who proclaimed their commitment to a tidy separation of judicial, legislative, and executive powers and resisted state and federal regulatory activities that would impede their corporate clients. They viewed the growth of administrative law as an alarming symptom of change. It not only threatened separation of powers and federalism but permitted the expansion of regulatory control into areas of economic activity traditionally free of public supervision. A "flood of administrative legislation" after March 4, 1933, upset the "balanced order" of separation of powers and jeopardized the judiciary which, according to an ABA committee, was "in danger of meeting a measure of the fate of the Merovingian kings." Administrative law was nothing less than a "fifth column," a "lurking omnipresence" imported by professors with alien, "continental" ideas.[2]

Virtually any assertion of government regulatory power from

Washington aroused fear of imminent cataclysm. Lawyers analogized from the Soviet Union under Stalin and Germany under Hitler to warn of impending dictatorship under Roosevelt. William Ransom, ABA president during 1935-36, referred to "diabolical plans" and "blue prints borrowed from old world dictatorship." A Los Angeles corporation lawyer predicted to John W. Davis that the continuation of the New Deal "may well result in the overthrow and destruction of American institutions and ideals, and indeed our very system of government. . . . The peril is as great, if not greater, than any war could bring."[3] Another lawyer warned of "an adroit, systematic and sinister effort to discredit and destroy the influence and the leadership of the stabilizing forces and institutions of American life"—the bar being conspicuous among them. During the New Deal years one bar association leader after another pleaded for the reassertion of leadership by lawyers who would defend "the fundamentals of the law from assault from without and intriguing falsities from within. . . ." Stabilization required defense of business values. "Our prime function is to implement the existing order," declared a Chicago lawyer. "Its sudden destruction . . . implies our own." Businessmen, wrote a New York attorney, "are still the most substantial and the most influential members of each community." Therefore, "it is the rehabilitation of the managers of the economic system with which the lawyer is primarily charged." Above all, lawyers must reassert their own leadership as a counterweight to popular excesses. A senior partner in Shearman and Sterling, expressing concern about the very survival of Western civilization, asked plaintively, shortly before Roosevelt's inauguration: "with the mob in control and a babel of mental confusion throughout the world, how can we be hopeful?"[4]

It would be difficult to exaggerate the sense of personal and professional dislocation that elite lawyers experienced with the accession of the Roosevelt administration. Their values were challenged; their clients were pilloried; the deference they were accustomed to was withdrawn; and their self-assumed role as defenders of American institutions was mocked. New Deal laws posed a moral no less than a legal challenge. As George Wharton Pepper, the eminent Philadelphia lawyer, told the Supreme Court at the conclusion of his oral argument in the *Butler* case, testing the constitutionality of the Agricultural Adjustment Act: "I am standing here to plead the cause of the America I have loved; and I pray Almighty God that not in my time may the land of the regimented be accepted as a worthy substitute for the land of the free."[5] Most painful, perhaps, was the evident flow of power to young lawyers from unacceptable social and ethnic backgrounds —lawyers whose training and values made them professionally comfortable within the administrative process. As they waved the banners of reform and Realism from their New Deal redoubts, they enraged their professional elders. Robert Swaine, a Cravath partner, described them as "cynical opportunists"; a partner in Strong and Cadwalader referred to "ill-educated 'brain-trusters' whose foreign blood continuously militates against their understanding of the Anglo-Saxon idea of self-government." Nobody, complained a Dallas attorney in 1935, paid attention any longer to what the American Bar Association did. A Pennsylvania lawyer, upset by the New Deal, wrote forlornly: "I don't know where to go."[6]

An older group of corporate lawyers was quite certain of its direction. The National Lawyers' Committee of the American Liberty League, an anti-New Deal group organized in 1934, enrolled two thousand attorneys committed to defending their version of the Constitution against the Roosevelt administration. The committee enlisted the cream of the corporate bar, including former Solicitors General John W. Davis and James M. Beck and former Attorney General George W. Wickersham. Their assault against the administration was prompted by political opposition, reinforced by professional displacement. Or, as Beck expressed it: when Congress, "demoralized by the delirium of emergency or inspired by class passion," enacted laws which "nullify" the Constitution, then lawyers must speak out.[7]

The most egregious statute was the National Labor Relations Act, a belated Congressional effort to resolve violent conflict by permitting the reallocation of power in labor-management relations. In its first report, nearly 150 pages long, issued over the signatures of fifty-eight lawyers, the National Lawyers' Committee declared the act unconstitutional. Earl F. Reed, counsel for the Weirton Steel Company and a draftsman of the brief, asserted that once a lawyer told his client that a statute was unconstitutional the law was a "nullity" which need not be obeyed.[8] Reed's statement, in conjunction with other committee activities, carried the Liberty League lawyers to—and perhaps over—the brink of ethical conduct as the profession had defined it. According to the Canons of Ethics it was improper for lawyers to engage in newspaper publication regarding pending litigation, to solicit legal business, or to stir litigation with offers of free advice and assistance. Under prevailing ethical norms lawyers were expected not to have indirect interests through retainers or business associations in the outcome of litigation. Yet the Liberty League lawyers had issued elaborate press releases, they had declared a law unconstitutional while a test case was pending, and Beck had made a radio offer of free legal defense for anyone whose rights were abridged. A swirl of criticism arose within the profession. A member of the Illinois law faculty, formerly a state supreme court judge, described the brief as a "brazen" attempt to "intimidate" the courts, while the usually staid *United States Law Review* expressed editorial incredulity that lawyers would declare a statute unconstitutional in the press before it had been judicially resolved.[9]

Upon complaint to the American Bar Association, its ethics committee met to consider the allegations of impropriety lodged against the Liberty League lawyers. Although even Beck harbored doubts about the propriety of public comments while litigation was pending, the committee exonerated the lawyers. Notwithstanding the canon proscribing newspaper publicity, it found no evidence of unethical behavior. Similarly, the radio offer of free legal assistance was deemed to be not an offer of free legal assistance or an attempt to solicit business but an offer to defend the constitutional rights of the indigent. One member of the Lawyers' Committee, pleased with ABA vindication, commented that "it was obvious that we would get it."[10] The unintended irony of his observation suggests an ethical double standard under which negligence lawyers were chastised as ambulance chasers but their corporate counterparts were exonerated. Justice Harlan Stone was prompted to complain that the handholding between the association and the Liberty League made him "despair of ever attaining better things" from the profession. A member of the ABA inner circle sadly confessed that he felt like "giving up and starting all over again" to organize a truly national association.[11]

The Liberty League experience wounded the professional elite. Mocked by unfavorable publicity, it sniped ineffectually at the New Deal until the beginning of Roosevelt's second term, when the President attempted to pack the Supreme Court with justices presumed to favor New Deal legislation. Coming on the heels of the protracted and unsettling sit-down strike in the automobile industry, the plan was received as a parallel assault upon American institutions, as the final unraveling of the fabric of law and order. Supreme Court justices might issue unpopular decisions, but the Court itself was too venerable a symbol to attack, especially at a time of national unease when labor-management turmoil was upsetting the traditional imbalance of power between workers and employers. The political winds shifted quickly. Lawyers were joined by other Americans who saw their country "in a state of absolute unrest, bordering on revolution," with respect for law and order disappearing.[12] John L. Lewis and Franklin D. Roosevelt were unlikely leaders of a proletarian revolt, but their apparent collaboration in the subversion of corporate and judicial autonomy offered the professional elite an opportunity to regain public favor and professional self-esteem.

John W. Davis was not alone in his declaration that the Court plan constituted one of the "gravest attacks" on government in his lifetime, if not in the life of the nation. All lawyers, Charles C. Burlingham reported from New York, were "greatly stirred." But the American Bar Association had learned important lessons from its identification with the Liberty League. Here was too vital an issue to jeopardize with the taint of public ABA opposition. Indeed, the president of the association was reminded by one of his predecessors that the organization was so inflexible and unimaginative that local associations should lead the counterattack. Lawyers publicly identified as anti–New Dealers were advised to remain in the background. Wise lawyers, Burlingham told Felix Frankfurter, were passing the same message to corporate lawyers: the Davises of the profession must be kept "conspicuously absent" from protest activities.[13]

Association leaders acted with restraint. A special ABA committee, which conducted a referendum of association members, also polled nearly 150,000 non-member lawyers. The results were overwhelming. Eighty-six percent of the ABA respondents, and 77 percent of the 50,000 non-members who returned ballots, opposed the court plan. The committee pointedly observed that the strongest opposition arose from smaller cities and county seats, not from the metropolitan corporate bar. One local bar association after another forwarded similar results to the White House: the Pennsylvania Bar Association reported that lawyers were opposed by a five-to-one margin; in Oregon 84 percent of the bar was against the plan; at a special meeting of the Association of the Bar of the City of New York, members concluded overwhelmingly that the President's bill was unsound and dangerous.[14] ABA leaders were jubilant. No longer, one of its officers concluded, would the association be dismissed as the mere mouthpiece of an urban corporate elite. It had shed its skin as "something suspect—and tainted—'a second Liberty League.' "[15]

Roosevelt's assault on the Court, and by extension on the rule of law, rankled for months after the plan was soundly defeated. "As lawyers, as an association, as individuals, we have been wronged by the President," ABA president Frederick A. Stinchfield wrote. "On very many occasions," he declared publicly, "the President has expressed what seems to amount to a hatred of the entire legal profession."[16] The Court fight offered anti–New Deal lawyers a rare opportunity to express resentment against the Roosevelt administration without incurring public censure. Grievances poured out against "constitutional perverts" (recent law-school graduates), trained by "flippant professors" who, as "Jewish panderers" to Roosevelt, bore responsibility for the Court proposal. An upstate attorney from New York told his associates in the New York bar association: "You gentlemen who graduated from law school prior to 1925 obtained a real education from experienced men. We who graduated since that time—what did we get? . . . We were told the courts were wrong and that the professor's theory was what the law should be. . . . We were to be his disciples and crusaders to change the laws that had been the backbone of the country since the Declaration of Independence."[17]

The Court fight summoned displaced lawyers to Armageddon to battle with their foes: teachers, Realists, New Deal administrators, brain-trusters, and Jewish radicals. Although it enabled the American Bar Association to identify with the popular side of a national issue and recover some lost prestige, it came too late to deter spreading professional discontent with the political conservatism of the association. Rumblings of criticism had been audible for years. Some Harvard faculty members were appalled when ABA president Earle W. Evans offered to their law students as a prescription for professional success the advice to attend church regularly, call people by their first names, and spread the word that they were public-spirited citizens. Other lawyers were dissatisfied with the tightly controlled process of officer selection (by an "inside clique," according to an ABA executive committee member) which produced professional leaders whose values seldom embraced anything beyond nineteenth-century rugged individualism or twentieth-century Babbittry.[18] Yet, as Thurman Arnold concluded, since the legal system channeled corporate lawyers to positions of professional leadership, the ABA would not change until the deference accorded corporation lawyers diminished. Dean Charles Clark of Yale told association members that they must choose between the impotence of a "social gathering of the older and financially successful lawyers" or the vigor that would accompany identification with contemporary professional and public trends.[19] Few professional leaders would accept Arnold's analysis or Clark's categories. Rapid social change only strengthened their resolve to cling to older values, especially when, in their view, the sanctity of Law, Constitution, and Court were under attack.[20]

Internal rigidity prompted external action. Late in 1936 professional and political impulses converged with the organization of the National Lawyers Guild, the first association formed to challenge ABA hegemony. The guild was a true child of the thirties. Professionally it expressed the resentment of aspiring minority-group lawyers over their marginal status, over restricted opportunities in an era of hard times, over an unrepresentative bar structure, and over parochial professional values. Politically the guild, in the words of one of its charter members, was "born in revolt —a revolt that embraced the entire intellectual life of the times." Drawing primarily upon legal Realism and New Deal reform, it spoke to lawyers who were committed to the liberal issues of the day: the right of workers to organize, civil liberties, and minority rights. It appealed to lawyers who were prepared to discard precedent in novel circumstances, who viewed law "as a living and flexible instrument which must be adapted to the needs of the people."[21]

Economic privation was encouraging various professionals to organize: newspapermen formed the American Newspaper Guild; actors joined the Screen Actors Guild. The Lawyers Guild never became a trade union, but its organization expressed the same impulses that impelled other white-collar professionals to act collectively. The attorney most familiar with the efforts of other professionals was Morris Ernst, a New York lawyer who had helped Heywood Broun to organize the Newspaper Guild in 1933. Ernst, the son of a Jewish retail merchant, had graduated from Williams College before World War I, sold furniture for a time, attended night law school, and joined a new firm whose partners, Williams classmates, enjoyed access to the prosperous German Jewish community of finance, commerce, and philanthropy. During the thirties he became progressively disenchanted with the organized bar. Speaking before the New York County Lawyers Association in 1934, he described the legal profession as "a vast army, retained in the main by those who have wealth to resist any social change." One of his listeners was sufficiently inspired to suggest the formation of a "democratic, socially-conscious, lawyer's guild," modeled after the Newspaper Guild, to offer lawyers outside Wall Street the opportunity to achieve professional satisfaction and success. Ernst despaired of reforming the American Bar Association. By 1936 he had come to see the need for a progressive national professional association that would defend the use of law as an instrument for implementing, rather than frustrating, popular wishes.[22]

Early in December a small group of liberal lawyers—including Ernst, Frank Walsh, Jerome Frank, and Karl Llewellyn—met at the City Club in New York to organize a lawyer's guild. They came with disparate motives and hopes: some were dismayed by the corporate law identity of the ABA; others were distressed by its active involvement in conservative politics; some wanted to rally lawyers to the New Deal; still others (Llewellyn especially) desired an organization committed to the provision of low-cost legal services. They were united only by the conviction that the ABA should no longer be permitted to speak for the legal profession. In "A Call to American Lawyers" the guild condemned that association, whose "concern for liberty has been secondary to its concern for property," and cited the "urgent need" for a national association of lawyers that would become "a truly progressive force in the life of the nation." If lawyers rediscovered their role as champions of liberty and justice, they could recapture their traditional position of community leadership. (Even this most progressive of associations looked to the past for its model.) The guild, Ernst told President Roosevelt early the following year, was "the first national answer to the Liberty League and the American Bar Association."[23]

Although the guild described itself as a grass-roots national movement, formed in response to the "spontaneous demand" of thousands of lawyers, it was always an incongruous, and occasionally tenuous, alliance of liberal lawyers with strong ties to the professional establishment, radical attorneys with equally strong commitments to the political Left, law teachers who were dismayed by the commercial tone of the organized bar, and low-status urban practitioners from ethnic minority groups who were frustrated in their quest for professional success. In 1937, for example, the guild executive board included teachers Thomas Emerson and Walter Gellhorn; Charles Houston, Jr. (dean of Howard Law School and chief litigator for the NAACP); practitioners Jerome Frank (by then one of Ernst's partners) and Osmond K. Fraenkel (a prominent civil liberties lawyer for the ACLU); New Dealer Abe Fortas; and radical lawyers Lee Pressman, Harry Sacher, Maurice Sugar, and Nathan Witt. The membership was drawn primarily from low-income, low-status urban practitioners along the East and West coasts. A young, struggling Irish Catholic lawyer from Washington, of "good but humble birth," responded with alacrity to the guild appeal. So did black lawyers, for whom the guild's policy of racial equality stood in sharp and welcome contrast to the all-white ABA. Although guild leaders were warned that their willingness to enroll black attorneys would keep the organization still-born in the South, the guild preserved its commitment to racial (and sexual) equality.[24]

But the precarious alliances between a rising elite and a submerged underclass, and between politics and professionalism, created serious organizational problems. With recruitment primarily from among low-income professionals, economic delinquency was rife, and financial difficulties plagued the guild from its inception. After dues were raised above a token dollar, nearly five hundred lawyers from New York alone allowed their membership to lapse. Although nearly three thousand lawyers joined the guild during its first six months, fewer than one thousand enrolled during the following year. Even more menacing than financial problems were the shifting winds of politics. The guild, launched amid the liberal euphoria that followed Roosevelt's re-election landslide, quickly became mired in the conservative backlash against CIO organizing drives, the General Motors sit-down strike, and the Court-packing plan. Guild endorsement of Roosevelt's Court proposal cost it substantial support and fed accusations that it was a radical organization. Within the guild sharp differences emerged over the proper functions of a professional association. Were public issues appropriate concerns? Did professionalism exclude politics altogether, or only certain politics? How were energies to be allocated between professional and political obligations? Should political beliefs be submerged to narrowly defined professional interests—or did professionalism draw strength from political engagement? Early in its life the National Lawyers Guild began to flounder on these issues.[25]

Virtually from its inception there was an undercurrent of political discord. In the guild, as in so many other liberal organizations, domestic and international politics merged: anti-Communists squared off against unaffiliated radicals, members of the Communist Party, and sympathizers with the Soviet Union. As the political climate soured, both at home and abroad, the spirit of the Popular Front between liberals and radicals waned. During the first month in the life of the guild Morris Ernst complained about the obstructive tactics of left-wing members. Soon a Milwaukee attorney was warning Frank Walsh, the guild's first president, that the price of organizational success was the exclusion of certain lawyers—"extremists"—from leadership. A New York lawyer complained to Ernst that political activists were diverting the guild from its proper course. Ernst, vexed over the intrusion of foreign-policy issues—particularly the Spanish Civil War—in guild debates, increasingly doubted the integrity of some guild leaders.[26]

The tension generated by financial difficulties, lagging enrollments, and, above all, conflict between left and center finally erupted at a meeting of the guild executive board in February 1939. Ernst, convinced that the growth of the guild was impeded because it was insufficiently committed to the Bill of Rights and to the democratic process, proposed the addition of a statement to the constitutional preamble declaring guild opposition "to dictatorship of any kind, whether left or right; whether Fascist, Nazi, or Communistic."[27] To subsequent generations, untutored in the rhetoric of ideological warfare that characterized the late 1930's, such a resolution might seem not only innocuous but commendable. Who, especially among liberals and progressives, would not vigorously oppose dictatorship? In context, however, the resolution was rife with political significance.

By 1939 liberals were badly battered by the conservative reaction against the New Deal, which played variations on the theme of Communist conspiracy and subversion in Washington, in trade unions, and in liberal associations. The anti-totalitarian pledge offered an escape. It confronted Soviet sympathizers with a wicked dilemma: they must equate the Soviet Union with Nazi Germany or suffer the political opprobrium assured by their failure to do so. Either way, liberals would avoid the taint of association with those whose politics were suspect. Ernst and his allies on the executive board were convinced that the guild could survive only if it defined itself as a strictly professional and social group; it must repudiate the impression that it was a Communist Party front organization. Their opponents, refusing to identify the Soviet Union with Nazi Germany, fought the intrusion of a political test. Some of them were deeply troubled by the libertarian implications of the proposed resolution: where did the freedoms of conscience and association end and the claim of public declarations of faith begin? Demanding an intellectually free organization, without the test of political identity, one member declared heatedly: "If [it is] going to be just another American Bar Association, I do not want to belong to it."[28]

Ernst insisted that it was consistent with the Bill of Rights to expose covert political movements. He carried the issue to the annual guild convention in 1939, where it precipitated another divisive quarrel. Increasingly, guild members felt compelled to define and declare their political convictions in a climate that was strongly supportive of anti-radicalism. Acrimony punctuated guild meetings; members issued charges and announced resignations; radical politics were ascribed to psychological disorders.[29] Even the most thoughtful members found balanced judgments difficult to sustain. Jerome Frank at first supported the Ernst resolution on the ground that denunciation of totalitarianism (and the implicit link between the Soviet Union and Nazi Germany) was vital to "the kind of liberal Americanism" that the guild properly espoused. Liberals, he concluded, should not refrain from stating "honestly and persistently their undeviating opposition to Communism," even though there was the risk that public condemnation of Communism would provide aid and comfort to the political Right. Yet five days later, enunciating the terms on which he would remain a guild member, Frank concluded that the guild could not properly ask members to declare their political faith as a condition of membership. Good professional standing, he insisted, should be the sole test.[30]

The politicization of the National Lawyers Guild became the salient feature of its organizational life. The clash between its liberal and radical factions diverted energies which just a few years earlier had flowed into the guild's creation as a progressive alternative to the American Bar Association, sensitive to the problems of marginal lawyers and to the need for responsive public law. Organized at a time when liberals and radicals cooperated without embarrassment or harassment, it foundered once the political props for such harmony collapsed. The guild drifted under a cloud of opprobrium as the era of the Popular Front waned. During the period of the Nazi-Soviet Pact it shared in the ostracism meted out to fellow travelers; after the wartime alliance between the Soviet Union and the United States was chilled by the Cold War, it was crippled by the federal government.[31] The guild never overcame its political vulnerability: the child of liberal euphoria, it was the victim of conservative reaction and liberal retreat.

The guild, in the eyes of some, was "wrecked on the red shoals."[32] But to focus exclusively on guild politics is to ignore a critical (and rarely explored) dimension of its organizational life and energies. Although internecine quarreling was a distinctive feature of its New York chapter, and, by 1938, of its national board of directors, other guild chapters, especially in Philadelphia and Chicago, launched innovative professional reforms.

There the guild developed pioneer programs for neighborhood law offices to provide legal services to low- and middle-income citizens, thus enabling it to tap the commitment of marginal and underemployed lawyers whose race, religion, age, and background had all but excluded them from professional opportunities in an economically shattered society.

Legal aid, the fig leaf that covered the modesty of professional efforts to remedy injustice inherent in service-for-a-fee professionalism, had since the late nineteenth century enabled lawyers to disregard the unmet legal needs of a vast lower- and middle-income populace. It took the Depression to provoke concern, less because people needed lawyers than because so many lawyers desperately needed clients. In their first reactions to the Depression, lawyers had reverted to the cry of "overcrowding," a euphemism (as Karl Llewellyn shrewdly observed) for too many low-income lawyers.[33] But under the impact of sustained economic hardship, to the point where needy lawyers began to register for work-relief, a tradition of professional indifference began to crumble.

Llewellyn, a member of the University of Chicago law faculty, became the intellectual godfather of a reform program predicated upon the idea of neighborhood legal services. In a profusion of articles, speeches, and letters between 1933 and 1938 he diagnosed the malady and proposed cures. The basic problem, he concluded, was the gap between "a canal-boat-and-buggy type of professional organization" and a modern urban industrial society. The legal profession still proceeded upon assumptions grounded in the homogeneous communities of the nineteenth century. Llewellyn pleaded for the adaptation of "this individually and individualistically organized profession to metropolitan conditions, and to conditions of modern business and technical organization."[34]

It was in the cities that the fabric of professional values had unraveled. Llewellyn's critique began with the Canons of Ethics, which, he claimed, were ill-suited to metropolitan life. "Turn these canons loose on a great city, and the results are devastating in proportion to its size," he wrote. Rules against solicitation and advertising, and condemnation of ambulance chasing, only benefited established lawyers and obstructed the provision of adequate legal services. The ambulance chaser, Llewellyn observed, brought legal services to people who needed them. If services were available at reasonable cost to a public that knew how and where to obtain them, illicit solicitation would disappear as a troublesome issue. Similarly, diagnoses that stressed "overcrowding" and prescribed a reduction in the number of lawyers were more closely attuned to the economic self-interest of practitioners than to the unmet legal needs of laymen. Complaints about the unauthorized practice of law—specifically that banks, title companies, collection agencies, and accountants poached upon the lawyer's preserve—masked the real "danger to the Bar's needed living being earned" by emphasizing the ostensible "danger to the Bar's needed service being rendered." Only the large firm, Llewellyn concluded, had adjusted to modern demands and adequately met the needs of its constituency. But the price was exorbitant: "we are left with two thirds of the bar and eighty per cent of the public who lack either the contact or the means of making it." Not only did the legal needs of a vast public go unserved; the fact that lawyers were compelled to scramble for clients increased the likelihood of unethical conduct.[35]

To remedy these deficiencies Llewellyn proposed drastic professional reorganization to develop new business for lawyers and simultaneously extend legal services to needy clients. He urged metropolitan bar associations to establish neighborhood law offices throughout their communities, to prepare lists of competent lawyers, to fix rates for standardized services on a sliding scale according to client's income, and to publicize the availability of lawyers who could provide these services. A prospective client could choose any lawyer from the professional list; the lawyer chosen would be required to take the case at a fixed fee. These arrangements would compel successful lawyers to take nonpaying business; they would help to provide income for unknown, but available, lawyers; and they would bring legal services to those who previously could not locate or afford them.[36] In private correspondence Llewellyn explored an even more radical possibility: a tax on the income earned by large firms to subsidize community legal services. Llewellyn hoped to deflect the bar from its endemic individualism and reorient it to community needs. He insisted that counseling, or "legal hygiene," complement the litigious approach of legal aid societies. But he hardly expected bar associations to respond with alacrity to his suggestions. They were, he concluded, "inert." Their leaders were "so highly specialized in corporate work that they are out of all touch with the little man's need." By 1938 bar associations occupied a conspicuously diminished place in his proposals. Neighborhood legal service bureaus would, he hoped, be staffed either by young unemployed lawyers, or, on a *pro bono* basis, by junior law firm members.[37]

Llewellyn's proposals spread ripples throughout the profession. Dean Clark of Yale, aided by funding from the Federal Emergency Relief Administration and by young lawyers on relief who conducted interviews, learned from a survey of Connecticut attorneys that "the public has undone legal business; the lawyers have free time." He suggested that bar associations scatter law offices throughout cities to enable the public to obtain legal advice from lawyers who could not find employment elsewhere. Professor Malcolm Sharp of Chicago advocated low-cost legal service bureaus to specialize in the economical handling of small claims on a cooperative basis. Attorneys prominent in the legal aid movement were jolted once again by the fear that unmet legal needs would precipitate social discord. Alarmed even more by the specter of socialized practice, they urged immediate action by the bar to undercut demands for government intervention. Reginald Heber Smith, conceding that "independent, self-respecting, hard-working, home-loving Americans" needed legal assistance, suggested that law offices sponsored by bar associations navigate between the traditional indifference of private firms and the restricted activities of legal aid societies. Smith still spoke from within conventional legal aid assumptions: he hoped that these offices would reject divorce cases, criminal cases (except for first offenders), or personal injury cases where the claim exceeded two hundred and fifty dollars.[38] Deference to professional autonomy remained the paramount consideration.

Young urban lawyers responded enthusiastically to the idea of neighborhood law offices. Writing to Llewellyn from Chicago, Atlanta, Denver, and Philadelphia, they expressed interest in the establishment of low-cost legal service bureaus in metropolitan neighborhoods far from the downtown clusters of corporate firms. These struggling attorneys, eager to locate potential clients, knew that expanded legal services would enhance their own professional opportunities while serving clients' needs. Professional reform and self-interest dovetailed perfectly. One law student told Llewellyn that law clinics would provide "a wonderful chance for both the young lawyer and also the people who can't afford to pay large legal fees." A Chicago attorney suggested that clinics would furnish "an excellent training ground for the younger lawyer as well as a basis for a fair income."[39]

Established bar associations were unresponsive to the needs of clients or lawyers. But the Lawyers Guild, uninhibited by traditional paralysis, acted. The influence of Llewellyn, who had attended the founding meeting of the guild (but never became a member), was evident. Early in its organizational life guild members explored the possibility of neighborhood legal service agencies whose appeal would rest upon easy access and low fees. The neighborhood law office was part of a more ambitious guild proposal to shift the professional ethos from traditional individualism that best served corporations and corporate lawyers to cooperative planning that met the needs of the bar's middle stratum and their potential clients. As Wisconsin Dean Lloyd K. Garrison told guild members at their second convention, "new kinds of organization, and new centers of cooperative activity" were required.[40]

Guild chapters in Chicago and Philadelphia were the first to implement the neighborhood law office proposal. Pursuant to a resolution adopted at the guild's first convention, which called for the establishment of legal clinics, Chicago members devised a centralized legal service bureau to serve low-income groups and to educate them to their legal needs. Although the plan presupposed the failure of individualized professional enterprise to meet community needs, it was tightly structured to conform to traditional professional values. Its governing board of trustees comprised lawyers and judges selected by bar associations, law schools, and the courts. The consuming public was expected to yield complete autonomy to professional expertise. Publicity was encouraged, but advertising was not. Relationships between staff members and clients were expected to follow attorney-client patterns of the larger firms. In substance, the Chicago plan offered a centralized office, controlled by the organized bar, whose lawyers would draft wills, leases and contracts, handle alimony, foreclosure, and workmen's compensation claims, and attend to wage disputes and probate problems on a standardized, mass-production basis. The bureau would replicate the legal aid service model, although development in novel directions was not prohibited.[41]

The Philadelphia guild chapter, by contrast, devised a decentralized network of autonomous neighborhood offices. The idea was expounded in the *Dickinson Law Review* by Robert D. Abrahams, formerly an assistant city solicitor in Philadelphia. Abrahams was especially concerned with the possible loss of legal business to lay agencies because potential clients were offended by high costs, inaccessible law offices, and disagreeable experiences with over-eager negligence attorneys. His decentralized legal service bureaus were designed to bring the law office closer to the client and to reduce costs. After Philadelphia bar association leaders displayed chilly indifference to the proposal, a group of interested lawyers, all of whom belonged to the guild, secured organizational approval for an experiment in preventive law based upon the novel supposition that a lawyer "should not be remote from his client either in geography or understanding." In the summer of 1939 nearly one hundred and fifty lawyers applied for positions in the neighborhood offices. Clients responded with evident enthusiasm to lawyers who displayed "friendliness and . . . accessibility," rather than austere professionalism. More than 80 percent of those who came during the first eighteen months had never before entered a law office. Although only a rare case brought in a fee that exceeded five dollars, the offices quickly became self-sustaining—so vast was the need and so eagerly were the law offices accepted as community institutions. A decade later the program still flourished. Eleven neighborhood offices served 4,200 clients annually, without subsidy and with "substantial profit" to most participating lawyers. The client's dignity, not the lawyer's, still received priority. Abrahams was especially concerned that traditional relationships between attorney and client should be preserved "without socialization or bureaucracy."[42]

The Chicago and Philadelphia chapters acted autonomously but with strong organizational support from the guild. Ironically, for an association rebuked for its excessive radicalism, the guild's sponsorship of neighborhood law offices reflected quite traditional professional impulses. "Individual incentive and private profit in the practice of the law" were reinforced, not repudiated. The offices remained under bar supervision. The nature of their business precluded the possibility of competition with well-established lawyers. Finally, by serving the economic needs of underemployed lawyers and the legal needs of underserviced clients, they may have reduced the likelihood of unethical legal practices.[43] Yet substantial professional energy was required to implement even these conventional ideas beyond established professional limits.

The impulse for change had little to do with political ideology; it originated in hard times. Innovation was encouraged by the guild's organizational appeal to professional outsiders, whose desperation for work encouraged experimentation. It is revealing, for example, that the thirteen-member executive committee of the Philadelphia guild chapter included eight Jews, three Catholics and one black.[44] The legacy of the National Lawyers Guild was not the internecine political quarreling so characteristic of its national executive board, but rather its diffusion of professional participation, its sensitivity to contemporary social and legal problems, and its commitment to innovative means toward fulfilling obligations traditionally ignored.[45]

The guild tapped professional restlessness in a troubled decade. But the concerns expressed in its constitutional preamble—for civil rights, for workers' rights, and for law as a flexible instrument of social change—were not its monopoly. During the 1930's civil rights law, labor law, and administrative law moved from the periphery to the center of professional life. Invested by the Depression with new importance, they attracted special groups of younger lawyers whose personal aspirations found expression in public law.

The Negro bar had limped through the 1920's in a state of segregated disarray. Although wartime migration patterns brought unprecedented numbers of rural blacks to cities, creating new possibilities for commercial practice, the potential remained largely unfulfilled. Black attorneys suffered from the predictable consequences of racism and poverty. They had less education, inferior training, fewer opportunities, and lower professional self-esteem than their white counterparts. Black business and commerce was too underdeveloped to sustain the black bar. Neither whites nor blacks who could afford the fees charged by white attorneys would retain blacks as counsel. Potential clients, regardless of color, doubtlessly assumed (all too correctly) that black lawyers were likely to be inferior lawyers whose color would jeopardize a client's claim in a legal system that absorbed the racism of American society. Confined to the least desirable areas of practice—criminal, domestic relations, personal injury, and small claims—they were caught in a vicious bind from which there was no escape. And, as one black lawyer pointedly recalled: all judges "wore black, but were white."[46]

From law school through law practice aspiring black attorneys shouldered a crippling burden of handicaps. Coming as so many did from impoverished families, they were compelled to forego college for remunerative employment. Those who lacked a college degree were excluded from national law schools, which admitted an infinitesimally small number of black students anyway, regardless of qualifications. Either they entered law school at a relatively advanced age, or they attended sporadically at night.

At Howard, a night school that did not require even two years of college until 1924, the average age of the graduating class was thirty-one. Graduates found few opportunities for clerkships or apprenticeships which might provide practical training. White firms would not hire them; black law offices seldom could afford to do so. Young black lawyers were shunted into solo practice, where they tried to survive amid the enmity of established black and white attorneys who were already competing for minuscule retainers from an economically disadvantaged clientele. With professional opportunities at a premium, black attorneys were compelled to supplement their incomes by non-legal work, which diminished their professional identity and self-esteem and contributed to a conspicuously high rate of failure. Indeed, failure often provided the only escape from sustained professional hardship.[47]

As grim as this picture was, and would long remain, there were discernible glimmers of hope in the postwar decade. During the race riots of 1919 black lawyers in Chicago and the District of Columbia formed defense committees to provide legal services for black defendants. Although these *ad hoc* committees did not survive that emergency, the precedent of black lawyers engaged in legal defense work in black communities endured. In 1925 the National Bar Association, an organization of black lawyers, was formed "to strengthen and elevate the Negro lawyer in his profession and in his relationship to his people. . . ." Enrollment lagged; there were only 221 members in 1930. But its leaders pledged their support to any organization committed to promoting the welfare of black Americans. The association, its first president declared, expressed "the development of the American Negro's belief in himself."[48]

Raymond Pace Alexander, who delivered that optimistic pronouncement, belonged to an emerging elite within the black bar. A graduate of Harvard Law School in 1923, he earned a reputation in civil rights cases that enabled him to build a lucrative practice in his own Philadelphia law firm. Among his Harvard contemporaries, Charles Houston, Jr., a class ahead of Alexander, had already begun an affiliation with Howard Law School which was to transform that institution into a laboratory for civil rights and a nursery for civil rights lawyers. William Hastie, a future federal judge who studied at Harvard several years later, earned accolades from Felix Frankfurter as one of the three or four best Harvard students during his first twenty years on the faculty. The idea of elite leadership was contagious. A young black student, applying to Harvard in 1928, urged the administrator of the student loan fund to consider that "the specialized training of a few individuals would do much to furnish the [black] masses leadership."[49] The "talented tenth" principle took hold within the black bar, as a coterie of superbly educated younger lawyers began to push against social and professional obstacles.

The decade of the twenties was crucial for the development of black cohesiveness. The horror of the postwar riots, the excitement of Marcus Garvey's black nationalist crusade, and the creativity of the Harlem Renaissance stirred pride of race in black urban communities. Black lawyers shared this rising consciousness. Alexander was among the first to advocate a new role for black lawyers. "We owe the law more than merely using it as a means of making a livelihood," he declared. "We owe to our people, who, more than any other people are in need of our services, a duty to see that there shall be a quick end to the discrimination and segregation they suffer in their everyday activity." In their desperate struggle for economic independence, black lawyers, he alleged, had grown aloof from their own people. A few even capitalized upon the financial opportunities in business practice, repeating the pattern of young, successful white lawyers. But the time had come for them to return to their own community, Alexander claimed. Other young black lawyers concurred. Coming of age in an era of black assertiveness, they spurned the prevalent model of success which bestowed rewards upon a black lawyer in proportion to his cultivation of a white clientele and the white bar.[50]

At this propitious moment the National Association for the Advancement of Colored People reassessed its legal strategy. In

1929, when veteran NAACP litigators Moorfield Storey and Louis Marshall died, the NAACP launched a more aggressive legal program. An NAACP legal advisory committee (including Clarence Darrow, Arthur Garfield Hays, and Morris Ernst) approached the American Fund for Public Service for funds to subsidize litigation to protect and expand the constitutional rights of black Americans. Ernst was also a member of the Fund's committee on Negro work, which funneled the NAACP request to the board of directors with a strong memorandum of support. Citing twelve million unorganized black workers as "the most significant and at present most ineffective bloc of the producing class," the committee requested a grant of $300,000 for a "large-scale, widespread, dramatic campaign" to secure the constitutional rights of blacks—a campaign that would intensify racial self-consciousness and self-respect and "inevitably tend to effect a revolution in the economic life of this country." It contemplated a broad attack upon segregation and discrimination in public education, transportation, property ownership, and the exercise of political rights. In mid-1930 the American Fund for Public Service awarded $100,000 to the NAACP to proceed with its work.[51]

At Frankfurter's suggestion, Nathan Margold was retained as the "ideal man" to direct the civil rights campaign. Margold, thirty-one years old, was the prototypical outsider whose social conscience and hunger for opportunity had propelled him into the legal profession. Raised in Brooklyn, he graduated from City College and from the Harvard Law School, where he was a law review editor. He sampled private practice, served under Emory Buckner as assistant United States attorney, taught for a year at Harvard, and then invested his energy and talent in the legal needs of minority groups, especially American Indians. For the NAACP Margold plunged into an exhaustive study of past litigation and future possibilities. His preliminary report set guidelines for the legal strategy that culminated twenty-five years later in *Brown v. Board of Education of Topeka*, the landmark Supreme Court school desegregation decision. Margold proposed an assault upon the separate-but-equal principle in public education in an effort to demonstrate that segregation invariably meant discrimination. Reluctant to attack segregation *per se* at that juncture, Margold was not yet prepared to "deprive Southern states of their acknowledged privilege of providing separate accommodations for the two races." Instead, he hoped to force them to comply with their responsibility to provide equal facilities as the first step toward fair treatment.[52] In time, the NAACP abandoned equality within segregation to press the point, finally accepted by the Supreme Court, that desegregation was the prerequisite for equality. But Margold's research and legal skills, Walter White recalled, enabled the association to launch "a broad frontal attack on the basic causes of discrimination. . . ."[53]

In 1933, when Margold left the association for a position in the Roosevelt administration, Charles Houston, Jr., the Harvard-trained son of a prominent black lawyer from the District of Columbia, replaced him. Houston's selection had momentous ramifications, most immediately for black lawyers and ultimately for civil rights law. Houston straddled two disparate legal worlds: one defined by his father's professional success, his own excellent record at Harvard, and the model of law as a form of social engineering which he had absorbed from Pound and Frankfurter in Cambridge; the other defined by the black community in Washington, Howard Law School, the struggle of his people for equality, and the aspirations of black lawyers. He was determined to upgrade Howard, to transform it from a marginal night school to a national institution attractive to black students in search of quality legal education relevant to their needs and to the needs of black people. As chief counsel for the NAACP, Houston retained black lawyers whenever possible—in sharp contrast to NAACP practice when Storey and Marshall were counsel. He also forged a relationship between legal education and civil rights litigation. Howard and the NAACP were his institutional bases; education and litigation were his instruments for social change.[54]

Once the NAACP had committed its energies to a strategy of litigation for civil rights, Houston, as chief counsel as well as dean at Howard, dominated the emerging field of civil rights law. "Not until Howard University in the thirties began producing numbers of Negro lawyers trained in civil rights did the race relations picture begin to change," observed one of his successors at the NAACP. Under Houston's tutelage, a talented, committed group of black lawyers moved through Howard to the NAACP to litigate the struggle for civil rights. For the first time NAACP officials felt comfortable with black counsel. Black lawyers, previously viewed as an organizational liability, now contributed important psychological dividends. Used in the past solely for the sake of appearance, they now were described as "competent, conscientious, and courageous" attorneys upon whom the struggle for legal equality depended. Thurgood Marshall, Houston's replacement as NAACP chief counsel, became the most prominent; it was Marshall who engineered the stunning *Brown* victory and sat as the first black member of the Supreme Court. After the *Brown* decision, Marshall said of his mentor: "Charlie Houston taught us all that we should be 'social engineers.' "[55] Houston, Hastie, Alexander, Marshall, and a handful of their contemporaries became the legal architects of a revolution in civil rights law and the privileged beneficiaries of their own success.

The generation of black civil rights lawyers who came of age in the 1930's forged an identity for themselves in the process of litigating equality for their race. Like Jewish lawyers who found opportunities in New Deal agencies that did not exist in private practice, these black lawyers fulfilled their aspirations with a commitment to the cause of civil rights. No longer was law practice a tenuous means of earning a living for black lawyers; it had become a mission of consuming social and personal significance. "Those lawyers who have been quick to take up cases of discrimination and segregation against Negroes have . . . developed a higher appreciation of the honorable profession which they practice," observed a prominent sociologist midway through the thirties. Black civil rights lawyers emerged from the Depression decade as the spearhead of a movement for social change in race relations that has yet to run its course. For other black attorneys, however, the struggle against racism had more direct political implications. Atlanta attorney Benjamin J. Davis, Jr., a Harvard Law School graduate, was retained by the International Labor Defense to defend Communist organizer Angelo Herndon against charges of insurrection brought by the state of Georgia. When the trial judge called Davis and his client "nigger" and "darky," he suddenly understood "the whole treatment of the Negro people in the South. . . . I felt at that particular moment that if there was anything I could do to fight against this thing and to identify myself fully with my own people . . . that I was determined to do it." The "best thing," Davis believed, was to join the Communist Party. He participated in the defense of the Scottsboro boys, became general counsel to the party in Alabama, and finally relinquished his practice to work for the party and the labor movement. In 1949 he was one of eleven Communist Party leaders found guilty of violating the Smith Act.[56]

Within the legal profession, as in the society at large, the struggle for racial equality had barely begun. In 1939 when William Hastie, by then a federal judge, was proposed for membership in the American Bar Association, a prominent civil liberties lawyer in the association questioned the wisdom of pressing for his admission at that time. Several years later the association was disrupted, as it had been in 1912, by the discovery that three black lawyers had applied for membership. Two of them, who publicly challenged ABA discrimination, were rejected; the third, who remained silent, was admitted. Southern members were profoundly upset. A former ABA president despairingly anticipated a time when the association might drop the color line within the profession. From Atlanta an attorney reported the futility of trying to enroll Georgians in the association if additional blacks were admitted. A Texas lawyer urged his Southern colleagues to remain in the association lest Eastern Jews come to dominate it. The entire subject, the former ABA president warned his future successor, was "fraught with great danger."[57] Indeed it was, but in ways not yet anticipated. As the Depression decade ended, an English teacher in Lansing, Michigan, told one of his black students, ". . . you've got to be realistic about being a nigger. A lawyer—

that's no realistic goal for a nigger." Twenty-five years later the student related that incident in his *Autobiography of Malcolm X*.[58]

Civil rights law was a delayed fuse that sputtered through the 1930's. Labor law exploded. In the twenties, when Professor Francis Sayre introduced a course on labor law at Harvard and published a casebook on the subject, it had only begun to be seen as more than "an insignificant off-shoot of the law of torts."[59] The mere handful of attorneys identified as labor lawyers reflected a climate in which workers' aspirations were perceived as either criminal acts or utopian fantasies. Some, like Clarence Darrow and George Vanderveer, were prominent criminal lawyers for whom labor law was a perversely logical extension of their practice. Darrow's experience prompted him to compare the allocation of legal talent in cases where a worker challenged a corporation to the distribution of force if a dwarf were to fight the heavyweight champion. Accepting lucrative retainers from the "despoiler" so that he could afford to defend the "despoiled," Darrow was determined "to get what I could out of the system and use it to *destroy* the system."[60] Vanderveer, an ambitious West Coast lawyer, was inadvertently drawn into defense work for the Industrial Workers of the World and forged new commitments and a new career as "counsel for the damned." Other labor lawyers, like Morris Hillquit and Louis Waldman, were Socialists who served as "passionate warriors in the class struggle." Waldman, the son of a Russian innkeeper, had learned his first lesson in labor relations when he was discharged for refusing to sign a statement exonerating his company for the injuries inflicted by its machinery on a co-worker. He became a Socialist first and then a lawyer, nurturing a glimmer of hope that the legal profession might be "a possible instrument for social and economic progress."[61]

The practice of labor law was precarious, economically and politically. In an era when there was little national protective labor legislation, and unions existed at the pleasure of employers, labor law practice involved an endless round of workmen's compensation claims, defensive tactics to circumvent injunctions, and an occasional courtroom spectacular in a criminal case. To represent a labor radical was to jeopardize one's career, and occasionally one's life. Law practices were ruined since clients, frightened by an attorney's reputation, looked elsewhere for counsel; and lawyers were threatened by the vigilante tactics of community and bar. Early in 1922 an attorney representing two Wobblies in Shreveport, Louisiana, was kidnapped, flogged, and expelled from the state. A Pennsylvania attorney who belonged to the IWW and defended wartime conscientious objectors was disbarred for his "utter lack of respect" for the laws of the land. Little wonder that before 1935, as one attorney subsequently recalled, "if there were ten lawyers known as union lawyers, I'd be surprised."[62]

The Depression, worker militancy, and New Deal legislation transformed labor law from an adjunct of tort and criminal law to a distinct and legitimate area of specialized practice. The National Labor Relations Act, which guaranteed to workers the right to organize and bargain collectively, revolutionized the law of labor relations by creating a framework in which grievances could be resolved by litigation rather than by violence. The National Labor Relations Board confined labor-management conflict to an adversary setting where briefs, oral arguments, and cross-examination, not Pinkerton spies, strikebreakers, and tear gas, were the most potent weapons.

But it was less the institutional setting than militant union zeal, combined with the cause of social justice for workers, that attracted the Depression generation of converts to labor law. With labor on the march during the 1930's, young lawyers eagerly joined the swelling ranks. Minority-group lawyers, especially Jews, flocked into a field of practice where it was possible to merge liberal reform or radical hope with professional fulfillment. Lee Pressman, for example, was counsel to the Steel Workers Organizing Committee and then to the CIO, where he became John L. Lewis's right-hand man. It was Pressman who gave the cause of the Flint sit-down strikers a powerful boost by the discovery that the judge who had issued an injunction against the strikers owned $200,000 worth of General Motors stock. Maurice Sugar did yeoman legal service for the fledgling United Automobile Workers and managed to find time to write "Sit-Down," a Depression folk-song classic. Nathan Witt left the Agricultural Adjustment Administration to become general counsel for the National Labor Relations Board, which was both a magnet for socially conscious young lawyers and a treasured employment agency. One young recruit described a job with the board as "a chance to live 'nobly' at the law"; another referred to it as "the glamour agency of the New Deal to young radical-minded lawyers."[63]

Labor law was *their* cause, just as civil rights belonged to black lawyers. It was, one of them recalled, "the glamour field for the young committed law students." But the glamor was not without its hazards. Edward Lamb, a Toledo corporation lawyer, defended the union during the Auto-Lite strike of 1934 and became a convert to the cause of labor because it provided "a thrill from accomplishing something socially valuable." Four years later Lamb, by then executive vice-president of the Lawyers Guild and regional counsel for the CIO, faced disbarment proceedings for his vigorous courtroom defense of striking shoe workers. Other labor lawyers encountered more direct intimidation: they were kidnapped and beaten by vigilantes in California's Imperial Valley, in Gallup, New Mexico, in Michigan, and in Harlan County, Kentucky.[64] For some, labor law was only the beginning of a professional career devoted to representing underprivileged and unpopular defendants; they retained their zeal long after some of their contemporaries had become indistinguishable from their adversaries on the corporate side of the bargaining table. Thirty-five years later Charles Garry recalled a time when workers were despised with the intensity subsequently reserved for Black Panthers, whom he also represented. "I remember walking into a court to represent certain labor unions and being looked down upon with more vehemence than I am today walking in with a Panther." Their labor experiences encouraged these lawyers to use law as "a weapon, as a tool in the service of the movement for social change"—a commitment that molded their lives and careers.[65]

Before the thirties ended, Morris Waldman recalled in his autobiography, law schools were yielding "their crop of earnest young men who came into my office and told me, 'I'd like to work with you . . . I want to go into labor law.'" That was a far cry, if not a long time, from the mid-twenties, when Waldman's office had been inhabited only by Waldman, a few chairs, a filing cabinet, and a typewriter. Long after labor militancy had subsided into business unionism, labor lawyers retained a distinctive identity shaped by their Depression experiences. Although the stigma attached to labor practice diminished with time, some labor lawyers were still distinguished by their close identification with their clientele and by their willingness to accept occasional compensation in the form of satisfaction with service to a social ideal. These veterans remained "partisans of the underdog."[66]

Yet the legitimation of labor law, which galvanized young lawyers in the thirties, soon mitigated the very zeal that it once had inspired. The New Deal institutional framework drained the passion from labor-management conflict in the mass-production industries. Unions living under the protective umbrella of New Deal statutes required lawyers who could draft, negotiate, and arbitrate. Specialization generated expertise, which in time assured respectability. As labor-management conflict moved from the picket line to the conference table, from confrontation to arbitration, labor lawyers and management lawyers came to share the same commitments to process and structure. As courses in labor law were absorbed into law-school curricula, their special allure diminished as their popularity grew. In 1935 only third-year students at Harvard took labor law, which merely explored the permissible purposes and conduct of strikes. Fifteen years later Harvard offered four labor law courses, which also covered protective labor legislation, the negotiation and administration of collective bargaining agreements, and arbitration, and enrolled three-quarters of the graduating class. Law-school graduates who entered the field perceived themselves as "legal technicians," not as "shock troops in class warfare."[67] Professionalism replaced passion. By 1962, when Arthur J. Goldberg became the first labor lawyer to be appointed to the Supreme Court, the process had run its course. Five years later, when Thurgood Marshall, Charles Houston's protegé, joined the Court as its first black member, civil rights lawyers received their symbolic reward. Children of the Depression decade, labor and civil rights lawyers had grown to robust professional maturity as their once-marginal fields left the fringes of professional life to become institutionalized specialty practices.

These developments coincided with a larger transformation in administrative law which the New Deal accelerated. The administrative process had emerged as an integral part of government in the modern state. Spurred by the reforms of the Progressive era, and by the exigencies of mobilization during World War I, administrative law filled the lacunae in governmental power in an industrial, technological, bureaucratic, federal society. By the 1920's courses in administrative law were taught in law school, the American Bar Association established a special committee on the subject, and lawyers were being drawn inexorably into its orbit. As governmental functions were transferred to boards, commissions, and bureaus at both the federal and state levels, professional opportunities multiplied geometrically. By the beginning of the Depression the largest bar in the country, proportionate to population, was concentrated in Washington, with the bulk of its energy and activity consisting of practice before regulatory agencies.[68]

Long before the New Deal, federal regulation had spread to transportation, banking, and insurance. The administrative process provided the flexibility and discretion required to meet the demands imposed upon modern government—and as yet little of the rigidity or the coziness between regulators and regulated that subsequently attracted criticism. But the profusion of New Deal agencies, especially in sensitive areas like securities regulation and labor relations, charged administrative law with political acrimony. Denunciation of the new alphabet agencies became a staple of bar association proceedings and the cutting edge of anti-Roosevelt criticism from lawyers whose clients were the targets of New Deal regulatory statutes. Administrative usurpation was denounced as an unsavory intrusion upon judicial autonomy. In translation, this meant that the Securities and Exchange Commission or "that God-damned labor board" enjoyed too much power to impose upon businessmen restraints that courts might otherwise have been expected to nullify. New Dealers responded with a vigorous defense of administrative innovation. "The administrative process," wrote James M. Landis, whose text on the subject became a bible to young New Deal lawyers, "is . . . our generation's answer to the inadequacy of the judicial and the legislative processes." Landis, recalled one of his young protegés, "spoke for all of us who had been deeply committed to the New Deal and who had been intimately associated with the administrative process." Felix Frankfurter, eager to soothe its sting of novelty, reminded lawyers that administrative law had not arrived "like a thief in the night." It was not a child of the New Deal, but father to it; awareness of its critical role in modern government was new, but the role was not. He warned an old friend not to permit his criticism of New Deal economics to become indiscriminate criticism of the administrative process, which inevitably accompanied government intervention in the economic life of a nation—before the New Deal as during it, outside the United States as in it. Frankfurter insisted that agencies, not courts, must resolve a wide range of controversies "if unchallenged social purposes [are] to be realized."[69]

New Dealers were the first coterie of legal experts trained to run the administrative process. They fought relentlessly with their antagonists in corporate law firms over the boundaries of administrative discretion. Each side was probing the limits of power to be wielded by their respective clients, the federal government and corporations. The longer they fought, however, the narrower grew the differences between them. Late in 1938, for example, after John Foster Dulles of Sullivan & Cromwell complained that litigants before administrative agencies encountered judges who were simultaneously prosecutors, and prosecutors who dictated the decisions of judges, a meeting was arranged between corporate attorneys and SEC lawyers. Frankfurter, an intermediary, reassured Dulles that "there ought not to be decisive differences among lawyers bred in the same tradition. . . ." Dulles was not entirely convinced. In an address at Harvard Law School early the following year, he conceded that administrative agencies expressed the valid urge for efficiency amid complexities of government that often rendered separation of powers obsolete. But he still saw "fundamental antagonism" between private attorneys and government lawyers because administrative agencies made law "mobile rather than static. . . . No longer is it possible for a lawyer to sit at his desk and by making logical deductions from past decisions advise his client with confidence as to his rights."[70]

Dulles' point was conceded by SEC chairman William O. Douglas, who described administrative law as "the most volatile aspect of contemporary government."[71] But the "fundamental antagonism" that concerned Dulles was bound to lessen once the novelty of New Deal innovations wore off. In time, corporate lawyers realized that not only could their clients live with the regulatory agencies, but that indeed, corporations could attain new heights of insulated prosperity under the protective aegis of the administrative process. Lawyers "bred in the same tradition" could, and indeed would, reach the necessary accommodations. Their shared commitments, traditions, and values were far more consequential, in the long run, than their temporary animosity amid the crisis politics of the early New Deal years.

At first the regulatory process seemed nothing more than an arbitrary weapon in the arsenal of government lawyers who were hostile to corporations. It did, certainly, invest them with more power than government had traditionally exercised in areas like securities regulation or collective bargaining. But the price exacted for this power was, over time, its exercise within limits that the regulated, even more than the regulators, would define. Acceptable solutions were those amenable to the process of negotiation and compromise—the arts at which lawyers excelled. Government lawyers were required to translate regulatory formulae into language that the regulated would understand and accept. Simultaneously a new breed of Washington lawyers, serving the regulated, began to function as interpreters between the government and their clients, "explaining to each the needs, desires and demands of the other."[72]

The common bond of professionalism ultimately prevailed over discordant politics. Once it did, lawyers could exert enormous leverage on behalf of their corporate clients, who emerged as the primary beneficiaries of the administrative process. The existence of regulatory agencies offered them unprecedented direct access to government decision-making. Their wealth assured them skilled legal talent. Before the end of the Depression decade, government lawyers were beginning their exodus from the New Deal to Wall Street and Washington firms. They had drafted the statutes, served in the regulatory agencies created by the statutes, argued the cases, and shaped the common law of regulation that emerged from agency decisions. Equipped with the expertise that their New Deal experience had provided in novel areas of law that were of consummate concern to corporations, they entered a seller's market in the private sector of law practice.[73] Corporate firms appreciated the benefits to their own clientele that New Deal lawyers could provide. Even ethnic animosities began to melt under the promise of lucrative retainers, and minority-group lawyers gained access to the Protestant professional establishment, either by joining older firms or by creating new ones in Washington. There was little danger from dropping the bars to entry, because, as Frankfurter had put it, the newcomers were "bred in the same tradition." They were trained to perpetuate it, not to subvert it. The New Deal, ironically, had certified its own lawyers for careers in the service of those very clients who were most hostile to the Roosevelt administration.

The New Deal spotlight riveted attention upon Frankfurter's young activist lawyers who darted in and out of alphabet agencies, applying lessons from their Harvard seminars to the task of conflict resolution in Washington. Most government lawyers, however, were not general counsel to alphabet agencies, nor were they necessarily responding to the allure of a reform administration. Yet even at more prosaic levels of government employment, remote from the glamor agencies, profound changes occurred during the Depression, changes which created the career of government lawyer. With the rapid expansion of federal power into new areas of economic and social life came an increase in the volume and complexity of work for lawyers in government. Although lawyers in established departments earned substantially less than the elite in New Deal agencies, or their counterparts in private practice, the promise of permanent employment, especially for those plagued by racial and religious handicaps, was compelling.[74] A civil service position had obvious appeal: income, tenure, and security. In 1932 nearly two hundred thousand applicants took the federal civil service examination; the number doubled in two years and redoubled by 1936. But few positions for lawyers were filled by competitive examination; most were patronage plums. Every legal position in the departments of State, Treasury, Justice, and Agriculture was unclassified; so, too, were all but a few of those in the newer agencies. Approximately one thousand government lawyers—perhaps one-quarter of the total —were civil service appointees; these were largely confined to the Interstate Commerce Commission and the Veterans Administration. By the end of the decade, however, strong momentum had built up for creation of a career service for government attorneys. An expanded federal bureaucracy, the persistence of depressed hiring in the private sector, and the call for service to the country in an emergency created the setting that made innovation possible.[75]

In 1941, in response to recommendations of a presidential committee on civil service improvement, chaired by Justice Stanley Reed and including Frankfurter and Robert H. Jackson, a board of legal examiners was created by executive order to develop a merit system for recruitment of attorneys into the classified civil service. Members of the Reed committee, "forcibly impressed by the pervasive role played by the lawyer in the administration of the American government," unanimously favored the establishment of a career legal service. Not only was this a sharp break with the patronage tradition; it revealed "a new conception of the role of the attorney in the government service." A government job, once viewed as a consolation prize for rejects, a political reward, or, at best, a stepping-stone, had become a career. Frankfurter's model of the British civil service finally had been replicated for American lawyers. Expectations ran high that this formalization of government service for lawyers would mitigate the hostility of the bar to the federal government and, perhaps as a consequence, wean lawyers from their attachment to laissez faire. "No place in our profession," Jackson insisted, "offers greater opportunity and urge to grow than the legal service of the government."[76]

The New Deal, it is evident in retrospect, was a lawyer's deal. Not in any preceding administration had there been such dependence upon lawyers' skills or such affinity for lawyers' values. The Depression emergency institutionalized the modern liberal activist state. Lawyers, trained to govern, enjoyed direct access to its newest and most critical levers of power and monopolized the instruments of governance. Before World War I Felix Frankfurter, joined by some of his professional colleagues, had proclaimed the public responsibilities implicit in professional legal training and the necessity for lawyers and law schools to serve the state. The New Deal represented fulfillment. Frankfurter pointedly observed that his job placement work for the Roosevelt administration was indistinguishable from his recruitment service for Wall Street firms, and therefore "didn't mean anything ideologically."[77] His disclaimer was intended to disarm conservative critics; to that extent he was quite correct. Yet, in a different sense, it meant everything ideologically: the ideology of legalism had become the dominant ideology of decision-making in government.

The virtues and vices of the legal approach to problem-solving were readily apparent in New Deal Washington. A commitment to flexibility, to instrumentalism, to skeptical realism, and to administrative discretion, applied by lawyers who were "bred to the facts" (in James Landis' words) and to a case-by-case approach, may have freed the Roosevelt administration from debilitating intellectual paralysis. Yet the lawyer's obsession with craft and process, which liberated his skills, also dominated his values and inhibited his social goals. Lawyers guided New Deal solutions between the bargaining extremes but invariably toward the existing balance of power between competing interest groups. Lawyers' skills (drafting, negotiation, compromise) and lawyers' values (process divorced from substance, means over ends) permitted New Deal achievements yet set New Deal boundaries. No substantive result was permitted to assume such transcendent importance as to rule out compromise, a value commitment that gave the New Deal its opportunistic, shallow side and made it all too willing to capitulate to private power holders. The prototypical New Dealer may well have been a freewheeling activist lawyer with considerable policy-making opportunity. Yet legal training (to say nothing of New Deal politics) set clear limits and narrow boundaries which became conspicuously apparent with the passage of time. Lawyers too easily assumed that any social problem was amenable to resolution by enacting a statute, creating an administrative agency, and staffing it with law review editors. They assumed that how results were reached, not what results were reached, was the only test of responsible government. They appreciated that substantive results without fair process were procedurally unconscionable. But they failed to understand that process isolated from substance transformed lawyers into mere technicians in the service of power. Responding to the existing allocation of power, most of them were not prepared to consider the wisdom of redistributing it.

New Deal lawyers did not make the world over; none was empowered to do so, and most were not so inclined. They controlled the pace of change more than its direction. But their training and socialization dictated how social problems would be perceived and resolved—or evaded. Their case-by-case approach to social issues was a source of concentrated energy but also a constriction upon vision. Policy considerations, Jerome Frank had insisted, should not dominate a lawyer's thought. Neither Frank nor his colleagues ever grasped how much of a policy legalism itself was —or how profoundly it shaped the identity of the Roosevelt administration.[78] The New Deal has recently been criticized for representing "a transfer of power from the man in the street to the man from the *Harvard Law Review*," a transfer that accelerated the creation of a "hierarchical, elitist society."[79] This is a half-truth, but a suggestive one nonetheless. Power hardly was held by the man in the street before 1933, unless the street was Wall rather than Main. It would be more accurate to say that the New Deal reshuffled elites within a hierarchical profession and society. Between 1933 and 1941 professional power in the arena of public policy shifted from a corporate elite, served by Wall Street lawyers, to a legal elite, dominated by New Deal lawyers who, in time, became a regulatory elite ensconced in their Washington law firms.

It was a momentous shift in professional direction for some lawyers to serve public institutions as passionately as their predecessors (and peers) served private power-holders. New areas of law emerged; new careers opened; government experience became a marketable professional commodity. Consequently new groups of lawyers jostled for power and for elite status. Although the strength of an elite usually is measured by its ability to set the terms of admission into its circle of influence, its survival may depend upon its ability to adjust to outside pressures and admit challengers.[80] During the 1930's the established professional elite retained its privileged sanctuaries in corporate law firms, in traditional bar associations, and in pressure groups like the Liberty League Lawyers' Committee. But it was powerless to halt the growth of parallel institutions that trained and certified a new elite drawn from different ethnic groups and social classes. Just as members of new-immigrant groups finally "made it" in the business world only by developing their own marginal areas of entrepreneurial activity—Hollywood entertainment, for example, or bootlegging and organized crime—so minority-group lawyers succeeded in government.[81] Rival elites coexisted in uneasy equilibrium during the New Deal years.

The creation of a New Deal counter-elite among lawyers was a special type of underclass rebellion. Lawyers with responsibility and power within the administration possessed impeccable professional credentials and were, for the most part, committed to quite traditional modes of professional behavior. Whether choice or circumstance deflected them from private practice, they found a novel opportunity to fuse personal ambition, social mobility, and liberal reform in government service. The New Deal encouraged mobility, but only for those lawyers already designated as the best graduates of the best schools. For this reason New Deal lawyers, once they were certified as securities or anti-trust experts, moved with relative ease into the Protestant corporate professional establishment, leaving behind the ideal of committed service in the public interest as their abandoned legacy. Professional democratization and reconsideration of the role of lawyers as "hired guns" (whether for a corporation or for the government) were indefinitely deferred.

Changing elites did not change professional values. The growth of a parallel elite did assure a necessary degree of social mobility within the legal profession. The ultimate assimilation of the new elite into the traditional structure strengthened professional values which might have been jeopardized by exclusion based solely upon social and ethnic factors. At the base of the professional pyramid little had changed. The battleground was at the apex, where old and new elites clashed. When the dust kicked up by their professional rivalry finally settled, the old structure was greatly strengthened by its newest inhabitants, who were, by their presence, its newest defenders. The economic underclass of the profession was left to fend for itself. Marginal lawyers remained the forgotten professionals; marginal clients, indeed many middle-class Americans, still could not afford the quality of legal services, or justice, that corporations and government commanded. In service to power, lawyers made government by a legal elite the culmination of New Deal liberalism.

Eight

Cold War Conformity

The Second World War, unlike the First, created minimal internal stress for the legal profession. National cohesion for the Allied cause spared lawyers a repetition of their previous conflict between professionalism and patriotism. With the Soviet Union a valued, if uneasy, ally there was no significant protest from the left against the war effort and, therefore, no hostility directed against political dissenters. Immigration restriction and the absorption of second-generation Americans had reduced ethnic-group animosity. The external foe was the only enemy; national unity bound lawyers to the nation without exacting any sacrifice of their professional obligations.

In the postwar years, however, old patterns and values reasserted their authority. The first familiar change, a harbinger of economic recovery, was the creation of new law firms and the reorganization of old ones as New Dealers scrambled to capitalize upon their government expertise in the newer fields of administrative regulation. Back in 1921 John W. Davis and Frank Polk had led the exodus of Wilsonian Democrats from Washington to Wall Street; twenty-five years later Roosevelt Democrats either repeated the journey or carved out new careers as Washington lawyers who represented corporate interests before the federal government. In New York Samuel Rosenman, Roosevelt's trusted associate, organized a new firm; the Paul Weiss firm expanded to include Lloyd K. Garrison, former chairman of the National Labor Relations Board, and Simon Rifkind, Senator Robert F. Wagner's legislative assistant. In Washington Thomas G. Corcoran had pointed the way even before the war. Disgruntled over his failure to be appointed solicitor general, he left the administration in 1940 to engage in free-wheeling practice before government boards and behind agency doors. After the war Thurman Arnold launched the most celebrated New Deal firm with his former Yale student, Abe Fortas. Its practice, Arnold wrote, was "devoted largely to the problems of corporate clients with the government"; its simple purpose, Fortas reminisced, "was to provide a means for its two partners to make a living."[1]

There was social significance to these private decisions. For the first time, minority-group lawyers in significant numbers gained access to the professional elite in private practice. Ethnic and religious lines still held fast in the older corporate firms. The partner roster at Davis Polk, for example, read like a page from the *Social Register*. (Not until 1961, six years after Davis died, was there a Jewish partner.[2]) But the expertise accumulated by lawyers in New Deal Washington increased their market value in private practice to newer corporations dependent upon the federal government for contracts and profits. These lawyers, skilled navigators of the federal regulatory labyrinth which they had themselves designed, quickly became insiders looking out as their days as professional pariahs receded. Professional mobility and elite circulation, set in motion during the 1930's, yielded their rewards abundantly, if belatedly, to those who followed the traditional path to success in corporate practice.

But for other lawyers the postwar era turned sour, then ominous, and finally disastrous. The Cold War brought a menacing reminder of an earlier postwar era: the Red Scare. Once again the organized bar discovered subversion in its midst and merged patriotism with legalism to compel conformity to elite political values. Just as bar leaders had used Americanization to cleanse the profession of undesirables after World War I, so they now capitalized upon Cold War hostility toward the Soviet Union to eradicate the New Deal legacy from their professional culture. In political terms that legacy meant the tolerant spirit of the Popular Front, when liberal and radical lawyers identified their own obligations as attorneys with political reform, civil liberties, minority rights, and the quest for social justice—especially in labor relations. Professionally, it signified an enlarged conception of the responsibility of lawyers to provide service to clients regardless of their economic resources. But the politics of the Cold War transformed these pursuits into subversive activities. The professional elite, encouraged and abetted by two attorneys general, by the House Un-American Activities Committee, and by politicians eager to ride anti-Communism to political power, attempted to purge the profession of lawyers whose political and professional commitments deviated from Cold War orthodoxy. To bar leaders, convinced that they confronted "a campaign directed from Moscow, aimed especially at our capitalistic system," nothing less was tolerable.[3]

Attorney General Tom Clark was the first to warn of impending danger and to propose palliatives. In a speech in 1946 to members of the Chicago Bar Association, reprinted in the American Bar Association *Journal*, Clark described a plot by Communists, "outside ideologists," and "small groups of radicals" to undermine the nation. He condemned the "revolutionary" who "uses every device in the legal category to further the interests of those who would destroy our government, by force if necessary." These lawyers, he suggested, should be taken by bar associations "to the legal woodshed for a definite and well-deserved admonition."[4] Clark's warning went unheeded for nearly two years. But as American relations with the Soviet Union deteriorated, the American Bar Association explored its woodshed for admonitory weapons.

At first there were only editorials, warning that "forces of disorder and disruption are gathering for strife that will come if and when the orders for it are issued from the faraway Kremlin." Occasionally an article criticized lawyers whose dogmatic adherence to constitutional rights impeded their response to the Communist threat.[5] At its annual convention in 1948 the association adopted a resolution declaring that any lawyer who provided assistance or support to the "world communist movement" was "unworthy" of ABA membership. The Board of Governors resolved that no member of the National Lawyers Guild should be accepted in the association. In an ironic twist that perfectly captured the new mood, the board referred a resolution condemning Communists to the Bill of Rights Committee (a beachhead of New Deal liberalism within the association) and sent a resolution calling for an educational effort on behalf of the Bill of Rights to the American Citizenship Committee (a bastion of reaction from the Red Scare days). The Bill of Rights Committee had already become a political weathervane within the ABA with its denunciation of those who claimed that inquiries into political beliefs and associations abridged constitutional rights.[6]

The National Lawyers Guild was a primary target for the government and for the American Bar Association. Wracked by internal political discord during 1939 and 1940 after the Nazi-Soviet pact, it survived the world war but not the cold war that followed. The Smith Act trial of Communist Party leaders, with the hysteria it both expressed and generated, was the turning point. Each of the defense attorneys cited for contempt by Judge Medina was a guild member; three were officers. With guilt by association endemic in postwar public life, vilification of the guild inevitably followed. The House Un-American Activities Committee issued a report denouncing it as "the foremost legal bulwark of the Communist Party, its front organizations, and controlled unions." Guild lawyers, it declared, "substitute insult for argument" and "resort to intimidation of judges." The committee recommended that the guild be placed on the attorney general's list of subversive organizations.[7] If intimidation of guild lawyers was not the intent, it certainly was the consequence. The guild quickly lost half its membership and was compelled to expend the major portion of its energies to save itself from annihilation.[8]

The Un-American Activities Committee opened the government assault against the guild, but Herbert Brownell, President Eisenhower's attorney general, turned the campaign into a crusade with an address to the American Bar Association in 1953. Brownell conceded that in the past the guild had attracted "some very well-known and completely loyal American citizens." But professional groups, he noted, were not exempt from Communist infiltration. Indeed, "at least since 1946 the leadership of the National Lawyers Guild has been in the hands of card-carrying Communists and prominent fellow travelers. . . . It has become more and more the legal mouthpiece for the Communist Party. . . . The evidence shows that . . . [it] is at present a Communist-dominated and controlled organization." Therefore, he concluded, the guild must show cause why it should not be placed on the attorney general's list of subversive organizations.[9] When the "Brownell bombshell" hit, reported a guild official, "the blow was paralyzing."[10] Not only was the guild forced to devote its energies to self-preservation; many members resigned, and others warned that they would also leave if the guild was listed. Many chapters withered, while most of the survivors were inactive or defunct. For five years the guild was engaged in protracted litigation to avoid listing; finally, in 1958, after it filed for dismissal of the suit, a new attorney general withdrew the recommendation to designate it as a subversive group.[11]

Brownell's allegations raised substantial issues of accuracy, propriety, and fairness. If the evidence demonstrated that the guild was Communist-dominated and controlled, as Brownell claimed, he offered no substantiation beyond his *ipse dixit*. Nor did he acknowledge that Communists were, after all, as entitled to legal defense as anyone else—and, during the Cold War, more likely to need it. He slighted one obvious explanation for the frequent defense of Communists by guild attorneys: the delinquency of so many other lawyers, which the anti-Communist pronouncements of government officials and bar leaders encouraged. It was, at the very least, of dubious propriety for an attorney general to present the initial declaration of his official intentions, which had legal consequences, in a public address to a rival professional association. Certainly the minimal components of due process entitled the guild to the courtesy of prior and private notification and an opportunity to respond before it was too late for any response to matter. The only precedent, claimed one shocked attorney, was a case cited by Lewis Carroll in 1857 in which the Queen had stated her preference for the verdict first and the trial afterwards.[12]

The political vulnerability of the guild during the Cold War was obvious. Less evident was the extent to which the anti-Communism of the bar camouflaged ABA efforts to undercut guild proposals for public funding of legal services for low-income groups. The broad dissemination of legal services had been a primary guild objective since its birth during the Depression. After World War II the guild, disappointed with the failure of charitable legal aid, advocated expanded government responsibility. Encouraged by the report of the Rushcliffe Committee in England, whose recommendations for government-subsidized legal aid were enacted in 1949, guild attorneys urged the appropriation of public funds for legal services for those who could not afford the fees charged by private attorneys. Privately funded legal aid, concluded a guild committee, had "never ripened from mere charitable indulgences into categorical constitutional imperatives." With two-thirds of the American population unable to afford legal services, the problem was ingrained and massive, not the "minor aberration in the functioning of our system of justice" that most lawyers claimed it was. Predictably, the guild proposal was denounced as socialistic; as guild attorneys wryly observed, however, federal judges received government salaries but no one ever complained about socialization of the bench.[13]

Leaders of the American Bar Association were appalled at the prospect of federally subsidized legal services. According to a member of the Board of Governors, federal legal services, like the "undermining influences of Communist infiltration" with which he equated them, would inevitably substitute state regimentation for professional independence. For proposing federal subsidies the guild earned the opprobrium otherwise reserved for Bolsheviks, another "organized minority with ruthless methods." Association president Harold J. Gallagher warned that "the entry of the government into the field of providing legal services is too dangerous to be permitted to come about in our free America." It was absurd, these lawyers suggested, to believe that a needy client should obtain assistance anywhere but in "the law office of an independent practitioner." The fact that only an economically privileged minority could obtain it there was not mentioned.[14]

The attack upon the National Lawyers Guild by government officials and bar leaders virtually destroyed it as an effective professional organization. Its notoriety is apparent from the most celebrated episode of the Army-McCarthy hearings in 1954, when the senator disclosed that one of attorney Joseph Welch's young associates in Hale & Dorr had once belonged to the guild. Welch's indignant response, charging McCarthy with reckless cruelty, was a rare and telling rebuke to the senator. But Welch, as his words of outrage indicated, accepted the prevailing view that mere attribution of guild membership constituted character "assassination." The accused lawyer, Welch predicted, would "always bear a scar needlessly inflicted" merely by public reference to his former guild membership.[15] The guild was triply vulnerable: for the politics of some of its members; for the willingness of its attorneys to defend clients and causes which other lawyers would not touch; and for its innovative recommendations for federally subsidized legal services. The socioethnic identity of its membership doubtlessly was an additional impediment, although one that is impossible to measure with precision. It is at least worth noting that the five guild lawyers sentenced for contempt after the Smith Act trial included three Jews, one black (who later became a Detroit judge), and one Irish Catholic—hardly representative of the most privileged professional groups. Clearly there was a reciprocal relationship between the social marginality that predisposed minority-group lawyers to guild membership and the attack of the professional elite on the guild for its political and professional identity.

The Cold War objectives of the American Bar Association went far beyond the elimination of a professional rival and the defense of free-enterprise legal services. "We are faced with a constant attack from within upon our form of government and

our American way of life by proponents of foreign ideologies,"
declared ABA president Cody Fowler in 1950.[16] The association
explored proscriptive and punitive measures of professional discipline, imploring other professional groups to follow its lead. First
it recommended that lawyers take a loyalty oath and file an affi-
davit declaring whether they had ever belonged to, or supported,
an organization that advocated the forcible overthrow of the government. It urged that lawyers who had once belonged to the
Communist Party should be investigated to determine their fitness
to remain in practice. In 1951 its Special Committee to Study
Communist Tactics, Strategy, and Objectives, which had worked
closely with the Un-American Activities Committee and the
American Legion, warned that lawyers who advised clients not to
answer questions which did not violate their constitutional rights
were guilty of a "deliberate effort to evade, confuse or distort."
(The committee took no notice of severe legal penalties which accompanied the decision to answer questions selectively.) The time
had come, it concluded, "to drive such lawyers from the profession" and to exclude lawyers found guilty of "embracing and
practicing" the doctrines of Communism or Marxism-Leninism.[17]
The House of Delegates recommended to all bar associations that
disciplinary action leading to expulsion commence against lawyers
who were party members or who advocated Marxism-Leninism.[18]

In 1952, when civil liberties lawyers were overwhelmed with
litigation arising under the First and Fifth Amendments, the ABA
Bill of Rights Committee concluded that it was unnecessary to
disseminate information about constitutional liberties and found it
"gratifying" that no violations of the Bill of Rights requiring
committee intervention had come to its attention. (One lawyer
did request an *amicus* brief in an attempt to stave off disbarment,
but the committee decided that it was "not . . . appropriate" to
provide it.[19]) The Communist Tactics committee decided, however, that "the time for ousting the communist attorney is at
hand." It was directed by the House of Delegates to cooperate
with the attorney general and with state and local bar associations,
to inquire into the conduct of attorneys who had been identified

as party members or who had claimed their Fifth Amendment rights before congressional committees, and to institute proceedings to determine their fitness for practice. The committee eagerly complied. Aided by the Justice Department, it compiled lists of culpable attorneys and concluded that party membership was "incompatible" with membership at the bar.[20]

Among various proposals for professional orthodoxy none elicited more criticism than the resolution approved by the association without debate in 1950 demanding that all lawyers attest to their loyalty with an anti-Communist oath.[21] Zechariah Chafee, Jr., of Harvard Law School, a patrician turned libertarian by World War I violations of the Bill of Rights, observed that the loyalty oath was adopted "with less discussion than would have been devoted to the menu at the next annual banquet." Chafee, author of the classic *Free Speech in the United States* and a charter member of the association's Bill of Rights Committee, decided that "if nobody else stood up on his hind legs and yelled, I was going to do so."[22] Chafee, characteristically, yelled gently but effectively. He spoke and wrote vigorously against the proposal, which in his opinion violated the First and Fifth Amendments, portended "a purge of the American bar," and was, he concluded, "the worst thing the American Bar Association has ever done."[23] Chafee directed his appeal to bar leaders within the association. Twenty-six prominent members—including Grenville Clark, Charles C. Burlingham, John W. Davis, Whitney North Seymour, and Harrison Tweed—co-signed a petition in opposition to the oath which denounced it as "unfounded in its implication of widespread disloyalty."[24]

The loyalty oath proposal was a powerful symbolic issue. Its adoption was praised as an antibody against the virus of subversion carried by lawyers who had infected the profession and their country. The names of three Harvard Law School alumni—Alger Hiss, Benjamin Davis (one of the Smith Act defendants in the 1949 trial), and Lee Pressman (who subsequently testified to his membership in a Communist cell in the Department of Agriculture during the early New Deal years)—appeared repeatedly in letters to Chafee from lawyers excoriating "scoundrely Reds" who believed in the resolution of issues by "violence, bloodshed, and death." Some of these Harvard graduates, warned one alumnus, "have been placed in key [government] positions, with appalling consequences to humanity in general and our country in particular."[25] Loyalty was a consuming national obsession during the postwar years. Although, as various critics noted, no test could insure loyalty—indeed, the term was rarely defined—society demanded the symbolic gesture of conformity that an oath provided.[26] Lawyers were not more hysterical than most Americans; neither were they less so. They did, however, permit "the hysterical men among [them] to exercise a disproportionate amount of influence."[27] At the state level there was little eagerness to follow the ABA lead, prompting Chafee to conclude in 1953 that the campaign was "a flop." The result, according to two scholars, was "a handful of oaths, . . . some half-concealed supplementary investigative techniques, and a few character committees with inquisitorial tendencies." In New York both the state and the prestigious city bar associations opposed the ABA proposal. Elsewhere, the demand for loyalty was confined to metropolitan areas, an expression of the traditional professional division between "the wicked city and the virtuous country."[28]

Although the loyalty oath proposal expired at the state level, it was only part of a sustained, invidious, and more successful effort to intimidate lawyers for unpopular defendants and to discipline those whose beliefs or associations were adjudged subversive. Once political conformity determined the boundaries of justice, many unpopular defendants could not find lawyers—and some unpopular lawyers could not retain clients. In one state after another bench and bar mobilized an array of professional weapons: contempt citations, disbarment, exclusion, and subtler forms of coercive socialization. A presumed tradition of professional independence and responsibility degenerated as the practice of law became "a hazardous occupation." There was, one scholar concluded, "an open season on lawyers who defend hated men."[29]

Overt repression began after the Smith Act trial in 1949. That trial was the key that unlocked Attorney General Clark's legal woodshed of discipline and harassment. It was long, frequently acrimonious, and laced with anti-Communist hysteria from the indictment stage through its "bizarre" culmination, when Medina found five defense attorneys guilty of criminal contempt and imposed prison sentences upon them without providing them with notice, a hearing, or the opportunity to defend themselves.[30] After Medina pronounced sentence, defense attorney Abraham Isserman cited Clark's "woodshed" recommendation for politically wayward lawyers, which had reappeared in another article by Clark during the closing months of the trial. Isserman, asserting that Medina had applied the punitive measures proposed by the attorney general, accused the judge of "an effort to intimidate members of the bar. . . ."[31] It may be difficult, as a special bar association committee on courtroom conduct recently concluded, "to decide whether the obstructive behavior at the trial arose out of the lawyers' individual, vigorous assertion of their clients' cause or was planned beforehand . . . , or whether their conduct was the result of a self-fulfilling prophecy, a consequence of Judge Medina's own behavior."[32] But there is no question that Medina conducted the trial with an attentive eye to the press and to his approving audience in Washington government circles. He solicited information about newspaper coverage from a friend in a Washington law firm; he received letters of praise from the attorney general and from FBI director J. Edgar Hoover; and he sent a copy of his charge to the jury to Senator Joseph R. McCarthy.[33]

Contentiousness between judge and defense lawyers, against a backdrop of national hysteria, inevitably redounded to the lawyers' disadvantage. Medina symbolized the majesty of law, meting out justice to Communist Party leaders accused of conspiring to advocate the forcible overthrow of the government. The defense attorneys were excoriated by lawyers who interpreted their conduct as an integral part of the conspiracy that was on trial.[34] Medina, haunted by the memory of a tumultuous Smith Act trial five years earlier during which the judge had died of a heart attack, shared that understanding of the lawyers' behavior. In his certificate of contempt he observed that before the trial had progressed very far he "was reluctantly forced to conclude" that the lawyers' actions "were the result of an agreement . . . deliberately entered into in a cold and calculating manner" to provoke incidents that would cause confusion and delay and impair his own health. Medina indicated that he would have overlooked the conduct of the lawyers, or merely reprimanded them for it, had it come from the heat of controversy or from zealous defense. But the "agreement . . . deliberately entered into" made all the difference.[35] Although Medina subsequently praised "courage and independence" at the bar, one of his critics observed that the judge "was promoted and the lawyers went to jail."[36]

The contempt citations triggered appeals, disciplinary proceedings, and collateral litigation which culminated in the first instances of disbarment for forensic misconduct. For the two lawyers most directly involved, Harry Sacher and Abraham Isserman, these proceedings, stretching over twelve years, can only be described as diabolic. They were veteran attorneys of acknowledged ability who had devoted substantial portions of their careers to civil liberties and labor defense efforts. During the Cold War they paid the price exacted by their profession for their commitments. But long before Medina's summary action reached its final judicial resolution the consequences of intimidation were apparent. Accelerated by each jolt in domestic politics—the Hiss and Rosenberg trials, McCarthyism, the Korean War—the momentum of professional repression increased.

On appeal the *Sacher* contempt case (involving all the defense lawyers) posed a dilemma for the Second Circuit. Medina's citation was challenged on the ground that summary punishment was proper only if it followed immediately upon commission of the contemptuous acts, not when it was administered weeks or months later, after the trial had ended. But reversal of Medina in the contempt case would substantiate the claim being pressed in the *Dennis* appeal (involving the Smith Act defendants) that the trial was unfair. This dilemma threatened to become an acute embarrassment to the government and to Medina when Judge Jerome Frank, "tentatively and most reluctantly" (he claimed), voted to reverse and remand for a hearing before another judge. Frank was convinced that the lawyers had acted contemptuously, but he also thought that the trial judge could not punish summarily after the trial. Since Judge Charles Clark had also concluded that the lawyers were entitled to a hearing and to an opportunity to defend themselves before a contempt citation was imposed, their votes were sufficient for reversal.[37]

Judge Augustus Hand, the third member of the Second Circuit panel, believed that Medina had acted properly. He also was distressed by the implications of reversal. He apparently persuaded Frank to change his vote, thereby upholding the contempt citations.[38] Hand, now joined by Frank, declared that the lawyers had committed "wilful obstruction" which need not be punished immediately (thereby risking an interruption of the trial), but only as speedily as circumstances permitted.[39] Yet the Hand-Frank majority, by resolving this problem, created new ones. By Medina's own admission in his contempt certificate, he would not have cited for contempt but for his belief that the lawyers had participated in a "plan" or "an agreement."[40] As Frank indicated, Medina's allegation "charges something in the nature of a conspiracy" —a notion which both Frank and Clark rejected.[41] So Frank, joining with Clark in voiding the "agreement" or conspiracy specification, joined with Hand in sustaining the other specifications— although Medina himself had conceded that but for the "plan" there would have been no contempt. This doubtlessly troubled Hand, who attempted to have his conspiracy cake yet not choke on it. Medina's statement about an agreement, Hand reasoned, meant nothing more than that the acts were "deliberate." It was, he concluded, quite unimportant whether Medina believed in a conspiracy.[42] If the reasoning was tortured the result was clear: contempt citations, justified by the trial judge solely on the presumption that conspiracy existed, were upheld on appeal notwithstanding the finding that no conspiracy existed.[43]

Frank, in his concurring opinion, dismissed fears expressed in *amicus* briefs that prospective lawyers for unpopular defendants would be intimidated by the contempt citations. "The fears are unfounded," he observed.[44] But they were quite well-founded indeed. Reports of difficulty in obtaining counsel soon reached the Supreme Court, which earlier had denied a petition for review of *Sacher*. Justice Robert Jackson, who concurred in the denial, changed his mind when he realized that "the difficulties of obtaining counsel in these cases has increased since we denied review and . . . it threatens to become impossible."[45] The Court heard the appeal and, by a 5-3 vote (with Jackson writing the majority opinion), upheld the power of a trial judge to defer summary contempt until the completion of the trial if he believed that the exigencies of the trial required it.[46] Justices Black and Frankfurter issued stinging dissents and strong rebukes to Medina for his evident demonstrations of hostility toward the defense lawyers. Black, referring to the proceedings, declared: "I cannot reconcile this summary blasting of legal careers with a fair system of justice. Such a procedure constitutes an overhanging menace to the security of every courtroom advocate in America. The menace is most ominous for lawyers who are obscure, unpopular or defenders of unpopular persons or unorthodox causes." Frankfurter added that Medina had "failed to exercise the moral authority of a court possessed of a great tradition." It was, he concluded, "a disservice to the law to sanction the imposition of punishment by a judge personally involved and therefore not unreasonably to be deemed to be seeking retribution."[47]

Sacher and Isserman went to jail (as did the other Smith Act trial defense attorneys), but their professional travail had barely begun. The Association of the Bar of the City of New York brought disbarment proceedings against Sacher, who already had been suspended for two years. The Second Circuit, again over Clark's dissent, upheld disbarment on appeal. Judge Augustus Hand, again for the majority, insisted that disbarment was necessary to protect the court and to increase respect for it by assuring that lawyers displayed "good professional character." Clark, however, observed that Sacher had a twenty-four-year record of "unblemished conduct" and that in appellate appearances on his own behalf, in trying circumstances, he had demonstrated "courteous and dignified" behavior and professional ability of "unusually high order." Clark expressed the hope that his disbarment would be rescinded when "the present atmosphere of hysteria has somewhat abated." His dissent concluded with the plaintive question: ". . . why must the most serious wounds to justice be self-inflicted?"[48] By 1954, when the disbarment appeal reached the Supreme Court, the hysteria cited by Clark had begun to abate. In a *per curiam* opinion reversing the Court of Appeals, the Court declared that permanent disbarment was "unnecessarily severe."[49]

Sacher's ordeal almost pales in comparison with Isserman's. Isserman, who had shared Sacher's two-year suspension in New York, was disbarred in New Jersey in 1952 for his "scandalous and inexcusable behavior."[50] After his disbarment there the United States Supreme Court ordered him to show cause why he should not be disbarred from practice before it, since the high court ordinarily followed the findings of state courts in disbarment proceedings. With the Court evenly split, disbarment resulted (a third disbarment, in New York, followed). The four dissenting justices observed that disbarment for contempt was rare, if not unprecedented. Indeed, they noted, Elihu Root and David Dudley Field, two earlier titans of the bar, had been cited for contempt after the trial of Boss Tweed without additional penalty—in fact Field was subsequently elected president of the American Bar Association.[51] In 1953 the Court denied a petition for certiorari appealing the New Jersey disbarment. But in 1954, the year that Sacher's disbarment was set aside, the tide turned. Isserman's Supreme Court disbarment was reversed under a new rule providing that no order of disbarment could be entered without the concurrence of a majority of the participating justices.[52] Five years later Isserman's New York disbarment was set aside as "discriminatorily severe," with the Court of Appeals noting that Isserman had endured *de facto* suspension for nearly a decade.[53] In 1961, twelve years after the first house of disciplinary cards was built upon its foundation of conspiratorial sand, Isserman's application for reinstatement to the New Jersey bar was granted. The state Supreme Court, writing the final page to a sordid chapter, concluded that a nine-year "stigma of disbarment . . . is more than enough."[54]

Sacher and Isserman were the first American lawyers known to have suffered the penalty of disbarment for forensic misconduct.[55] But if the severity of their penalties was unique, their punishment was only the prologue to a period of sustained professional harassment of politically vulnerable attorneys. Lawyers who obeyed the ethical canon requiring them to engage in advocacy with zeal and courage, without regard for the unpopularity of the client or his cause, found that they had plunged through a trap door to disciplinary proceedings. Additionally, prospective lawyers confronted the paradoxical situation of asserting their own constitutional rights only to discover that this assertion might be sufficient to disqualify them from practice. The number of reported cases—barely a dozen reached the Supreme Court in a decade—does not accurately measure the damage that Cold War professionalism inflicted. It is impossible to calculate with precision the impact of a few conspicuous prosecutions or proceedings on the entire profession, or on those whose caution or fear may have prompted silent acquiescence and consequent anonymity. It cannot be determined how many prospective attorneys for unpopular defendants were deterred by the threat of professional discipline or the risk of economic adversity. Nor can it be estimated how many concealed political rebuffs were issued to clients in need of an attorney's services. Justice William O. Douglas referred in 1952 to a "black silence of fear" in the legal profession.[56] Some sounds remain audible; some dark recesses can be illuminated. Doubtlessly for all but a relatively few lawyers it was business as usual. But for those who were hounded as political pariahs (and for their professional hunters) the Cold War represented an unprecedented episode of professional repression.

Within months after the Smith Act trial concluded, its ripple effect was apparent. In California two defense lawyers for Harry Bridges, the left-wing union leader who was a venerable target of government prosecution, were summarily cited for contempt and served jail sentences for courtroom conduct which, in language strikingly reminiscent of the Medina contretemps, was described as "designed and calculated to contemptuously provoke the court."[57] (One of the lawyers, Vincent Hallinan, became the vice-presidential nominee of the Progressive Party in 1952.) In Pennsylvania lawyer Hyman Schlesinger was defending a client in a trespass case when the trial judge interrupted the proceedings to inquire if he had ever belonged to the Communist Party or to its front organizations. Schlesinger, refusing to answer, was pronounced "morally unfit" and held in contempt. (The judge, who a year earlier had wrongfully dismissed a grand juror whom he had concluded was a Communist, was rebuked and reversed for conducting "arbitrary and unjudicial proceedings."[58]) In separate proceedings Schlesinger was charged with "professional misconduct" by a bar disciplinary group. For a period of eight months he was unable to find defense counsel, "a lamentable commentary" according to the Pennsylvania Supreme Court which set aside his disbarment—a decade later.[59]

In Florida a lawyer was disbarred for claiming his Fifth Amendment right against self-incrimination in declining to answer questions regarding his alleged membership in the Communist Party. The appellate court reversed, noting that, although a lawyer who was a party member forfeited his right to practice (because he was "guilty of a species of treason"), mere refusal to answer questions about party membership was insufficient for disbarment.[60] In Maryland a lawyer convicted under the Smith Act of conspiring to advocate the forcible overthrow of the government was disbarred upon termination of his three-year jail sentence for a crime involving moral turpitude and for being "a subversive person."[61] (Eighteen years later, when he sought reinstatement, a three-judge panel agreed that his conviction had been "largely political in nature."[62]) In Hawaii a lawyer for Smith Act defendants addressed a public meeting while the trial was in progress at which she claimed that "shocking and horrible things" were occurring in court and that there could not be a fair trial in a Smith Act case.

The trial judge invited the Hawaii Bar Association to scrutinize her professional conduct. A one-year suspension resulted which the Court of Appeals upheld on the ground that she had made "a wilful oral attack" upon the administration of justice. In a 5-4 decision the Supreme Court reversed on the narrow ground that the evidence was insufficient to support the conclusion that her speech impugned the integrity of the judge. Only four justices could agree that lawyers "are free to criticize the state of the law."[63]

There was less punitive harassment. In Michigan two lawyers were called before a local ethics committee to account for their criticism of the Rosenberg trial. Lawyers appearing before legislative investigating committees and grand juries frequently were questioned about their political beliefs and associations and those of their clients.[64] An American Bar Association committee concluded in 1953 that even lawyers who were conspicuously identified as anti-Communists were subject to "severe personal vilification and abuse" if they defended Communists or suspected Communists.[65] A Los Angeles attorney, asked to assist in the defense of Communist Party leaders in a Smith Act prosecution, was warned by friends that he would inherit their guilt. Deciding not to jeopardize his practice, he quoted a fee that he knew was too high to be met. But, angered by the high bail set for the defendants, he reconsidered and entered the case. "In less time than it took [him] to reach his decision, his law practice vanished."[66] Lawyers who were spared disciplinary proceedings by their local bar association suffered economic reprisals and damaged reputations. It was, Justice Douglas concluded, "a dark tragedy."[67]

Lawyers were not the only victims of professional reprisals. Defendants in political cases found it difficult in some instances, and virtually impossible in others, to obtain counsel. The deterrent effect of disciplinary harassment and economic risk was apparent when the government began its prosecution of second-string Communist Party officials, who approached one hundred and fifty lawyers before they could obtain defense counsel. According to their claim a former Cabinet member and a former Supreme Court justice were among those who declined. Twenty-eight law firms were contacted: twelve did not reply; sixteen refused either to interview the defendants or to take the case. Finally, a panel of willing lawyers was obtained through the combined efforts of the National Lawyers Guild and two bar associations.[68] Steve Nelson, defendant in a Pennsylvania sedition trial, was compelled to represent himself after he unsuccessfully approached local bar associations, visited twenty-five lawyers in Pittsburgh, and wrote to fifty others in Philadelphia, New York, and Chicago—none of whom would defend him. Other Smith Act defendants, in Baltimore and St. Louis, encountered identical difficulties.[69] For political nonconformists the right to counsel disappeared into constitutional limbo.

The pool of lawyers willing to defend unpopular defendants, reduced by expulsion and intimidation, was further depleted by the diligent efforts of bar associations to exclude prospective applicants who did not satisfy prevailing political tests. A reliable old weapon—good moral character—was refurbished for Cold War service. In Texas the Houston Bar Association issued an adverse recommendation against an applicant who had participated in activities on behalf of Julius and Ethel Rosenberg and had associated with a lawyer suspected to be a party member. The federal district court affirmed, suggesting that admission to the bar required that "private and personal character shall be unexceptional." The Supreme Court reversed.[70] In New Mexico an applicant who belonged to the Communist Party between 1932 and 1940, used aliases to escape anti-Semitism in employment, and was arrested (but never tried or convicted) for organizing workers and recruiting volunteers for the Spanish Loyalist cause was prevented from taking the bar examination because he lacked the requisite moral character. The Supreme Court again reversed, holding that a state could not exclude a person from practice for reasons that contravened the due process or equal protection clauses. This disqualification, it suggested, had no rational connection to the applicant's fitness for practice. Neither unorthodox ideas, nor party membership, wrote Justice Black, demonstrated questionable moral character.[71]

In a companion case the Supreme Court reversed a California decision which upheld the exclusion of an applicant who refused, on First Amendment grounds, to answer questions regarding his past or present political beliefs and associations. His examiners claimed that he had failed to demonstrate good moral character; the Supreme Court reversed on the ground that the record did not support reasonable doubt about his character. Without reaching the issue of whether the failure to answer was an independent ground for exclusion the Court, in another opinion by Black, declared that "good moral character" was "unusually ambiguous" and that any definition would reflect "the attitudes, experiences, and prejudices of the definer." It was, therefore, "a dangerous instrument for arbitrary and discriminatory denial of the right to practice law."[72]

The New Mexico and California decisions, an attorney suggested in 1957, represented a "return to reason" by the judiciary.[73] Disclosures of past party membership apparently were insufficient to sustain inferences of bad character. That interpretation gained credence when the Supreme Court also vacated an Oregon decision under which an expelled party member was excluded from the bar on the ground that his disclaimer of belief in the forcible overthrow of the government was a lie that demonstrated bad character. The Court asked Oregon to reconsider in the light of its recent rulings.[74] Bar associations, rebuked by the Supreme Court in four consecutive cases that turned on definitions of good character, seemed to have exceeded their exclusionary limits.

But optimistic assessments were premature. Bar associations did not relent, and the Supreme Court ultimately acquiesced in their exclusionary efforts. First, the Oregon Supreme Court, again ruling that the expelled party member swore falsely when he declared his belief in the party's (and his own) non-violent objectives, upheld his exclusion on character grounds. The Supreme Court denied certiorari.[75] The Konigsberg case in California returned to the Supreme Court in 1961. Again the applicant had been asked about party membership; again he refused to answer; again the bar examiners refused to certify him on the ground that his silence obstructed a full inquiry into his qualifications. The first time around the Court had declined to reach the issue of whether refusal to answer was ground for exclusion; but this time, by a 5-4 vote, it affirmed, holding that a state could deny admission if an applicant refused to answer questions relevant to his qualifications.[76]

The second *Konigsberg* opinion, written by Justice Harlan, was as tortured in its reasoning as the earlier *Sacher* opinion in the Court of Appeals had been. In *Sacher*, contempt charges resting necessarily upon a conspiracy were upheld even though the conspiracy claim was rejected. In the first *Konigsberg* case the Court had dismissed the contention that the available evidence pertaining to Konigsberg's past, or his refusal to answer, supported adverse character inferences. Now, however, it held that Konigsberg's refusal to answer left the record in sufficient doubt—even though his acknowledgment of party membership would *not* justify exclusion. If Konigsberg conceded party membership his admission to the bar was incontestable; but his refusal to answer enabled the Court to uphold his exclusion. The Court declined to look beyond the reason offered by the bar examiners, even though it was apparent that Konigsberg's exclusion rested upon their belief that he had been a party member. Konigsberg, wrote Black in dissent, was "but another victim of the prevailing fashion of destroying men for the views it is suspected they might entertain."[77]

In another case decided the same day the Court upheld exclusionary proceedings in Illinois that made the California bar seem almost libertarian by comparison. Back in 1950 the Chicago bar committee on character and fitness, reviewing the application of George Anastaplo, discovered that the applicant had quoted from the Declaration of Independence about the right of the people to abolish their government if it was destructive of the ends for which it was established. The specter of Jefferson disturbed committee members, who questioned Anastaplo about membership in the Communist Party and about his belief in a Supreme Being. When he refused to answer on the ground that such questions constituted illegitimate inquiries into his political beliefs, the committee refused to certify him. The Illinois Supreme Court upheld the relevance of the inquiry for determination of Anastaplo's qualities of citizenship and affirmed. Anastaplo's appeal to the United States Supreme Court was dismissed.[78]

In 1957, encouraged by the ray of hope cast by the first *Konigsberg* decision, Anastaplo sought a rehearing on his application. The character committee refused, but the state Supreme Court reversed and directed a rehearing. The committee again declined to certify, claiming that Anastaplo's refusal to answer was tantamount to a failure to demonstrate good character. Anastaplo argued his own case before the Illinois Supreme Court, which agreed with the bar committee that his "recalcitrance" prevented him from demonstrating good character, notwithstanding uncontroverted evidence of good character in numerous testimonials from lawyers and faculty members at the University of Chicago, where Anastaplo was teaching and studying for his doctorate.[79] Anastaplo carried his case to the Supreme Court, again arguing his own appeal. By another 5-4 vote the Court affirmed his exclusion. Justice Harlan, relying upon *Konigsberg*, declared that the state's interest in learning about party membership outweighed "any deterrent effect" upon the freedoms of speech and association.[80]

To Justice Black, again in dissent, the case illustrated the consequences of permitting bar members to deny the protection of the First Amendment to applicants for admission. "To force the Bar to become a group of thoroughly orthodox, time-serving, government-fearing individuals is to humiliate and degrade it," Black wrote. "But that," he concluded, "is the present trend."[81] Throughout the proceedings, ironically, Anastaplo had concealed his desire to retain his air force reserve commission so that he might fight against the Soviet Union if war came. Some years later Chicago law professor Malcolm Sharp would refer to Anastaplo as "the staunchest anti-Communist" he knew. After Anastaplo was expelled from the Soviet Union for taking too many photographs, Sharp observed: "The similarity between the Russian police and the Character and Fitness Committee of the Chicago Bar must strike a detached observer."[82]

The power of bar associations to police the profession was not unrestricted. Although the ABA, in the words of one of its officers, "has taken the leadership in attempting to remove the communist lawyers from the practice of law," its followers often lagged behind.[83] As the loyalty oath campaign demonstrated, the association could recommend but not always command. Furthermore, the power to exclude received greater judicial protection than the power to expel. Yet the professional elite succeeded even as it failed. Before the affirmative Supreme Court decisions, intense hostility toward nonconformity pervaded the profession as bar leaders tried to extirpate lawyers whose primary offense was the unpopularity of their political beliefs and associations. These embattled lawyers were also vulnerable for other reasons, which were deeply embedded in the history of the professional culture. It may only be coincidental that an ethnic profile of the lawyer-defendants in admission and expulsion cases which were appealed to the Supreme Court reveals that, of a total of thirteen lawyers, there were eight Jews, two Irish Catholics, one Greek-American, and two probable Protestants—one of whom was female. Then again, it may not have been coincidental. The professional elite found the deviants that its own ethnic and political exclusiveness had created. Armed with the full disciplinary powers of the profession, it stripped lawyers of their livelihoods, clients of their lawyers, and the Bill of Rights of several vital provisions.

There was no shortage of rhetorical obeisance to the lawyer's responsibility. John W. Davis, the venerable leader of the appellate bar, declared that it was the "supreme function" of lawyers to serve as "sleepless sentinels on the ramparts of human liberty and there to sound the alarm whenever an enemy appears."[84] But Davis, like so many other professional leaders, slumbered soundly on the ramparts. It was, of course, an article of professional faith that when constitutional freedoms were jeopardized established lawyers would rise to the responsibility of defending the unpopular and beleaguered. Thus Charles Evans Hughes had protested against the expulsion of Socialist members of the New York legislature during the Red Scare; indeed, Davis himself had once carried to the Supreme Court an appeal on behalf of a theologian who claimed that selective conscientious objection should not disqualify him from citizenship. Yet there lurks the suspicion, overabundantly documented during the Cold War, that incidents like these were celebrated less because they were typical than because they were exceptional. As early as 1951 Justice Felix Frankfurter concluded: "Our own profession is giving a miserable account of itself these days."[85]

It was Davis too, nearing the end of a brilliant career which had earned him the accolade "lawyer's lawyer," who demonstrated how wide was the gap between rhetoric and reality. During the Cold War years he refused to sign an *amicus* brief attacking federal loyalty programs; to argue "the 'Commie' case" before the Supreme Court; or to sign an *amicus* brief on behalf of the lawyers cited by Judge Medina for contempt. Although he always claimed that he would take any case that came into his office, when Gus Hall, a Communist convicted under the Smith Act, approached him Davis responded that he was too busy. Considering Davis' volubility during the 1930's, when New Deal statutes elicited repeated statements of outrage, his Cold War silence strongly suggests a double standard which permitted him to believe that power was repressive, and Liberty Leagues were necessary, only when New Dealers regulated corporations.[86]

The point is not that Davis was exceptional. Twenty-three other lawyers also refused to provide Gus Hall with counsel—but no one ever lauded their devotion to professional ideals. Indeed, Davis-Polk partners probably did more than most attorneys to furnish counsel to politically unpopular defendants who could pay their fees.[87] But the dismal record of the professional elite and the organized bar is beyond dispute. Grenville Clark, senior partner in the Root, Clark firm, with a long record of concern for the Bill of Rights, attributed the erosion of civil liberties to "the supineness of the organized Bar," which "appalled" him.[88] The historian of the Association of the Bar of the City of New York, which is generally acknowledged to have been among the most responsible bar associations during this period, concluded that those lawyers "who had the courage [to defend freedom] had not the stature; and of the few who had the stature, apparently not one had the courage."[89] A prominent Washington lawyer not only refused to take loyalty cases but declined to recommend lawyers who might do so. "I wouldn't be caught dead sending them on to another lawyer—for fear he would think I think he's a Communist, or something. I know that's bad, but most lawyers feel the same way."[90] By the end of the Cold War decade elite irresponsibility had sifted down to law schools: among one group of law students who conceded overwhelmingly that a probable Communist was entitled to defense, only half said that they would take the case themselves. Eighty percent of those who said they would decline attributed their delinquency to the fear of unfavorable public opinion.[91] The students had learned what their professional leaders taught.

Thus lawyers who provided counsel to unpopular defendants did not honor any professional tradition of service; they violated a tradition of indifference. Cold War politics and law-firm economics kept their numbers small. By the early fifties few lawyers could have been unaware that the price of guilt by association might be a prison term, suspension, disbarment, or a shattered practice. Those lawyers who could most easily afford long, unpopular cases were the ones least likely to be interested in civil liberties violations. Lawyers who sought uneasily to balance their professional responsibility against their personal fear occasionally accompanied their appearance for an unpopular defendant with a statement of disaffiliation from their client's beliefs and associations in an attempt to shield themselves from the taint of their client's politics. A double standard of professional service plagued defense lawyers: they received unfavorable notoriety for efforts on behalf of political dissenters which no lawyer questioned when such efforts were made in defense of property and privilege. Elite lawyers, who split politics from professionalism whenever corporate practice was the issue, asserted the principle of guilt by association when political radicals were on trial.[92]

Evidence bearing on the motivation of lawyers who did defy professional complicity in Cold War repression is elusive and fragmentary, but it does suggest that, courage aside, professional and political marginality, reinforced by personal associations, were crucial determinants. The law firm most prominent for its defense work in loyalty cases was Arnold, Fortas & Porter, the new firm of New Deal lawyers. Its involvement may have been inadvertent; the victims of McCarthyism it represented were often Depression liberals who were referred to the firm by New Deal acquaintances of the partners. Yet there was empathy, generated by "the background and ideological identification of the members of the firm." As Fortas said: "We were 'liberals.' We were 'New Dealers.' "[93] Renunciation of the New Deal was a primary component of Cold War politics; predictably, the most concerned law firm also was the most conspicuously New Deal firm.

But the most vigilant lawyers seldom came from prominent Washington or New York firms. They came disproportionately from among the children of politically radical parents and entered the legal profession during the Depression. They were drawn to labor law, practicing either for a CIO union or for the National Labor Relations Board. Their network of personal relationships provided contacts with labor radicals who, as targets of McCarthyism, turned to lawyer-friends for assistance. Responsive lawyers were quickly inundated; as one of them said, "if you take one or two of these cases you find you're not handling any others." These lawyers, receptive to the politics of the harassed radicals, were an alienated and disaffected professional group whose members shared similar career patterns, friendship bonds, political values, and Lawyers Guild membership. "They perceived their function in the [loyalty-security] litigation as defenders of the political left against government harassment."[94] For some, the fusion of political commitment and professional activity offered personal fulfillment which they had never before experienced. As

George Crockett told Judge Medina, in reference to his defense efforts, after his citation for contempt at the conclusion of the Smith Act trial: "For the first time in the 15 years that I have been practicing law I have had an opportunity to practice law as an American lawyer and not as a Negro lawyer. I have enjoyed that brief trip into the realm of freedom."[95]

Occasionally a lawyer who did not fit into any of these categories emerged from obscurity to offer his services. Royal W. France had left a successful New York corporate practice in 1929 to teach economics in a Florida college. "I had become a servant, even if a highly paid one, of big business and I did not feel at home with myself." He taught for twenty years and became active in Socialist Party politics. In 1951, nearing the age of seventy, he relinquished his second career to resume his first. Disturbed by the political climate, by the vulnerability of unpopular lawyers and defendants, and by the timidity of the bar, he returned to practice. He met with Harry Sacher and was drawn into the defense of second-string Communist Party leaders. He joined the Lawyers Guild and served on its executive board. A self-described "old-fashioned type of liberal who believes that the First Amendment means what it says," he explained: "I could not be at peace with myself until I had genuinely and without reserve offered myself, at a crucial moment in history, to defend the principles which lay at the basis of my philosophy of life. To do so required defending Communists."[96]

Measured by need there were too few New Deal firms, radical lawyers, mavericks like France, or vigilant defense organizations. The professional tone was set not by the few who responded but by the many who were silent. The bar was not distinctively supine during the Cold War; other institutions also capitulated. But the bar had special responsibilities—to promote the administration of justice and to provide equal justice under law. Once professional leaders demonstrated that they would not only tolerate repression, but abet it, the legal and judicial processes tilted toward "political justice": the use of legal procedures for political ends.[97] In its eagerness to demonstrate loyalty to prevailing political values the professional elite sacrificed important standards of professional conduct and basic constitutional rights. The disciplinary measures of the bar, which the Supreme Court tolerated, created a "dual status" theory of constitutional rights for prospective lawyers. Entitled to certain rights of expression and association as citizens, they were excluded from the practice of law if they asserted them. One legal scholar properly wondered how fundamental rights were protected "by making their exercise ground for disbarment."[98] Not only the First and Fifth Amendments but the Sixth—the right to the assistance of counsel—was jeopardized by anti-Communist zeal within the profession. The Constitution aside (which was where lawyers often placed it), the Canons of Ethics offered contradictory admonitions which the Cold War exposed. One canon urged lawyers to defend the accused regardless of their personal opinion as to guilt; but another, permitting lawyers to reject distasteful clients, diluted that responsibility and may have contributed to the eagerness of lawyers to publicly disavow their clients' beliefs. The obligation to provide vigorous advocacy must have sounded hollow to lawyers for unpopular defendants after the Smith Act trial in 1949.[99]

Even as the Cold War thawed, the political values of the professional elite remained frozen. Indeed, just as the Supreme Court showed signs of rousing itself from acquiescence in Cold War politics, influential voices on the bench and in the leading law schools articulated an ideology of professional craft and reason which added intellectual gloss to Cold War values. Their outcry was triggered by the hostile reaction to a series of decisions that reached a climax on "Red Monday" (June 17, 1957), when the Supreme Court began to reject claims that national security justified any incursion on individual rights. The first *Konigsberg* decision, in May, had marked the prelude; then in *Yates, Service, Watkins,* and *Sweezy* the Court toppled important props supporting loyalty-security proceedings, sweeping forays by the House Un-American Activities Committee, and the prosecutorial zeal of the Department of Justice under the Smith Act. These decisions elicited stinging criticism from Congressional conservatives and some ABA leaders and vigorous efforts to restrict the appellate jurisdiction of the Supreme Court.[100] As the Warren Court began assertively to fashion jurisprudence as an instrument for the protection of individual rights (and racial equality), voices of professional foreboding grew strident.

A group of influential scholars, galvanized by the critics, attempted to deflect the Supreme Court from its activist libertarian course. These spokesmen for judicial restraint, neutral principles, craft, and reasoned jurisprudence were in a quandary. They remembered the attacks against the pre–New Deal Court for its conservative activism in defense of property and privilege; now they found it difficult to accept the proposition that the Court should defer to legislatures in matters of economic regulation, but aggressively protect fundamental constitutional freedoms from legislative abridgement. Their advocacy of undiscriminating judicial passivity wrenched an argument against judicial activism from its historical context and elevated it to the status of an eternal verity. But behind the mask of their commitment to transcendent values of process and reason was capitulation to the politics of the status quo.

Learned Hand, delivering the prestigious Holmes lectures at Harvard Law School in 1958, raised the first flag of surrender. Asserting canons of judicial self-restraint which Thayer and Holmes had articulated at a much earlier time when the Court had persistently nullified social legislation, Hand was critical of broad judicial review and apprehensive about judicial choice. He advocated definitions of First Amendment freedoms and due process which did nothing more than reflect "an honest effort to embody that compromise or adjustment that will secure the widest acceptance and most avoid resentment. . . ." A judge should be "the mouthpiece of a public will, conceived as the resultant of many conflicting strains that have come, at least provisionally, to a consensus."[101] If the public will wished to dismantle the Bill of Rights, presumably the highest obligation of the judge was to express that preference and thereby legitimize it. Consensus replaced Constitution as the arbiter of judicial decisions.

Professor Herbert Wechsler of Columbia, who delivered the Holmes lecture the following year, shared Hand's yearning for "neutral principles" of adjudication. Wechsler wanted criteria "that can be framed and tested as an exercise of reason and not merely as an act of willfulness or will." Rejecting *"ad hoc* evaluation," which measured decisions against "the interests or the values" of personal choice, he insisted that courts must make principled decisions which rested upon "reasons that in their generality and their neutrality transcend any immediate result that is involved."[102] Professor Henry M. Hart, Jr., of Harvard added his authoritative voice. Calling for judicial opinions "which are grounded in reason and not on mere fiat or precedent," for opinions which articulated "impersonal and durable principles," Hart, co-author of an influential text on the legal process, pleaded with the Supreme Court to be "a voice of reason." Reason, he insisted, was "the life of the law."[103]

A school of jurisprudence had emerged that was perfectly attuned to the end-of-ideology politics of the Cold War. Its passion for judicial passivity represented, at the very least, "simply another road to judicial surrender."[104] At a deeper level it expressed what Thurman Arnold, in a rebuttal to Hart, described as the new professional "theology." Hart had complained that a sharply divided Supreme Court, insufficiently committed to reasoned decisions, was "threatening to undermine the professional respect of first-rate lawyers."[105] To Arnold this merely meant that corporate lawyers in the American Bar Association were critical of the Court's decisions protecting individual rights—as indeed they were—and that it was Hart's desire that the Court regain the admiration of the corporate bar. At this level, therefore, "Professor Hart's theology" not only represented an attack upon the competence of libertarian judges but an appeal to the politics of the professional elite as the valid standard of judicial decision-making.[106]

The political implications of the Hand-Wechsler-Hart doctrines, which Arnold and others spotted, were less significant than the camouflage of neutrality that disguised political preferences as reason and craft. Political disagreements were debatable. But when a debate over political values was transformed into a choice between politics and craft, when that very transformation was concealed by pleas for "neutrality" and "reason," when indeed "neutrality" and "reason" were labels assigned to certain political choices, then thought and meaning were subverted. When a standard of professionalism emerged that incorporated political judgments yet denied the incorporation, when that standard became the sole acceptable criterion of professionalism, and when it was promulgated as an unquestioned truth to lawyers, law students, and the public, so that people actually *believed* that craft and process were value-free, then the consequences for critical thought were devastating.

The advocates of judicial self-restraint, neutral principles, and reasoned elaboration either did not comprehend, or could not acknowledge, that their jurisprudential preferences incorporated deeply conservative, even anti-libertarian, political values. This was not the first time in modern professional history that lawyers offered neutral principles to achieve particular results—and to preserve the powerful, magical image of their own professional skill as a mysterious science incomprehensible to unreasoning laymen.[107] The defense of legal functionalism—that law was best when it expressed dominant community values—was deceptively alluring. It capitalized upon hostile memories from an earlier historical era when judicial activism had been asserted in defense of corporate privilege and in opposition to legislative regulation in the public interest. But the analogy between two quite different historical issues, and the corresponding judicial responses, was deeply flawed; for in Bill of Rights cases, if courts did not reject the public will then civil liberties would wither and die at the mercy of hostile majorities. If judges did not make law but only reasoned their way to it according to canons of modesty that invariably assured deference to the legislative will, then the sociological sensitivities and libertarian values of justices like Earl Warren and William O. Douglas could be exposed and denigrated as examples of sloppy craftsmanship. The new jurisprudence of process, reason, and craft concealed policy preferences for the status quo and for a deferential Supreme Court, even at the sacrifice of First Amendment freedoms.[108] The craftsmen would not concede that a legitimate argument against pre-1937 activist jurisprudence did not, either logically or historically, justify judicial passivity twenty years later when the issues had changed. The movement "toward legality and away from purpose," a dominant theme in legal thought at mid-century, expressed the final capitulation of the professional elite to the passions of the Cold War.[109]

The Cold War experience represented a variation on themes that resonated through the modern history of the professional culture, themes which became audible whenever the values and status of elite lawyers were jeopardized by social change. After World War I, against the backdrop of the Bolshevik revolution, professional patriotism and boosterism expressed hostility toward the changing demographic base of the profession. After World War II, in the chill of the Cold War, the quest for professional loyalty and conformity expressed the desire to repudiate the reform politics of the New Deal, legal Realism, liberal professionalism, and the activist jurisprudence of the Warren Court. The professional elite tried to disguise its acceptance of Cold War values with a formal theory which, like a magic wand, would transform the pumpkin of politics into the golden coach of craft. Yet just as the theory began to harden into dogma its internal contradictions were wrenched to the surface by a decade of turmoil. Not only the theory, but the legal and social culture that nourished it, even —and most ominously—the very rule of law that sustained it, began to crumble.

Nine

The Disintegration
of Legal Authority

Nothing in the modern history of the bar presaged the sustained crisis of professionalism that began during the 1960's and culminated in Watergate. The legal profession, as a social institution, had always absorbed cultural values while asserting its own autonomy. As long as the public retained faith in the integrity of the legal process, the bar preserved its precarious compromise between the politics of professionalism and the rule of law. But amid the most severe national crisis since the Civil War, faith in legal authority disintegrated; and once it did, the revealed role of lawyers in preserving a discriminatory rule of law, followed by evidence of the complicity of lawyers in lawlessness, demolished claims of professional neutrality. The elaborate structure of ethics and values which had defined professional responsibility for nearly a century collapsed in a shambles.

Ever since the beginning of the urban industrial era the bar had adopted a series of compromises to reconcile public responsibility with professional self-interest—and to conceal the distance between them. These compromises unraveled once the Cold War thawed and public attention shifted to neglected domestic concerns. The civil rights struggle, followed by the brief war against poverty, exposed standards of professional behavior which preserved the glaring inadequacy of legal services for citizens who were black, poor, or both. For a brief moment, until the war in

Vietnam corroded faith in institutional legitimacy, democratic currents swirled through the profession. Then, as confidence waned in the fairness of legal and judicial processes, resulting in the eruption of protest and violence, demands for law and order pervaded professional and political life. Watergate marked the final demolition of credence in legal authority. It revealed that law and order was a mask for illicit repression; that those sworn to uphold the law had conspired to subvert it; that lawyers, including the chief law enforcement officers of the nation, were deeply implicated in lawlessness; that double standards of professional conduct protected the wealthy and powerful while destroying the promise and possibility of equal justice under law.

Perhaps disintegration began as early as 1955 when Mrs. Rosa Parks, a black seamstress, refused to move to the back of a bus in Montgomery, Alabama. That moment, which marked the resumption by black Americans of their struggle for civil rights and racial equality, made explicit a conflict between the obligations of law and of conscience—between the commands of law and the claims of justice. As the civil rights crusade spread through the South, the relentless prejudice of a Jim Crow bar surfaced. The complicity of the legal profession in preserving unequal justice under law, for whites only, raised questions about professionalism and its social responsibilities. For at least a decade after the *Brown* desegregation decision, Southern lawyers persisted in their defense of "Nordic, White Protestant, Anglo-Saxon Christian values." Few dared to defend advocates of racial equality. Daring was costly: it prompted harassment by courts, legislatures, vigilantes, and fellow professionals (while the American Bar Association shrugged aside the problem as a "political" issue beyond its purview). In Mississippi, the president of the ABA (along with forty other lawyers) refused to represent a civil rights advocate. One white lawyer in the state, one of a mere handful who demonstrated the courage of his professional convictions, was disbarred. A black lawyer engaged in desegregation litigation was harassed by a federal judge whose behavior, according to an appellate court, contributed to the "humiliation, anxiety, and possible intimidation of . . . a reputable member of the bar." The result, not only in Mississippi but throughout the South, was "timid lawyers and neglected clients."[1]

If local white lawyers were unwilling to defend those who challenged the racial status quo, few black lawyers were available to do so. Racism suffused professional life. In 1944 Gunnar Myrdal had speculated ironically that great opportunities should exist for black lawyers, so considerable were the wrongs inflicted upon their people. Two decades later, however, professional opportunities remained virtually nonexistent; the brief promise of the interwar decades was unfulfilled. Although the litigation strategy of the NAACP had succeeded brilliantly by 1954, any hope that the demonstrated excellence of a black professional elite would encourage the acceptance of black attorneys on their merits remained chimerical. The overwhelming majority remained second-class professionals who were confined to solo practice, an impoverished black clientele, and a narrow range of legal business. Constricted by their inability as blacks to secure justice (or even respect) in most courts, they languished in a "starvation profession." Integrated firms were rare; a Department of Labor study in 1963 revealed that there were only thirty-five in the entire country, a fact that prompted Secretary of Labor Willard Wirtz to declare that "the legal profession has got a lot to answer for on this particular score." Black Americans, "sharply aware of . . . the general ineffectiveness of the Negro lawyer to handle 'the white man's law,'" were deprived of a legal system that could command their loyalty and respect.[2]

The refusal of white Southern lawyers to defend black clients, combined with the unavailability of black lawyers, prompted another invasion of carpetbaggers across the Mason-Dixon line: Northern lawyers committed to the objectives of the civil rights movement. Once it became evident that the defense of black clients would fall either to outsiders or by the wayside, hundreds of lawyers and law students went south to offer free legal services to those whose color or politics prevented them from obtaining local counsel.[3] Northern attorneys quickly discovered the hazards of desegregation litigation. Although the Supreme Court, in a 1963 decision, had overturned efforts by the state of Virginia to apply professional canons of ethics to impede litigation by the NAACP, visiting attorneys encountered constant harassment for unauthorized practice and solicitation. In Louisiana, for example, it required Justice Department intervention, and a favorable federal court ruling, merely to enable a young volunteer to defend a black client. (A lengthy brief documented the interlocking racism of state law schools and the bar: few aspiring black students were admitted to law school; between 1927 and 1947 every black applicant to the Louisiana bar was rejected; by the 1960's only 1 percent of the bar, compared to 32 percent of the population, was black.)[4]

The civil rights movement jarred the bar into some concern with the consequences of professional prejudice, but even during the most euphoric days of the struggle, when white hostility wavered before the nonviolent disobedience of black resistance, the ideological constraints of professionalism retained their force. Everything in a lawyer's training argued for the separation of client from cause (although the corporate lawyer as assuredly represented the cause of private property as a civil rights lawyer represented the cause of equality), and for the halting process of case-by-case litigation. Professional inhibitions were most evident in the Department of Justice which was, paradoxically, the federal agency most strongly committed to racial equality during the Kennedy presidency. But whether at the policy level, dominated by the Burke Marshalls, Byron Whites, and Nicholas Katzenbachs, or down the halls where droves of recent law school graduates clustered, the legal mind was hemmed in by the ingrained professional preference for process and order, moderation and compromise. It was uncomfortable with the choice posed so starkly by the civil rights movement: between the law of segregation and the justice of equality.

A professional "code of the Ivy League Gentlemen" constricted Justice Department lawyers. It dictated their preference for negotiation and settlement; their belief that reasonable men (the law's everyman always was reasonable) could, and should, achieve a satisfactory compromise; their endless patience with endless litigation. They thought like the corporate lawyers they had all been trained to be, demonstrating "caution, moderation, respect for precedent, patience in the face of protracted litigation, commitment to the system as is. . . ." Yet their "passion for and expertise at confrontation-avoidance," in encounters with Southern governors like George Wallace and Ross Barnett, inevitably meant "peaceful co-existence with present injustice." It was the President's assumption, and their conceit, that "the paradigmatic Wall Street lawyer . . . is equipped to solve any problem."[5] But federalism, individualism, legalism, and professionalism—their primary loyalties—were insufficient to resolve the civil rights problem, as ensuing racial chaos tragically demonstrated. A regional problem spread into a national catastrophe: from Little Rock, Birmingham, and Selma to Watts, Detroit, and New York. One consequence of permitting government lawyers trained for Wall Street to impose their solutions as the only reasonable ones was the perpetuation of racial injustice long beyond the moment when a golden opportunity existed to apply federal power to ameliorate it.

The racist tilt of legal institutions, and the repressive violence that sustained it, so shocked some white attorneys that their conventional definitions of professionalism were shattered. An unknown New York lawyer named William Kunstler, who observed his first lunch-counter sit-in in 1961, discovered from his defense work "that there is considerably more to the practice of law than learning precedents or interpreting statutes." The Freedom Riders taught him that he was "a human being first and a lawyer second"; that professional austerity must yield to personal engagement. Charles Morgan, Jr., a Birmingham attorney who was bored with "the antiseptic legal world of administrative proceedings, municipal bonds, proxy contests and corporate problems," learned an identical lesson. He opened his own office, took civil rights cases, and discovered that he had "fundamentally changed" his perspective: "I had given up the luxury of detachment from the causes I represented." Civil rights lawyers began to wrestle self-consciously with the question of whether their client was, as traditional professional precepts dictated, a solitary party to a discrete case—or, as some lawyers were coming to believe, a cause larger than any single client. Perhaps, argued Harvard law professor Mark Howe, the "movement" lawyer "must see his responsibility as different . . . from what it has been in other times and other settings. If he is to be true to his profession he must, perhaps, see himself as engaged not by a client but by a cause."[6] For the first time in a troubled decade of domestic turmoil, but not for the last, lawyers contemplated the contributions of professionalism to social injustice.

Once the civil rights movement gathered momentum it overflowed both its regional and racial channels. As it moved north, attention shifted beyond racial injustice to economic inequality and to the link between race and poverty in American society. More self-consciously than ever before, the legal profession had to confront the implications of wealth and poverty for the distribution of legal services. The federal government, declaring war on poverty, committed unprecedented energy to assure legal services to those who needed them as a matter of right, not privilege. Its program precipitated within the bar a divisive struggle which exposed the tenacious grasp, and debilitating consequences, of the traditional free-enterprise professional ideology.

Nearly half a century earlier Reginald Heber Smith had documented the existence of a dual system of justice in the United States: one for the affluent, another for the impoverished. But the flurry of concern with legal aid that his study inspired did not survive the postwar emergency that produced it. After World War II new proposals for some assumption of responsibility by the government were aborted by the Cold War. The American Bar Association, convinced that "the lines of class and wealth have largely disappeared" from American society, applauded the adequacy of legal aid and turned its energies to referral programs for the middle class. Even this interest was rooted in Cold War politics: "in addition to combating the threat of socialism," an ABA committee reported, "with one stroke . . . we dispose forever of the communist theme that in capitalist countries the lawyer is the slave of the wealthy class."[7] But the Cold War anti-Communism of the bar was not a solution. Legal aid remained a charity, not a right.

Once poverty was rediscovered early in the 1960's, comfortable assurances about the adequacy of legal aid dissolved amid abundant evidence of the maldistribution of legal services along economic lines. The civil rights movement contributed to this reawakening. So did the Supreme Court's *Gideon* decision, which, in assuring counsel to indigent defendants in felony proceedings for the first time as a matter of constitutional right, testified to the privileges that money had traditionally purchased within the criminal process. In 1964 there was a federal conference on the extension of legal services to the poor; Congress passed the Economic Opportunity Act; Edgar and Jean Cahn, veterans of the civil rights movement, published their seminal proposal for neighborhood law firms; and Attorney General Robert Kennedy delivered a biting Law Day address in which he declared that "lawyers must bear the responsibility for permitting the growth and continuance of two systems of law—one for the rich, one for the poor."[8]

While poverty legislation was still being drafted, a legal services proposal was considered but dismissed. Then, prodded by the Cahns, the new Office of Economic Opportunity, directed by Sargent Shriver, belatedly grasped the potential of legal services in the war against poverty. The Cahns (Edgar was Shriver's administrative assistant) proposed "a university affiliated, neighborhood law firm" to serve as an advocate for the poor by making "public officials, private service agencies, and local business interests more responsive to the needs and grievances of the neighborhood." Shriver approved the plan and appointed Jean Cahn to administer it. Thus, largely as an afterthought, and without consultation with bar leaders, the first federal legal services program was launched.[9]

The plan was an amalgam of rather shopworn ideas—suggestions

for federally subsidized legal services and neighborhood law offices were at least twenty-five years old; its novelty lay in their
fusion at a propitious moment. But successful implementation required more expansive conceptions of the lawyer's role and responsibility than the bar had ever tolerated. Austere detachment,
accompanied by protestations of political neutrality, might suit
Wall Street practice but could hardly inspire confidence in those
for whom the new program was designed. The Cahns understood
this: a lawyer, they asserted, "need not be apologetic for being
partisan, for identifying. That is his function." As warriors against
poverty, lawyers were expected to be "activists for the poor" just
as civil rights lawyers had been activists for a submerged black
constituency (and corporate lawyers were activists for the affluent). Attorney General Katzenbach, perhaps galvanized by his
civil rights experiences, urged lawyers "to go out to the poor
rather than to wait. . . . To be reduced to inaction by ethical
prohibitions . . . is to let the canons of lawyers serve the cause
of injustice." OEO guidelines for lawyers went beyond representation in court to encourage advocacy of reform; beyond legal aid to what one of its attorneys described as "legal action designed to change the structure of the world in which poor people
live."[10]

The legal services program, launched independently of the bar,
elicited an uneasy response from the ABA hierarchy, caught between fear of government intervention, concern with its obstruc-
tionist reputation, and desire for federal subsidies for its own projects. President Lewis Powell, Jr., authorized cooperation with
OEO contingent upon its adherence to prevailing ethical norms.
Early in 1965 the Cahns met with ABA leaders to arrange a *quid
pro quo:* bar support for the federal program; federal funding for
bar programs. Soon thereafter the ABA House of Delegates
pledged cooperation on Powell's terms: if OEO utilized the "expertise and facilities of the organized bar" to the maximum extent
feasible and if it relied upon lawyers to provide service "in ac-
cordance with ethical standards of the legal profession."[11]

Powell was concerned lest OEO depart from prevailing ethical norms or establish autonomy from professional organizations. His insistence upon fidelity to traditional ethics threatened to undercut the basic purposes of the program; for strict adherence to the inhibitions incorporated in the Canons would contribute to the very problems that OEO was created to resolve. The choice was clear: either OEO pursued a course of innovative activism and reform, to the peril of professional mores, or it yielded to the dictates of the Canons, thereby sacrificing the commitments that justified its existence.[12]

Early in 1965, just months after the birth of the program, these unresolved conflicts precipitated a political struggle for control over agency policy. Shriver, under pressure from Powell and other bar leaders, removed the Cahns and appointed E. Clinton Bamberger, Jr., who enjoyed ABA confidence, as director. Not until its own influence was exerted within OEO did the American Bar Association defend it against even more hostile professional critics.[13] Despite the constraint of professional suspicion, OEO was able to launch bold initiatives which threatened basic norms of professionalism. At its furthest reach it contemplated the use of legal means to reallocate political and economic power in American society. If the legal system could abet the maldistribution of power, OEO attorneys argued, then it might also hasten redistribution. OEO substituted purposeful litigation for random case-and-client individualism. It rejected middle-class premises of professionalism which confined an attorney to his office until a knowledgeable and affluent client appeared. It repudiated the bureaucratic paternalism of legal aid, which dispensed charity to clients and displayed deference to the bar. It not only provided some poor people with effective representation; it encouraged client participation in decision-making. And it pursued a new style of advocacy: instigatory, not passive; experimental, not traditional. Its commitment went beyond single clients to a class of clients—and to their cause.[14]

Sargent Shriver, announcing the first legal services grants, tried to provide reassurance that OEO would not attempt "to subvert or oppose the organized Bar." Instead, OEO would demonstrate to ghetto residents that the legal system could redress their grievances, thereby making them "less likely to turn to violence and rioting."[15] Yet part of its commitment was to "change the basic structure of property rights and the labyrinthine governmental procedures that have traditionally supported business and middle-class interests in our society."[16] OEO attorneys in urban ghettos challenged landlords who refused to comply with housing codes, state officials who imposed residency rules to disqualify welfare recipients, and lending agencies that exacted usurious loans. An especially effective Rural Legal Assistance Office in California sued to overturn the bracero policy of the Department of Labor and to secure economic justice for migrant workers. Class action suits were brought on behalf of entire groups of victims of private or public exploitation. Hundreds of thousands of Americans were assured legal remedies, for the first time as a matter of right, not wealth.[17] The ripples of change spread quickly: to courtrooms, where OEO attorneys engaged in zealous advocacy on behalf of neglected clients; to legislative halls, where they lobbied; to law schools, where new courses were introduced on a range of poverty law subjects and students swarmed into clinical work. The effort to provide legal services to the poor, declared Justice Abe Fortas, had become "the legal frontier of our time."[18]

But the pioneers on that frontier encountered relentless professional and political opposition. Although Shriver pledged that OEO would not subvert the bar, he failed to secure a reciprocal pledge from the bar toward OEO. Lawyers began to issue warnings about the dangers of "socialized law," the dire consequences of lay interference with professional prerogatives, and the subversive infiltration of neighborhood law firms by political radicals. The president of the National Legal Aid and Defender Association, the organization that monopolized (and restricted) professional legal services until OEO appeared, reminded Lewis Powell about the dangers of community control. Powell, despite his public posture as a professional friend of OEO, worked behind the scenes to keep the program tightly reined in by the organized bar.[19]

The most intense professional opposition to OEO erupted at the local level, where OEO community efforts triggered the animosity of solo and small-firm lawyers who feared intrusion upon their business. The prospect of losing clients who might qualify for free services, and of confronting skilled salaried attorneys in litigation, filled them with foreboding. Ironically these plaintiff's lawyers, usually the professional liberals, tenaciously defended the status quo, while elite lawyers, traditionally the professional stalwarts, were most tolerant of federal legal services—perhaps because they were least likely to lose a case or a client to an OEO attorney. Some local bar associations merely refused to support OEO programs; others (in Texas, Pennsylvania, Florida, and California) brought suits to enjoin OEO from violating ethical canons. In Tennessee the president of the state bar association issued the dire warning that lawyers confronted "the most serious threat ever posed to the independency and individuality and strength of the profession."[20]

Although hostility was most intense at the local level among economically marginal lawyers, it cut through the various professional strata. Solo lawyers voiced economic grievances, although few of them had a substantial clientele whose low income might entitle them to OEO benefits. (OEO clients always were more likely to be rescued from professional neglect than spirited away from private attorneys.) At loftier professional levels the grievances were political (expressed in ethical terms), not economic. Although elite lawyers, with least to lose economically from OEO, could afford to vote their support of federal legal services at ABA conventions, the association never was comfortable with OEO reform objectives or with its indifference to traditional ethical admonitions. Allegations of improper advertising, soliciting, stirring of litigation, and group practice were designed to blunt the reform function of OEO. As laudable as the widespread provision of legal services might be in principle, praise halted where the involvement of OEO attorneys in rent strikes or class-action suits began.[21]

The intensity of professional opposition reflected the reluctance of the bar to modify those traditional precepts which sustained inequity. Given the presumption that the only proper role for an attorney was to represent the immediate interests of a client, it followed that lawyers overstepped their function when they used the legal process instrumentally to achieve social change. Lawyers were trained to perceive random individuals who would enjoy legal equality once they secured professional assistance, not groups or classes which required special initiatives to rectify the common plight of their members. Nothing in their conception of a legal system had prepared them to tolerate aggressive law reform as a proper component of the adversary process.[22] OEO could not escape the tentacles of professional constraint. Lawyers dominated OEO boards of directors and exerted additional influence in bar associations, which interpreted ethical norms narrowly and litigated to defend their definitions. Courts usually sustained these restrictive definitions of the permissible bounds of professional activity. In New York, for example, the Appellate Division rejected a petition for approval from poverty law firms whose solicitation, group representation, and lobbying, in the court's judgment, elevated law reform above client service.[23]

Early in the life of OEO political hostility converged with professional mistrust. Conflict erupted first in California, where the Reagan administration attacked the California Rural Legal Assistance program, a vigorous advocate for Chicanos, migrants, and other exploited victims of the state's sprawling farm factories. Powerful farm interests in the state were outraged that a federal agency should sue a state government to secure equal rights and fair treatment for field workers. (As the lawyer who presided over OEO programs in California declared: "What we've created in CRLA is an economic leverage equal to that of large corporations. Clearly that should not be.") In 1967 Senator George Murphy introduced legislation, defeated by the Senate, providing that no legal action could be brought against any state or federal agency—a potentially lethal blow to OEO litigation. One year later Congress passed an amendment proposed by Senator Edward Gurney which prohibited OEO attorneys from engaging in criminal representation—ostensibly to protect the preserve of public defenders but more probably because rioters and other "inappropriate" people were being defended by OEO attorneys in the wake of ghetto eruptions. In 1969 Murphy returned with a proposal to empower governors with a veto over OEO projects; it passed the Senate but met defeat in the House.[24] By the beginning of the Nixon administration OEO confronted not only the suspicion of professional associations and difficulties in court, but Congressional sniping and the hostility of a President who was determined to emasculate the program.

The surrender in the war against the legal consequences of poverty offered a corollary to the demise of the civil rights movement: if it was unlikely that blacks and whites would receive equal treatment, so it was improbable that rich and poor could enjoy equal justice. But civil rights and poverty lawyers had exposed the racial and economic bias of the legal system. Their work generated expanded definitions of professional responsibility which incorporated personal commitment and explicitly political values. The consequences were momentous for a generation of law students who were simultaneously attracted to the promises of the legal profession and repelled by the performance of legal institutions.

Students from the silent generation of the 1950's had spurned career choices which, in the expectant words of *Harvard Law Review* editors, did not offer "the faint glimmer of those enormous earnings which top the pyramid of legal remuneration in the world of the 'corporation' lawyer."[25] Within a decade, however, as a new generation was ripped away from the conventions of its predecessors by its encounter with racism, poverty, and the Vietnam war, law schools reeled under challenges to their educational purpose and to the authority of the legal order they defended.[26] The schools, staid and secure, were unquestioningly committed to the values of craft and process, to the virtues of the case method, and to a curriculum weighted toward protection of the interests of the wealthy and powerful. Except for a flurry of criticism during the Depression, when clinical training won a sympathetic hearing as a supplement to Socratic dialogues, legal education remained frozen in its Langdellian mold until the 1960's. Then, with startling suddenness, its fundamental precepts were subjected to withering scrutiny and disparagement. Young lawyers, who discovered hidden political preferences lurking beneath the claim of value-free craftsmanship, rebelled against the demand that they sacrifice their personal autonomy to professional austerity, which seemed to elevate legal order above social justice.

It was the proudest boast of law schools to train students to think like lawyers. This process entailed a highly stylized mode of intellectual activity that rewarded inductive reasoning, analytical precision, and verbal felicity. But relentless doctrinal analysis of appellate opinions severely restricted the range and depth of inquiry. Craft was rewarded over choice and process over purpose. Thinking like a lawyer required the application of technical skills to the task of problem-resolution within the confines of the adversary system. Justice, it was assumed, emerged inevitably from the confrontation of skilled advocates in an adversary setting. Legalism instilled "laissez-faire morality" in its converts: the presumption that proper decisions were the decisions that the legal system produced.[27]

Critics increasingly disputed the claim that legal education and professional craft were value-free. Ralph Nader charged law schools with the failure "to articulate a theory and practice of just deployment of legal manpower"; he described legal education as "a highly sophisticated form of mind control that trades breadth of vision . . . for freedom to roam in an intellectual cage." Edgar and Jean Cahn complained that it "operates to enervate moral indignation and to inculcate intellectual and moral timidity." Law schools, according to the Cahns, "reward removal from the chaotic world of emotion, events, passions and people to the rarified stratosphere of metaphysical debate." The curriculum, with its emphasis on the problems of the wealthy, was an obvious target: "a course on bankruptcy would focus on business failures and railroad reorganization, rather than on wage-earner bankruptcies. . . . In real property little time would be spent on the most common and vital real estate transaction for many members of our society—the renting of a residential apartment. And needless to say, the course in corporations would not consider the use of the corporate structure for non-profit goals." As one law review editor complained, law school "taught us . . . to think less of social horizons than of bread and butter."[28]

As the legal system began to lose its aura of neutrality, the very process of thinking like a lawyer became suspect. Students objected to professional training which seemed designed to preserve wealth at the expense of competing social purposes. Inspired by the civil rights struggle, appalled by the rampant inequities of racism and poverty in an egalitarian and affluent society, they found it difficult to believe that justice emerged inevitably from the normal functioning of the legal system. They began to insist upon professional training that would at least acknowledge, if not cultivate, the exercise of conscience and commitment by individual attorneys; that would embrace goals other than client-caretaking—especially when affluent whites were the predominant clientele. As one graduate complained: "law school teaches students to deal with every conceivable loss, that of an arm, a leg, five dollars, a wife—every one, that is, but the most important, the loss of one's self."[29]

Technical skills were necessary, but to a new generation of students they were insufficient. Institutional neutrality and professional detachment ceased to command allegiance as believable precepts. Indeed, the law school was criticized for being "too much a part of the society around it." Students spurned craft at the expense of social concern. Craft, they knew, was "a tool that can serve any master"—and they were reluctant to enslave themselves to its commands.[30] The lawyer as "hired gun," who was (in William Sloane Coffin's deprecatory description of James D. St. Clair at the time of the Benjamin Spock trial) "all case and no cause," commanded diminished respect. A young Harvard graduate who left private practice for public interest law declared his impatience with "the conventional lawyer's willingness to take either side."[31]

Every generation responds with special intensity to the tension between its inherited cultural values and the formative events of its time. During the 1960's observers detected "a new generation of law students and recent graduates more conscious of the urgency of social reform than any past generation of lawyers." There was, for example, a sudden, sharp turn away from private practice as an initial career choice. Young lawyers in significant numbers repudiated private gain for public service. For some, early in the decade, it was civil rights litigation. Then, as professional alternatives multiplied, neighborhood law offices, social action agencies, public interest firms, and even law communes attracted young lawyers who searched for opportunities to combine professionalism with social activism in the public interest. Even those who flirted with Wall Street firms exacted their own terms, often demanding some measure of firm accountability for the social consequences of their clients' activities and released time for their own *pro bono* work.[32]

Private firms saw their buyers' market of the fifties tumble. Led by the Cravath firm, Wall Street offices raised the salaries of first-year associates from $10,500 to $15,000 in an effort to retain their flow of graduates. But it was evident that for many students the type of practice, not the salary scale, had become the paramount consideration. Simon Rifkind of the Paul Weiss firm complained that "an extraordinary amount of bright young men no longer seem interested in practicing law," adding somewhat plaintively that "the old-fashioned practice of law ought not to be considered beneath anyone." Law firms expended considerable energy to recruit, only to discover that their representatives might be badgered, picketed, held accountable for their clients—or ignored. "Lawyers should be responsible for their actions," a student asserted in a representative comment of his generation, "not merely 'hired guns.' "[33]

The erosion of interest in private practice considerably exceeded the low-water mark of the Depression years. The percentage of graduates entering private firms declined from 54 to 38 at Harvard between 1964 and 1968; midway through the academic year 1969-70 not a single law review editor from the graduating class planned to join a Wall Street firm. Similar shifts occurred in other schools: a decline of 10 percent at Yale; 9 percent at Virginia; and nearly two-thirds at Michigan within a single year. "It is clear," wrote Harvard dean Derek Bok, "that the desires and aspirations of many students are presently out of joint with the practice of law as it has been known in large metropolitan law firms."[34] This represented a striking change, which affected the pride of established firms. Twenty years earlier, Abe Fortas reminisced, "there was no suggestion that lawyers were not justified in conducting a purely commercial practice." When Fortas joined Arnold and Porter no one had proposed that a firm should assess "the social implications of the activities of those seeking its services."[35] But a new generation of lawyers demanded precisely that assessment, understanding that the failure to evaluate social consequences did not automatically imply professional neutrality.

In any period of social dislocation the lawyer's role in preserving the status quo had invariably incurred sharp criticism, even from within the profession. Brandeis had appealed for social responsibility in the Progressive era; and during the Depression Harlan F. Stone sharply rebuked lawyers who acted as hired men. Even a Cravath firm partner had acknowledged, in response to such criticism, that "probably too frequently" the lawyer's advice was limited to the "technical validity" of a proposed course of action, rather than encompassing its social implications.[36] Amid the social turmoil of the sixties, many lawyers struggled to discover more satisfying definitions of professionalism than those they had inherited. It became difficult for them to act in their clients' interest "without any regard to society's needs or anyone else's needs."[37] Once again lawyers proclaimed their moral and professional obligation to assess the consequences of their clients' behavior—and their own.

The demand for individual responsibility and social accountability left most lawyers uneasy. Little in professional values, ethics, or education prepared lawyers to confront that issue. Instead, the laissez-faire model of professionalism encouraged lawyers to evade it. Abe Fortas, writing in praise of the traditional style of private practice, condemned "project-lawyering, rather than client representation." He fondly recalled the time when "the social implications of the position to be taken on the client's behalf were submerged by the lawyer's dedication . . . to the obligations of advocacy in an adversary system." Fortas lauded "the disciplined isolation of personal passion from judgment which is an essential ingredient of the professional."[38] Lloyd Cutler, another prominent Washington attorney, decried the allegation that law firms incorporated any ideological pattern of client representation. He asserted the virtue of variety: lawyers should do "whatever work they found honorable, constructive and rewarding." He said that his own firm tried "not to urge any position for which there is no reasonable public policy basis." In disputes between government and business (which defined Washington practice) there usually was, Cutler asserted, "much to be said for both sides"—a convenient denial of responsibility for choices that any law firm made about the kind of business it conducted. For these lawyers the morality of the adversary process must prevail: "judgments of right and wrong," Cutler insisted, "are to be made after the process is completed, not before it begins."[39]

The profession was unsettled by any assertion that the public interest had any meaning beyond client service or that a lawyer, as Brandeis had insisted, should weigh the private interest of a client against a larger social interest. Not only was the adversary process presumed to be fair; it defined fairness. Merely by serving a client a lawyer served justice. Inaccessibility and inequality of counsel were excluded from consideration no matter how substantially they flawed the quality of justice that the adversary system produced. The adversary process presumed the equality of lawyers who often were unequal; and the profession had an obvious interest in preserving this presumption, for lawyers "simply could not accept their function in the system without some assurance that they are, in fact, instruments of justice rather than injustice."[40]

The professional elite never professed indifference to the "public interest." But its definition depended on the presumed fairness of the adversary process, which in the 1960's was no longer as self-evident as it once had been. In its Darwinian model, individuals and groups competed openly and fairly; consequently the process, not any substantive result, defined the public interest.[41] But the model had critical flaws, which the public events of the decade amply revealed to lawyers less indoctrinated in the traditional virtues. Survival of the fittest made little sense as a description of fairness beyond the privileged circle of Wall Street and Washington, nor did it take account of the social implications of activities that Wall Street and Washington lawyers defended. The adversary process hardly was value-free when access to counsel was determined by wealth, or when the interests of those likely to be affected by the outcome of negotiation or litigation were unrepresented in the neat two-sided division of adversary proceedings. With knowledge of the need for legal advice unevenly distributed, the posture of professional neutrality and passivity still served best those who knew that they confronted a legal problem and could afford to retain a lawyer to resolve it. The assumption that justice was automatically done within an adversary setting was valid only if the competitors were evenly matched, if all who wished to compete could do so, and if those who were affected by the result participated in the process. But adversary proceedings tolerated unbalanced legal resources and made a virtue of unquestioning service to clients whose activities, measured by any test external to the process, might cause substantial social harm.

Once the supreme virtue of client loyalty was placed in its social context, its imperatives seemed to rest as firmly upon economic self-interest as upon detached professionalism. In a society where only corporate clients and wealthy individuals were assured of access to lawyers, the principle of client loyalty could hardly be separated from the identity of the privileged clientele.[42] Consequently the loudest demand for public accountability was directed at Washington and Wall Street lawyers, in whose practice the disparity was greatest between the claim of professional neutrality and the reality of service to the privileged. Critics demanded that elite lawyers acknowledge their policy role and expand it to embrace a more equitable allocation of legal resources and a more inclusive agenda of social purpose.

Dissident lawyers insisted that corporate attorneys in fact engaged in project-lawyering as relentlessly as did any Movement lawyer. But their project, the preservation of private wealth, could easily be disguised as client loyalty. It was understandably difficult for lawyers to acknowledge that they were "as a group . . . no more dedicated to justice or public service than a private utility is dedicated to giving light."[43] Elite lawyers wanted it both ways; so, yearning for "the prestige of the profession, and the profits of business," they transformed the pursuit of self-interest into the virtue of public service.[44]

The professional rewards for serving any larger community interest, instead of private profit, were negligible. Consequently law firms approached public interest work—representing the unrepresented—"in terms of the acceptable levels of dispensed grace." *Pro bono* work, however highly touted as a professional innovation, never was more substantial than an elite gesture. Even in the most responsive firms, which collectively recorded 23,000 hours of *pro bono* volunteer time within a single year, the average was dismally low: only five hours annually for each lawyer, approximately 0.4 percent of the firms' billable time. This was "hardly a massive redirection of lawyer effort." Long after the debate began over an enlarged conception of professional responsibility, "price still remain[ed] the major mechanism for allocating legal services."[45]

For public interest law to survive, therefore, it had to escape elite confines. Encouraged by the OEO legal services program, public interest firms proliferated. But they floated precariously on a sea of professional indifference and financial uncertainty. Heavily dependent upon private foundations for funding once OEO was crippled by political opposition, they were vulnerable to the dictates of philanthropic groups who held them to rigorously traditional standards of professionalism. According to the Ford

Foundation, a primary benefactor, it was not for public interest lawyers to balance competing social interests; that role was reserved (as always) for the adversary process. Determination of "where the public welfare lies . . . is the job of the court."[46]

The most innovative, if fragile, expression of alternative practice was the law commune. Participating lawyers and workers openly shared radical political values along with the belief that lawyers should be more than "neutral legal technicians." Commune members identified with the political objectives of their clients (the first commune, in New York, arose in the wake of the Columbia student upheaval), whom they described as "partners in struggle." As one commune lawyer explained: "They're part of us and we're part of them." It was difficult, however, to be both a lawyer and a radical. Tension erupted between lawyers and non-lawyers; competition for cases flared up; the burden of lengthy, debilitating trials for impoverished clients could be crushing. Communal lawyers sought fulfillment from commitment and cooperation rather than from income, status, or proximity to power.[47] But it was virtually impossible to graft a communal institution onto a competitive profession. Communes disintegrated under pressure from the very social values that had generated them.

Many lawyers in the sixties acted upon the precept, enunciated by Oliver Wendell Holmes, Jr., that one must share in the actions and passions of his time at the peril of otherwise being judged not to have lived. They picketed, demonstrated, and protested, dropping the mantle of professional detachment to proclaim openly their commitments. They pursued the obligations of conscience, asserting the legitimacy—even the necessity—of value judgments. With an unerring eye for the contradictions in professionalism that sustained the maldistribution of justice, they pursued ways to relate their professional careers to their concern for a more equitable society. They were unabashedly committed to using law as an instrument of change in the interest of disadvantaged groups. Although established practitioners accused them of undermining public respect for the legal order, their activities demonstrated that even within the legal profession a substantial crisis of confidence in the legal system already existed. They displayed shock and anger at confronting a legal system which reflected the values of a society that preached equality and practiced racism, that promised abundance and tolerated poverty, and that waged war in the name of peace. The disparity between rhetoric and reality bred cynicism and revolt; a young Columbia graduate recalled his sudden awareness of "the unforgivable irrelevance of my legal education to what was happening in my head, in the courtroom and in the streets of our cities."[48]

For veteran radical attorneys, the 1960's provided an opportunity to resume New Deal commitments that had been thwarted by the Cold War. Two decades of political activism, separated by twenty years, were fused. Morton Stavis remembered the frustration of Wall Street anti-Semitism during the 1930's, which turned him toward labor union practice and a career of defending the unpopular and underprivileged. Charles Garry, born into an Armenian immigrant family, also had been prodded to professional dissent by the sting of ethnic discrimination. In the 1960's he found in the Black Panthers "a kind of cohesion of all the things that the labor movement originally started out in the '30's fighting for." He learned the satisfaction of working with people for whom he could confess "unqualified respect and love"; never, he declared, had he "felt as professionally fulfilled." The events of the sixties bred their own converts. Ramsey Clark, who experienced one of the most striking professional conversion experiences of the decade after his tenure as attorney general, found himself inhibited by the traditional obligation to "make the best of a case, good or bad." Wanting "greater freedom" than private practice provided, he left the Paul Weiss firm "to devote all my time to matters that I thought were important, and that's a little incompatible with the general practice of law."[49]

In a politically conservative profession such choices generated "perpetual tension or contradiction." Defenders of the disadvantaged were simultaneously officers of the court who were bound by the very legal system they criticized. Committed to the legal process, they were inexorably drawn to represent those who were victimized by the social values it embraced. Opposed to the existing political structure, they strengthened its processes by their participation. They knew that law oppressed, but they retained hope that it might protect.[50] Ever since the beginning of the urban industrial era, elite lawyers had engaged in "partisanship combined with technical excellence."[51] But partisanship for the underprivileged, combined with technical excellence, was an unsettling combination. New definitions of professional responsibility, which raised the prospect of a more equitable distribution of legal services, confronted the bar with a choice between its elitist tradition of service to the privileged and the more uncertain future of equal justice.

Amid growing criticism of the bar for neglect of its public responsibilities, the American Bar Association re-examined professional ethics for the first time in more than half a century. In 1964, at the request of President Lewis Powell, it established a special committee to propose changes that might accommodate the Canons to contemporary conditions. An additional spur to reconsideration, potentially even more unsettling than public criticism, came from the Supreme Court. In a series of decisions that began in that year the Court chipped away at professional prerogatives which preserved the self-regulating autonomy of the bar, often at the expense of the public interest in expanded legal services. In a Virginia case the Court overturned an injunction sought by the state bar to terminate a union plan for advising members of their need for legal services and recommending lawyers who could provide them. Professional claims of illegal solicitation and unauthorized practice, the Court ruled, could not infringe the right to be fairly represented in a law suit. In a subsequent decision, delivered while the Canons were under revision, the Court upheld the right of a union to employ a salaried attorney to represent its members over the insistence of the Illinois Bar Association that this also constituted unauthorized practice.[52] With OEO generating its own pressure for expanded professional responsibility, the specter of group legal services (which the bar had tenaciously opposed for thirty years) increased professional unease.

The bar could not consider the possibility that single-minded client loyalty, within a framework of individualized practice, was the source of maldistributed legal services, not the solution for it. To assure the availability of counsel at reduced cost, some private groups (particularly trade unions) had retained their own salaried attorneys to advise members about their legal problems. Bar association hostility to this practice surfaced during reconsideration of the Canons. The subject of group legal services provoked a debate that was reminiscent of the antagonism toward contingent fees at the time the original Canons were adopted.

In a new Code of Professional Responsibility, approved in 1969, the American Bar Association tried to balance the economic self-interest of the bar against the undisputed existence of a vast neglected public for whom legal services were unavailable. The Code went beyond the old Canons in recognizing that every person should enjoy access to a lawyer's services; it conceded that the circumstances of life in a complex urban society had destroyed the effectiveness of the passive selection process reflected in the Canons; and it asserted, for the first time, the affirmative duty of the profession to make legal counsel available.[53] But even as it repudiated the more outmoded laissez-faire precepts of the Canons, it remained committed to the very principles of individualized responsibility that contributed so substantially to the economically distorted allocation of legal services. Conceding that the efforts of individual attorneys often were insufficient (especially for the indigent), it nonetheless asserted that the obligation to provide legal services remained a matter of individual responsibility.

Thus the presumption that the social obligations of the profession could be entirely discharged by the random results of individual effort underlay the Code just as it had the Canons. Once again the association went no further to assure the provision of legal services than the minimum obligations of law required. Just as approval of contingent fees had been limited to those already "sanctioned by law," so group services by salaried attorneys were prohibited except insofar as Supreme Court decisions required them. In 1969, as in 1909, the social interest in expanded legal services was sacrificed to the self-interest of the established bar. Short of declaring its refusal to abide by court decisions, the American Bar Association could hardly have done less to respond to the pressing need for group legal services. Its new rule, complained an ABA dissenter who urged a more expansive provision, went "grudgingly as far as the Supreme Court requires but not one step beyond. . . . How appalling, as a statement of the policy of a *service profession!*" The result, professional critics complained, was a "conservative, timid rule" which was "unrealistic, inadequate, irresponsible, and unprofessional."[54] It suggested that the new Code, like its predecessor, was designed to be more cosmetic than remedial.

Beyond the inadequacy of services, another deficiency was restricted information. Although the Code did not mention solicitation, and even asserted that lawyers should help laymen to recognize their legal problems, it retained the virtue of secrecy as an essential ingredient of professionalism. Availability of legal services depended, in part, upon knowledge of their existence (and in larger part, of course, upon their actual existence). But constraints on publicity and advertising assured that knowledge would remain unevenly, and unfairly, distributed. Even under the new Code, "virtually all channels of communication which reach large segments of the general public are closed to an attorney." Lawyers could advertise their specialties in "reputable" law lists (as certified by the ABA), which were circulated to banks, corporations, and insurance companies, but they could not do so in the public media which reached a wider and less privileged clientele. Ethical advertising remained only that advertising which was directed to corporate clients; anything else, as persistent bar association efforts to impede advertising by public interest firms demonstrated, was unethical. A professional double standard was sustained by rules "which aid some in protecting their right of access to the legal system while throwing obstacles in the path of others."[55]

With its decisions to impede the development of group legal services and contain efforts of lawyers to communicate with a larger public, the ABA retained the ethical precepts which restricted access to legal services to a privileged clientele and monopolized legal business for the established bar. The Code failed to provide for prepaid legal services or for any other form of legal insurance that might lower costs and increase accessibility. Like the Canons it replaced, it concentrated its energies upon the preservation of a professional monopoly, not the provision of legal services.[56]

By 1969, when the Code was adopted, the professional mood, like the mood of the country, was turning sour and vindictive. For nearly a decade the legal profession had been subjected to some of the most severe criticism in its history. The civil rights movement exposed the complicity of law and lawyers in racial discrimination. The war against poverty demonstrated the contribution of legal institutions to economic injustice. Students challenged the legal establishment to repudiate its elitist values, to expand its public commitments, and to redefine its social obligations. Minority group members demanded access to the restricted sanctuaries of law schools and law firms. Young lawyers engaged in new styles of practice and promulgated expanded definitions of professional obligation. For years, city streets and university campuses (even law schools and, increasingly, some courtrooms) had seethed with demonstrations against economic and racial injustice at home and military intervention abroad. The American public was sated beyond endurance with protest, no matter how legitimate the grievances that inspired it or how violent the resistance applied to repress it. "Law and order" was the new rallying cry; absorbed by the bar, it encouraged lawyers to resist movements toward professional democracy.

At times during the decade, before the law-and-order reaction set in, it had seemed possible that the walls of professional elitism and privilege might finally crumble. As the war in Vietnam spread its corrosive disillusionment through national institutions, the political function of the legal profession as an adjunct of privilege moved sharply into focus. Some ripples of democratization nurtured hope for socially responsible professionalism. But the hope was chimerical; the elitist social structure of the bar swayed, it did not break. Under the stress of severe social turmoil it lost its neutral veneer but little of its resilience or retaliatory power.

The bar responded to the political discord of the sixties by refurbishing the instruments of discipline and intimidation left from earlier professional crises during the Red Scare and McCarthy eras. Convinced that occasional professional protest represented systematic subversion, it took its cues from Chief Justice Warren Burger, an outspoken proponent of law and order on the bench, who warned of "a tiny fragment of reckless, irresponsible lawyers" who "seem bent on destroying the system." Burger praised the virtue of civility, suggested that "insolence and arrogance are confused with zealous advocacy," and decried "adrenal-fueled lawyers" and "overzealous advocates." Attorney General John Mitchell also chastized activist lawyers before an approving ABA audience.[57] Encouraged by the law-and-order rhetoric of the Nixon administration, and by public intolerance with "permissiveness," the bar launched another campaign for professional conformity.

By 1970 it was once again "professionally and legally dangerous to be a lawyer representing the poor, minorities, and the politically unpopular." Contempt citations and disciplinary proceedings expressed the hostility of bench and bar toward professional dissidence; bar associations resumed their familiar role as "prosecutor rather than protector" of lawyers who defended unpopular clients. The lawyer who served as the scapegoat for professional hostility was William Kunstler, a conventional practitioner of commercial law during the 1950's whose defense of the Mississippi Freedom Riders was the first step on his journey to political activism. The trial of the Chicago Eight, the first and most dramatic of the political trials designed to coerce law and order during the

Nixon administration, was his Rubicon. Judge Julius Hoffman's partisan conduct of the proceedings, which drew professional praise at the time and appellate rebukes subsequently, dissolved Kunstler's loyalty to those forms and conventions of the judicial process that had become masks for prejudice and injustice. As Kunstler began to identify with the flamboyant protest of his clients, and as his clientele became more radical, he emerged as the prime target of professional opprobrium. Kunstler became the lawyer who personified the counter-culture values that jeopardized the authority of the legal system.

Cited for contempt at the conclusion of the trial, Kunstler was chastized in professional circles for his unprofessional behavior. But those who denounced him as a Robespierre at the bar overlooked the fact that his moderate liberalism, with its affirmations of faith in the inevitability of equality, had persisted until the late 1960's. (They also slighted his skillful application of federal removal proceedings as a device to deter harassment of civil rights workers by Southern state judges.) By the end of the decade, however, Kunstler's political and professional consciousness had changed under the impact of events. As his professional austerity melted, his patterns of advocacy (and the identity of his clientele) began to disturb the guardians of professional propriety. Kunstler was berated in the American Bar Association *Journal* for declaring that he was "not a lawyer for hire" but would "only defend those I love." An editorial denounced his "increasing flow of inflammatory remarks," his "cult of admirers," and his "approach to human problems that is anti-intellectual, frantically impatient, defiantly non-objective and intolerant, and eager for violence." The ABA, suddenly discovering the virtue of competent counsel for all, observed: "We know from long collective experience that many will go without legal defense or representation if they must depend upon finding a lawyer who 'loves' them."[58] Still, that experience had never much impressed the ABA until Kunstler and other activist lawyers engaged in their own form of amorous selectivity. The ABA conceded that "all lawyers who can afford it" were selective, preferring prosperous clients and popular causes.

But the profession, it asserted, had moved significantly closer to the provision of competent counsel for all; therefore client selectivity, however venerated by tradition, was now discredited. Once the ABA discovered that political dissidents might enjoy the benefits of preferential advocacy, it showed uncommon concern with the unrepresented.

Yet there were revealing similarities between Kunstler and his critics. He might wear an armband in court, or scorn the preferred mode of professional reticence, but he was hardly the first lawyer to identify with his clients, to emulate their life style, or to absorb their political values. (John W. Davis, for example, had done no less.) Kunstler's politics made him vulnerable: the grievance committee of the Association of the Bar of the City of New York was so eager to bring disciplinary charges against him for his conduct during the Chicago trial that it violated its own procedures, which prohibited such action until judicial proceedings were exhausted.[59]

Kunstler was not the only lawyer to feel the sting of professional discipline. If trials could be initiated to harass dissidents, so disciplinary proceedings could be instituted to intimidate their attorneys, whose zealous defense was translatable into "misconduct" and "moral turpitude." Good character re-emerged as a flexible test to exclude aspirants whose political activism displeased admissions committees. In California the bar examiners refused to certify an applicant who had been convicted for civil disobedience during a civil rights protest. (A California court ruled that civil disobedience did not demonstrate moral turpitude.) The Arizona bar denied a petition for admission of an applicant who refused to state whether she had ever belonged to the Communist Party. In Ohio applicants were required to list every organization they had joined after the age of sixteen. The Arizona and Ohio cases reached the Supreme Court, where Justice Black, for a bare majority, declared that "views and beliefs are immune from bar association inquisitions designed to lay a foundation for barring an applicant from the practice of law." (Justice Blackmun, in a dissenting aside, displayed the prevailing professional obsession with an "overabundance of courtroom spectacle" that was used to justify professional discipline.) For a time, the American Bar Association even considered an examination for all law students that would "identify those significant elements of character that may predictably give rise to misconduct in violation of professional responsibilities." It urged bench and bar to intensify their efforts "to root out the known . . . character risks already engaged in the practice of law."[60]

Professional obsession with courtroom disorder, however convenient for disciplinary purposes, had little foundation in fact. As the chairman of a New York City bar association committee concluded, after the hysteria subsided, the bar "misconstrued . . . the dimensions and causes of courtroom disorders." Acting as though courts had suddenly been seized "by an organized group of radical lawyers," the bar, panic-stricken, had "confused zeal in the defense of clients with revolution, and thus moved in the direction . . . of intimidating defense counsel."[61] Discipline still reflected the ethnic and economic divisions within the bar: those who administered discipline remained a class apart from those who received it. Professional associations still represented a "white, wealthy, and politically conservative" constituency that was determined to resist the political activism of those who were often not white, usually not wealthy, and invariably not politically conservative.[62]

Discipline provided only one example of the structural tenacity of professional discrimination. The most persistent and glaring structural defects were inequitable access to professional opportunity and to legal services. These issues were inextricably connected: elite lawyers promulgated and defended a set of professional values that served the interests of the ethnic, social, and economic groups to which they and their clients belonged. Any change in the social structure of the profession depended upon prior changes in law school admissions; there was virtually no possibility for the interests of a neglected clientele to receive a fair hearing within the legal system until the political base of professional power was substantially broadened.

Law schools, spurred by the civil rights movement, initiated action to remedy the glaring underrepresentation of black and other minority group lawyers. As late as 1964-65 less than 2 percent of the students in approved law schools (or lawyers in the profession) were black. Pressure for equality produced a sudden increase in minority recruitment. As law schools changed their admissions policies, the enrollment of black students increased from 700 to 4,800.[63] But this shift was temporary in duration and modest in results. In 1972 only Yale, from a sample of private and state law schools, enrolled a student body that was more than 10 percent non-Caucasian. Notwithstanding the modesty of the change, minority recruitment provoked other ethnic and religious groups to angry protests, which culminated in the challenge of the *De Funis* case to preferential admissions. Once some schools began to give preference to racial minority group members, in an effort to reverse decades of prior discrimination, they encountered charges of racism from those whose places in law school were protected by prevailing admissions criteria. But if law schools did not modify their traditional practices, they would inevitably perpetuate professional discrimination. The outcry against quotas had specious allure; but nominally objective criteria of merit were hardly unbiased, as one advocate of preferential admission observed, when they preserved the bar as "an almost totally white man's or Anglo's province." As long as the proportion of minority students remained far below their proportion to the population, any argument from social policy pointed in favor of a legal profession that was accessible to all. For a time this might require preferential admissions, but that hardly was a professional innovation. It was only when blacks became the beneficiaries that the virtues of color blindness suddenly prevailed. But "least of all professions," as one lawyer wrote, "can the legal profession condone the continued exclusion of minority groups." The prestigious schools, however, continued to recruit from "elite family backgrounds"; their students still represented a privileged economic, racial, and religious sample.[64]

Law schools constituted the first, but not the only, hurdle to professional access. Black graduates confronted bar examiners who remained as determined as their predecessors had been to preserve the racial purity of the profession. The evidence pointed to nothing less than a national scandal, and none of it suggested improvement. Although three-quarters of white applicants nationally passed bar examinations, only half that proportion of black applicants did so. In Philadelphia, where lawyers had pioneered in devising racial exclusion practices in the 1920's, bar examiners photographed applicants and seated black candidates consecutively in the same row to facilitate the grading of their examinations. Understandably, blacks failed in disproportionate numbers; there was, a bar association study concluded, "lacking any available hypothesis other than race by which to explain these proportions." No black applicant passed the Delaware bar examination between 1957 and 1974. In Ohio, where four of every five white applicants passed, one of three blacks was successful. Between 1969 and 1974 in South Carolina, 2 percent of white applicants and 50 percent of blacks failed. In Georgia, where more than half of all applicants passed the examination, no black applicant did (including—or, more properly, excluding—two graduates from Yale, one from Harvard, and three from Columbia). And in California, where there was one white lawyer for every 450 whites, there was one black lawyer for every 3,000 blacks (and one Chicano lawyer for every 16,000 Chicanos). Racial bias suits were brought in no fewer than ten states in an effort to end the discriminatory practices of bar examiners.[65]

Racial origin, as a primary determinant of law school admissions and bar examination success, inevitably pervaded law firm hiring, despite the ample publicity given to scattered examples of minority recruitment. In 1973 there were fewer black lawyers in the nation's twenty largest law firms (a total of 30 lawyers out of 3,200) than there had been five years earlier. In Chicago, where 32 percent of the population and 3.5 percent of the legal profession was black, only .37 percent of black lawyers practiced in firms—striking proof of their continuing professional marginality. The figures were as dismal elsewhere. Among Midwestern law firms, "the formerly closed corporate door [was] only slightly ajar" by 1971. According to an American Bar Foundation study, "the latent racism which has accumulated in all major American social institutions also pervades the legal profession"—after a decade of the most extensive efforts to undo racial discrimination in the history of the bar. Little wonder that one black lawyer complained: "I would like to learn what Anglo-Saxon law is all about so that I'll know how to dismantle it, because I do not think it is relevant for black people."[66]

Nor was the picture appreciably brighter for other minority groups (with Jews, having already served their long apprenticeship as professional undesirables, the conspicuous exception). In New Mexico and Arizona, where 135,000 Indians lived, there was not a single Indian lawyer; no Indian had ever received a law degree from university law schools in those states. In New York, where Italian-Americans comprised 18 percent of the population, there were two Italian-Americans among the 628 partners in the twenty largest firms. For women, that unique minority group that actually constitutes a majority of the population, sex discrimination was still rampant. For decades elite law schools had excluded them altogether. The dean of Columbia, under suffragette pressure during World War I, had predicted that, if women were admitted, his school soon would be swarming with "freaks or cranks." Although Columbia survived their presence, Harvard did not yield until after World War II. In 1963 women comprised 2.7 percent of the profession; five years later they were still being selectively recruited into low-status fields of specialty practice that were regarded as suitably feminine: domestic relations, juvenile, and probate work. Although law schools substantially expanded their enrollment of women during the decade (to the point of enrolling almost half the number to which their proportion in the population might entitle them), prevailing attitudes toward the proper role for women in society—which the law schools, being overwhelmingly male, naturally shared—kept the law a precarious and often demeaning profession for its female practitioners.[67]

Not only did the social structure of the profession prove resistant to change; as the sixties ended, the career choices of young lawyers, a reliable barometer of professional values and opportunities, shifted once again from public service to corporate practice. Droves of young activist lawyers were quickly superseded by students who might have admired Ralph Nader but decided not to emulate him. By 1970 social activism was a casualty of the law-and-order crusade, the diminishing opportunity for legal work in the public sector, and economic retrenchment. In major law schools the proportion of graduates entering private practice increased sharply between 1969 and 1971. A Harvard graduate from the activist generation concluded that law students no longer constituted a "new breed"; instead they had rejoined "the same old herd."[68]

Career decisions shifted to the rhythms of social change, but student dissatisfaction with legal education persisted long after political activism in the law schools subsided. Despite a flood of applications for admission, students demonstrated continuing confusion about the role of lawyers in society and concern over the impact of professional training upon their personal values. Their unease suggested that legal education, despite a decade of self-scrutiny and curricular innovation, was still mired in the contradictions generated by its professed preference for means of process over ends of substance. As the authors of a new Carnegie Commission study of legal education observed: "the rational study of values, or ends, has been viewed as off limits to most scholars." This same avoidance suggests why the authors, both of them law teachers, could conclude that legal education was "a good thing," yet cite the persistence of student "malaise and discontent."[69]

Law schools (like the firms that once again attracted their best students) still proclaimed their institutional neutrality and asserted their preference for technical competence unrelated to social goals. But technical competence had never been unrelated to social goals; in denying connections that were so evident to so many law students, law schools managed only to generate deep cynicism about the values to which the legal profession was ostensibly committed. Process divorced from substance, as a young Harvard graduate observed, represented a "false equation between moral abdication and intellectual independence."[70] Student anxiety and frustration suggested that legalism raised more questions for the newest professionals than it answered.

Indeed, as Judge J. Skelly Wright observed caustically, "self-appointed scholastic mandarins" on law school faculties, who praised "principled" and "reasoned" adjudication by the Supreme Court, supplied the most glaring model of the politics of professional detachment. "How are we to decide," Wright asked, "without first making value choices?" Suspicious of " 'value-free' values," he suggested that the neutral, scientific approach favored in the law schools rested upon implicitly political—and politically conservative—definitions of the good society.[71] Law schools might proclaim the neutrality of craft, and law students might resume conventional careers, but faith in the moral authority of the old legal order was diminished within the very institutions that bore responsibility for instilling it.

While the elitist social structure of the profession hardened, efforts to democratize the delivery of legal services disintegrated under political pressure. The OEO program, at peak strength, had scattered nearly 2,500 lawyers in three hundred communities to provide representation to more than one million clients, most of whom were "black and brown and red and yellow" citizens who could not otherwise have afforded legal assistance. It became politically fashionable to accuse OEO attorneys of harboring revolutionary intentions, but their commitment (like their statutory obligation) was to reform. Many poverty lawyers preferred to practice law in the traditional way, "except with poor people as clients." But poor people not only confronted mistreatment as individuals; as members of economically and racially disadvantaged groups they also suffered from institutional injustice which the snail's pace of case-by-case adversary proceedings could not overcome. Class action suits had enabled OEO to secure substantial benefits for entire groups of indigent migrants, welfare recipients, prisoners, tenants, and assorted victims of racial and economic discrimination. That was more than sufficient to make OEO legal services politically vulnerable.[72]

Crippled even before 1969 by gubernatorial vetoes, restrictions on criminal representation, attempts to decentralize its activities, and uncertain funding, the legal services program was finally scuttled by the Nixon administration. Representing a silent majority that had little tolerance for equal justice, and capitalizing upon opposition within the legal profession to OEO, the administration accused OEO lawyers of placing "causes ahead of cases" and using clients "as mere vehicles to promote sweeping social and political change." Once again traditional professional values were applied in defense of the political status quo—and to perpetuate the maldistribution of legal services. In 1971 the administration proposed to eliminate all criminal representation (in effect, to deprive indigent suspects of counsel), prohibit funding of public interest firms, impose prior review of appeals, restrict attorneys' activities, eliminate back-up centers (the critical institutions in the preparation of class-action suits), and transfer the entire program to an "independent" corporation whose directors would be appointed by the President. Congress approved an independent corporation, but Nixon vetoed the bill because it restricted his appointment power. Within two years three directors resigned or were discharged under political pressure. In the waning months of the Nixon administration, while the President availed himself of a somewhat different federal legal services program in his own defense against impeachment, Congress passed another bill, which established a permanent legal services program but gave the President the power to appoint all directors of the new corporation. In return for Nixon's promise not to veto the proposal the Senate, at the last moment, denied corporation funding to back-up centers to finance class-action litigation, thereby depriving legal services of the most effective means yet devised for securing the rights of poor people.[73]

The OEO program, an afterthought of Lyndon Johnson's war against poverty, was a specific target of the Nixon administration's war against justice. Professional groups, which offered support at the end when only the hollow shell of OEO was left to defend, had contributed substantially to its demise. Their belated recognition of the problem, and their incessant sniping at the program until its remedial potential vanished, belied their protestations of concern. The organized bar was more interested in winning the confidence of the poor "in our legal establishment" than in implementing a program that would actually secure the rights of poor people. The ABA insisted that OEO lawyers observe traditional ethical precepts and modes of advocacy, even if these contributed to the very maldistribution of legal services that OEO was committed to rectify. OEO, once it chose to take only those cases that "affect the poor as a class," placed itself squarely in conflict with the values of an individualistic profession. Although the bar insisted that these traditional values sufficed, OEO acted upon the more venturesome (and more accurate) assumption that injustice for the poor, in the words of one of its attorneys, "is endemic in American law because the structure of the legal system is not designed for the indigent"; and, committed to changing that structure, OEO was deprived of its effectiveness—which, after all, only proved the point. A legal profession that remained divided along ethnic and class lines could hardly serve the substantial segment of the American population that inhabited the underside of an identically divided society. Equal justice, like equal access, was incompatible with an elitist profession that remained distinguished by its "scandalous failure . . . to serve those who need it most."[74]

The legal profession had always accommodated itself to the broad social and political currents of American public life. Solemn reference to The Law might imply transcendent obligations, but the bar was a social institution which absorbed the demands of time and place and the values of the American political culture. Throughout the twentieth century it hovered between the poles of aristocracy and democracy defined by Tocqueville and Lincoln a century earlier. During brief periods of social ferment it responded reluctantly to public demands for accountability; during the more sustained periods of political retrenchment that invariably followed, it pursued its elitist objectives. At one level, professional life in the sixties merely constituted another episode in that rhythm, a turbulent interlude between the Cold War conformity of the McCarthy era and the law-and-order lawlessness of the Nixon administration. At another level, however, the professional activism of the decade represented the most substantial effort in modern history to transcend the constricting values which defined the traditional limits of professional propriety, service, and responsibility. Ultimately, however, it failed.

For more than a decade, between the hope of the New Frontier and the revulsion of Watergate, American society experienced a prolonged national trauma, culminating in the resignation of a President (who happened to be the twenty-third lawyer to occupy the office he vacated) to avoid impeachment and conviction for nothing less than the subversion of constitutional government. One sordid event after another tore the nation apart: assassinations; riots; war; domestic rebellion and repression; substantial evidence proving that law-enforcement officers and the highest government officials were law-breakers. The cumulative impact of these events, and the dissembling by public officials that so often accompanied them, educated a generation of Americans to the enormous disjunction between the beneficent promise of American life and its everyday performance. The relentless evidence of these contradictions not only demystified institutional authority; as the gap between myth and reality widened, social cohesion and political loyalty disappeared. In the end, nothing less than the legitimacy of legal authority was in jeopardy.

The assertion by civil rights and antiwar protesters of the justification for civil disobedience had resounded with disturbing frequency throughout the decade. It did not mean, as some too hastily concluded, that anarchy was abroad in the land; but it did suggest how fragile, and how inadequate, legal authority was once the foundation of consensus in its moral integrity crumbled. As early as 1964 a prominent liberal attorney complained that "our legal culture is under new and dire assault. Our people have been encouraged to lose faith in the use of reason, quiet courtrooms, and adversary advocacy." Six years later the Association of the Bar of the City of New York, eager to celebrate its centennial anniversary, instead held a symposium to explore whether "faith in law . . . was still valid." Eugene V. Rostow observed in his introduction to the published proceedings (entitled *Is Law Dead?*) that "our credo and our institutions are being tested with a vehemence the nation has not known since the Civil War. . . . Some deny the idea of law itself as the compass of our social system."[75] Yet professional values and behavior had themselves done much to abet this crisis of confidence.

Watergate was the most severe jolt to the integrity of legal authority. The mask that disguised lawlessness as law and order disappeared. The law-enforcers, lawyers all, were the law-breakers. "The record of the lawyers around Richard Nixon," wrote Anthony Lewis in the *New York Times*, "is one of the most appalling aspects of his Presidency." Administration lawyers, Lewis concluded, "have made their names symbols of contempt for law."[76] "How in God's name could so many lawyers get involved in something like this?" asked John Dean.[77] As the roster of lawyers implicated in burglary, perjury, and obstruction of justice lengthened, other attorneys echoed his bewildered question. Bar leaders quickly promised more stringent discipline; law teachers proposed additional training in legal ethics. An optimistic *New York Times* editorial cited "sophisticated observers" who believed that the bar, "if sufficiently motivated, has the requisite financial, intellectual and moral resources to improve significantly upon past performance."[78] But a long history of selective enforcement of discriminatory precepts offered little encouragement. It suggested the existence of structural defects in the professional culture which money, energy, and intelligence were insufficient to overcome as long as these defects were maintained by the existing framework of professional values. The question was not whether ethics should be taught (they already were), but which ethic should be taught: the ethic of the marketplace and client loyalty, or the ethic of equal justice.

Implicit in reform proposals was the notion that the legal profession could, and should, regulate itself. For obvious reasons lawyers insisted that legal ethics involved complex issues that only lawyers could comprehend. But the preservation of a professional monopoly was the problem, not the solution. Although Watergate hardly proved that an entire profession was unethical, the criminal behavior and professional malfeasance of so many lawyers suggested glaring defects in professional mores—despite the new Code of Professional Responsibility. As long, however, as lawyers were permitted to monopolize solutions to problems that their monopoly of solutions had created, the problems would endure; and Watergate offered little support for the proposition that responsibility for professional ethics should remain vested in the very professionals whose ethics were in question.[79]

While the profession struggled to retain its monopoly, lawyers tried to absolve themselves of responsibility for the derelictions of their professional colleagues. An entire profession, some asserted, should not be judged by the transgressions of a few. What Watergate lawyers did, explained the counsel to the grievance committee of the Association of the Bar of the City of New York, "usually was not in their capacity as lawyers."[80] The most sweeping exculpatory defense came from Alexander Bickel of Yale Law School. Watergate lawyers, he wrote, had indeed elevated "ideological imperatives and personal loyalty" above "the norms and commands of the legal order." But so, too, had all those protesters who opposed racial discrimination and the war in Vietnam; so had the Warren Court. All of them had enlisted in "the armies of conscience and of ideology" which displayed attitudes "that at least contributed to Watergate." They all had tried to "coerce the legal order by destroying . . . the procedures by which it conducts its business." The protest of the sixties, combined with the judicial activism of the Supreme Court, was "prologue"; Watergate was little more than a "replica" of these earlier "transgressions." The legal order, Bickel asserted, "heaved and groaned for years under a prodigality of moral causes" as "the derogators of procedure and of technicalities . . . rode high, on the bench as well as off." Civil disobedience was a habit-forming drug: "the habit it forms is destructive of the legal order." Watergate lawyers were only the last in a succession of addicts, hooked by the same intensity of moral passion that beset their predecessors. Bickel had nothing against excessive morality. The test of a legal order, he acknowledged, was its moral authority. But the legal order had its own superior morality which could not survive "the continuous assault of moral imperatives" from other sources. Its fundamental premise was the "high moral value, in most circumstances the highest," of its own stability. The "highest morality," Bickel concluded, "almost always is the morality of process."[81]

This was legalism with a vengeance: an ideology that superseded all ideologies; a morality that transcended all moralities; an omnipotent deity to be worshipped by a cult of monotheistic professional zealots. Yet the very "norms and commands of the legal order," that Bickel had absorbed so thoroughly and praised so fulsomely, held clues to Watergate and to the distintegration of legal authority that it accelerated. Bickel rested his argument upon the ostensible separation of process from substance and the presumed independence of law from policy and power—the precepts which defined professional thought in the twentieth century. They enabled him to equate the civil disobedience of black citizens who protested against laws that deprived them of their constitutional rights with the criminal acts of government officials sworn to uphold the law. It did not seem to matter to Bickel whether law served the cause of equality or of discrimination, whether courts were active in defense of indigents' rights or of unregulated corporate property—or whether the biases of the law elicited those very acts of disobedience that Bickel so casually lumped with the crimes of Watergate. It was precisely this failure to relate process to purpose, or to comprehend how tenaciously the legal order had aligned itself with certain moral imperatives at the expense of others, that had prompted a generation of Americans to repudiate his kind of legal authority. The notion that process could be divorced from substance was a fiction proclaimed by the philosophers of legalism.

The breakdown of legal authority did not occur because aggrieved citizens engaged in civil disobedience. Citizens protested because once legal authority ceased to embrace equity and justice it lost a substantial measure of its legitimacy. Unlike lawyers, they would not permit the authority of the legal order to exist independently of the moral values which the society proclaimed; for once function superseded purpose, the legal order drifted rudderless in a sea of moral relativism, inevitably blown by the strongest winds of wealth, power, and politics. Watergate demonstrated that the legal order no longer commanded the respect even of those who were elected or appointed to preside over it. The morality of process could only be justified as the highest morality if the citizenry was convinced of its fairness; but too many events in the sixties and early seventies destroyed that belief. For too long the morality of process and power had superseded the morality of justice. The stability of the legal order, deprived of external nourishment to give credence to its claim of fairness, dissolved.

The response of legal institutions to Watergate was as destructive to confidence in the legal order as was the criminal activity itself. The procedures used to re-establish the primacy of law, one observer wrote, "appear . . . to have perilously subverted it." The discretionary inequities built into the legal process received such blatant exposure that few people even bothered to defend the principle of even-handed justice; the advantages of discretionary justice were enthusiastically praised. The confessions that pried open the Watergate conspiracy were secured only with a judge's threats of harsh and punitive sentences. Plea-bargaining, endemic to the criminal process, became a national scandal after Vice-President Spiro Agnew was permitted to plead no contest to a minimal charge of tax evasion in return for his resignation and former Attorney General Richard Kleindienst was allowed to plead guilty to misleading the Senate to avoid facing the disbarment proceedings that would inevitably have followed his conviction for perjury. The Watergate prosecutions revealed a special standard of justice reserved for those who knew how "to buy or bargain their way out of trouble—and routinely denied to poor, black, ignorant Americans" who lacked power, influence, or money. One obvious lesson was that "justice in America is unequal, and the 'old boys' continue to get favored treatment."[82]

The morality of process, battered by Watergate lawyers and their prosecutors, was bludgeoned by President Nixon's counsel, whose conduct vividly displayed the perils of undeviating client loyalty at the expense of any competing consideration of the public interest. James St. Clair, a prominent Boston trial attorney, had defended both the antiwar dissent of William Sloane Coffin and the racist practices of the Boston School Committee with undiminished professional zeal. He always insisted that a lawyer's "private opinions of the people and principles in any lawsuit are not material." His posture defined proper advocacy within the adversary system. But it was precisely his limitless defense efforts in Nixon's interest (at least until his own professional reputation was jeopardized) that enabled an entire nation to observe the deleterious effects of absolute client loyalty. When the release of edited tapes revealed the President's willingness to pay hush money to conceal Watergate crimes, St. Clair declared that "not once does it appear that the President was engaged in a criminal plot to obstruct justice." When additional transcripts revealed Nixon's expressed desire for a "cover-up or anything else" that might save his presidency, St. Clair claimed that the passage had been deleted from White House transcripts because it was "not that relevant." St. Clair never checked Nixon's claims against Nixon's tapes—whose witholding he defended without knowledge of their contents. His defense of the President's refusal to release evidence finally incurred the wrath of federal judge Gerhard Gesell, who was compelled to instruct him: "It's wrong. You know it's wrong. . . . It borders on obstruction. It's offensive."[83] The final demise of the morality of process came after Nixon's resignation. St. Clair had persistently asserted the immunity of the President to the legal process, a proposition widely ridiculed when Nixon occupied the White House but confirmed after his departure by President Ford, whose pardon aborted the legal process before it had even begun to function.

As the legal profession nursed its Watergate wounds, most of them self-inflicted, it confronted the most serious crisis of public confidence in the rule of law in American history. The structure, the values, even the most cherished principles and basic processes of professionalism were perceived as instruments of injustice. It was on that note that the modern century of professional history closed. It had begun in the 1870's, when the founding of the American Bar Association marked the emergence of an organized bar and heralded the appearance of a new professional elite whose interests would mold the identity of the profession and the quality of the services it provided. It ended in the 1970's with discredited professional values and precarious faith in the legitimacy of legal authority. Renewal of faith depended upon the unlikely prospect that the legal profession would become a truly public profession, with a broader and deeper definition of social responsibility than the bar had ever tolerated.

Afterword

Historians are trained to comprehend the past, not to contemplate the future. But my interest in the modern history of the legal profession emboldens me to venture briefly into forbidden terrain. After so much description, it is hard to resist some prescription. Many years ago Karl Llewellyn, one of the wisest legal scholars of this century, observed that "the best talent of the bar will always muster to keep Ins in and to man the barricade against the Outs"; though it is not law but society, Llewellyn insisted, that "puts the screws on in favor of the Ins." Lawyers "mirror undistorted" the very society that accuses them of social irresponsibility.[1] Nothing in the past forty-five years has diminished the acuity of Llewellyn's perceptions. The professional elite still protects the Ins; in doing so, it still reflects (but also substantially reinforces) the disparity between democratic ideals and hierarchical realities that haunts American society and accounts for so much of its recent turmoil.

Back in 1916 a young corporate lawyer described an "unusual experience" in his professional career: Louis Brandeis, before taking a case, had inquired into "the justness of our position. . . . I had occasion to retain other lawyers," the attorney testified, "and no one ever raised that question."[2] Surely the time has come to raise it again, to raise it constantly, and to incorporate it in a system of professional values that has persistently evaded it. Lawyers will repress painful memories of Watergate with comforting (but deceptive) assurances that the self-correcting mechanisms of an essentially sound legal system proved its strength. Before they do, however, their profession should be held to an unprecedented measure of public scrutiny and accountability. If the legal system is to recover some measure of the legitimacy and authority that it has lost, difficult questions must be not only confronted but answered in radically different ways. Should the availability of legal services depend upon wealth rather than need? Should an adversary system that depends upon economic means be permitted to define justice? Should the morality of process and craft obliterate the social responsibility of lawyers?

If the practice of law is to become a public profession, not remain a private club, new values and voices are necessary. Justice should be defined not only by process but by product: is the result, measured by the interests of clients and the needs of society, fair? Legal services should exist by right to all citizens, not as a privilege to some. Substantial federal subsidies, supplemented by an excess profits tax on corporate law firms, can make this possible. But the prerequisite to reform is public regulation of the legal profession in the public interest. Otherwise, equal justice under law will remain subservient to unequal justice under lawyers.

Notes

The locations of manuscript collections appear in the Bibliographical Essay. Reference in the Notes indicates location within a divided collection.

INTRODUCTION

1. Geraint Perry, *Political Elites* (New York, 1969), 32-33.
2. Grant Gilmore, "Law, Anarchy and History," University of Chicago Law School *Record*, 14 (Autumn 1966), 2.
3. J. Willard Hurst, *The Growth of American Law: The Law Makers* (Boston, 1950), 4, 6.
4. Morton J. Horwitz, "The Conservative Tradition in the Writing of American Legal History," *American Journal of Legal History*, 17 (July 1973), 281, 283.
5. Hurst, *Growth*, 251, 254-55.

CHAPTER ONE

1. Alexis de Tocqueville, *Democracy in America*, trans. by Henry Reeve (3rd ed., London, 1838), II, 102-12.
2. R. Kent Newmyer, "Daniel Webster as Tocqueville's Lawyer: The *Dartmouth College* Case Again," *American Journal of Legal History*, 11 (1967), 146-47.
3. James M. Sheean, "The Country Lawyer," Illinois Bar Association, *Proc.*, 24 (1900), 194-95; L. V. Hocker, "The Country Lawyer," Missouri Bar Association, *Proc.* (1908), 216; John T. Barker, *Missouri Lawyer* (Phila-

delphia, 1949), 13, 17, 21; Arthur G. Powell, *I Can Go Home Again* (Chapel Hill, N.C., 1943), 58-59, 116, 135; *Law Student's Helper*, 12 (September 1904), 309. These are only samples from a voluminous literature; references to Lincoln as a professional prototype are far too numerous to cite.

4. David Dudley Field, "The Study and Practice of the Law," *Democratic Review*, 14 (1844), 345.

5. Roscoe Pound, "The Administration of Justice in the Modern City," *Harvard Law Review*, 26 (1912-13), 310-11.

6. James Bryce, *The American Commonwealth* (New York, 1888), II, 490-91, 622, 624, 626.

7. Arnold M. Paul, *Conservative Crisis and the Rule of Law: Attitudes of Bar and Bench, 1887-1895* (Ithaca, N.Y., 1960), 2-5, 131-32, 221-29.

8. See David J. Rothman, *The Discovery of the Asylum* (Boston, 1971), 243, 253-54, 257, 261-62, 283, 286, 290-91; Anthony Platt, *The Child-Savers: The Invention of Delinquency* (Chicago, 1969), 41-42, 65-66, 98, 177; Alan P. Grimes, *The Puritan Ethic and Woman Suffrage* (New York, 1967), *passim;* Joseph R. Gusfield, *Symbolic Crusade* (Glencoe, Ill., 1966), *passim;* James Weinstein, *The Corporate Ideal in the Liberal State: 1900-1918* (Boston, 1968), 108-10; Paul S. Boyer, *Purity in Print* (New York, 1968), chs. 1-2; Michael B. Katz, *Class, Bureaucracy, and Schools* (New York, 1971), 115-16.

9. Moses Rischin (ed.), *The American Gospel of Success* (Chicago, 1965), 8-9.

10. *Ibid.*, 9-10.

11. William Miller, "American Lawyers in Business and in Politics," *Yale Law Journal*, 60 (January 1951), 66-76.

12. For an ambivalent statement of the accessibility myth, see Lawrence M. Friedman, *A History of American Law* (New York, 1973), 266-67, 276, 550. See also Rischin, *Gospel of Success*, 10; Christopher Lasch, *The Agony of the American Left* (New York, 1969), 137. For the persistence of this pattern, see Erwin O. Smigel, *The Wall Street Lawyer* (New York, 1964), chs. 3-5.

13. William Miller, "American Historians and the Business Elite," in Miller (ed.), *Men in Business* (New York, 1962), 328.

14. Robert Stevens, "Aging Mistress: The Law School in America," *Change* (January-February 1970), 34.

15. See Robert H. Wiebe, *The Search for Order 1877-1920* (New York, 1967), 111-13, 116-17.

16. Arthur H. Dean, *William Nelson Cromwell, 1854-1948* (New York, 1957), 52.

17. Robert T. Swaine, *The Cravath Firm and Its Predecessors 1819-1947* (2 vols.: New York, 1946), I, 370; David J. Danelski and Joseph S. Tulchin (eds.), *The Autobiographical Notes of Charles Evans Hughes* (Cambridge, Mass., 1973), 88, 89, 96, 114-15.

18. "A Well-Known Lawyer Honored," *American Lawyer*, 9 (March 1901), 105.

19. Swaine, *Cravath Firm*, I, 370, 573-75, 587, 658; II, 3-4, 7, 10. See Walter K. Earle, *Mr. Shearman and Mr. Sterling and How They Grew* (privately printed, 1963), 178; *Davis Polk Wardwell Gardiner & Reed: Some of the Antecedents* (privately printed, 1935), 16-17, 27; Henry W. Taft, *A Century and a Half at the New York Bar* (New York, 1938), 174-75.

20. Richard Ames, "Suggestions From Law School Graduates as to Where and How to Begin Practice," *Harvard Law Review*, 27 (January 1914), 261-62, 265.

21. Elihu Root, "Some Duties of American Lawyers to American Law," *Yale Law Journal*, 14 (December 1904), 64.

22. "What Great Lawyers Receive for Their Work," *American Legal News*, 27 (September 1916), 21-26; "The Legal Profession and Its Opportunities," *Law Student's Helper*, 10 (January 1902), 8.

23. Danelski and Tulchin (eds.), *The Autobiographical Notes of Charles Evans Hughes*, 76.

24. H. Louis Jacobson to Louis Marshall, September 4, 1900, Louis Marshall MSS; Dulles, Foreword to Dean, *Cromwell*, i; Brandeis to Morris Jacob Wessel, May 13, 1912, Brandeis MSS, copy in possession of Professor Melvin Urofsky; Joseph M. Proskauer, *A Segment of My Times* (New York, 1950), 30. Established German and Sephardic Jews had an easier time, as Edward S. Greenbaum discovered when his father placed him in a firm whose partners were Benjamin Cardozo's cousins. See Edward S. Greenbaum, *A Lawyer's Job* (New York, 1967), 28.

25. Samuel Untermeyer to William Armstrong, December 24, 1909, Marshall MSS.

26. Swaine, *Cravath Firm*, I, 554-55.

27. "Graveyard of Reputations," *Law Student's Helper*, 8 (April 1900), 116.

28. Felix Frankfurter to Philip L. Miller, October 31, 1911, Felix Frankfurter MSS, Library of Congress. Frankfurter observed of his New York classmates: "The New York contingent has settled down on the whole to drab uniformity."

29. Elting E. Morison, *Turmoil and Tradition* (New York, 1964), 60, 77; Henry L. Stimson and McGeorge Bundy, *On Active Service in Peace and War* (New York, 1947), 3-7, 17.

30. Frederic C. Howe, *The Confessions of a Reformer* (Chicago, 1967), 199. Newton D. Baker also spurned an offer in 1901 to join a leading Cleveland firm in order to accept a position with Mayor Tom Johnson. Hoyt L. Warner, *Progressivism in Ohio 1897-1917* (Columbus, 1964), 64.

31. William Elder to Stimson, April 13, 1909; Cordenio A. Severance to Stimson, April 22, 1909, Box 40, Stimson MSS.

32. "James M. Beck's Promotion," *Law Student's Helper*, 11 (July 1903), 206.

33. Eugene Wambaugh to Frankfurter, November 10, 1911; Winifred T. Denison to Frankfurter, November 16, 1911, Frankfurter MSS, LC.

34. See Robert Stevens, "Two Cheers for 1870: The American Law School," 432-35, 443n, and Jerold S. Auerbach, "Enmity and Amity: Law Teachers and Practitioners, 1900-1922," in Donald Fleming and Bernard Bailyn, *Law in American History* (Boston, 1971), 551-62.

35. *Felix Frankfurter Reminisces* (New York, 1960), 26-27. There is evidence (from old examination books) which suggests that professors did not then grade "blindly," i.e., without knowledge of the identity of the student whose paper they graded.

36. *Ibid.*, 26.

37. Christopher Jencks and David Riesman, *The Academic Revolution* (New York, 1968), 77.

38. *Ibid.*, 90; Laurence R. Veysey, *The Emergence of the American University* (Chicago, 1965), 440.

39. Hurst, *Growth*, 307, 309.

40. Beck to Louis Marshall, March 29, 1903, Marshall MSS. See Albert Boyden, *Ropes-Gray 1865-1940* (Boston, 1942), 93-97.

41. Untermeyer to Marshall, April 29, 1903, Marshall MSS; Thomas B. Gay, *The Hunton Williams Firm and Its Predecessors 1877-1954* (Richmond, Va., 1971), I, 52, 54.

42. Otto Kirchner, "Starting in Law," *Law Student's Helper*, 10 (July 1902), 196.

43. "The Practice of Law in Large Cities," *Law Student's Helper*, 10 (January 1902), 4. See also "Fighting His Way to Success in Law," *Law Student's Helper*, 13 (April 1905), 126-29; James G. Rogers to Arthur Corbishley, August 8, 1911, Rogers MSS.

44. Charles F. Chamberlayne, "The Soul of the Profession," *Green Bag*, 18 (July 1906), 397. See also Howard Leslie Smith, "The Lawyer's Vocation," Illinois State Bar Association, *Proc.*, 24 (1900), 186-93. Smith declared (p. 189): "Some metropolitan law offices affect an organization like a department store and borrow much of its machinery. . . ."

45. Thorstein Veblen, *The Theory of the Leisure Class* (New York, 1934 ed.), 231.

46. Herbert Croly, *The Promise of American Life* (Cambridge, Mass., 1965 ed.), 134-46.

47. Theodore Roosevelt, *Presidential Addresses and State Papers May 10, 1905, to April 12, 1906* (New York, 1910), IV, 419-20.

48. John R. Dos Passos, *The American Lawyer* (New York, 1907), 25.

49. *Ibid.*, 25, 33, 34.

50. Theron G. Strong, *Landmarks of a Lawyer's Lifetime* (New York, 1914), 347, 354, 377-78.

51. James Hamilton Lewis, "The End of Lawyers," *American Lawyer*, 13 (March 1905), 115-17.

52. Edward M. Shepard, "Annual Address," New Hampshire Bar Association, *Proc.*, 2 (1906), 273, 276-77. Other lawyers agreed. See J. Aspinwall Hodge, "The Bench and the Bar in Their Relation to the People and the Corporations," *Albany Law Journal*, 69 (February 1907), 56.

53. Harlan F. Stone, *Law and Its Administration* (New York, 1915), 165-66, 176.

54. Woodrow Wilson, "The Lawyer and the Community," *North American Review*, 192 (November 1910), 609-10, 614, 619, 620. In an address the following year Wilson declared that "the country is full of men . . . whose practice is so specialized that they have become parts of the great business machine and are no longer parts of a great profession." Wilson, "The Lawyer in Politics," Kentucky State Bar Association, *Proc.*, 10 (1911), 111.

55. Louis D. Brandeis, "The Opportunity in the Law," *American Law Review*, 39 (July-August 1905), 559-61.

56. Champ S. Andrews, "The Law—A Business or a Profession?" *Yale Law Journal*, 17 (June 1908), 609-10.

57. James B. Dill, "The Business Lawyer of To-Day," *Albany Law Journal*, 65 (April 1903), 112; "Specialism in the Law," *Central Law Journal*, 53 (September 27, 1901), 255.

58. For the criticism see, e.g., "The Law as a Business Versus True Advocacy," *Central Law Journal*, 57 (October 16, 1903), 301.

59. Henry Wollman, "The Era of the Commercial Lawyer," *Green Bag*, 25 (September 1913), 383-84.

60. Levy Mayer, "Has Commercialism Impaired the Lawyer of To-Day?" Illinois Bar Association, *Proc.*, 24 (1900), 102, 103, 105.

61. Charles W. Needham, "The Commerce Lawyer," *Case and Comment*, 20 (September 1913), 247-48.

62. Henry W. Sackett, "The Modern Lawyer's Test," offprint of Address at Cornell University College of Law, February 26, 1915.

63. Edward P. White, "Changed Conditions in the Practice of the Law," New York State Bar Association, *Reports*, 27 (1904), 115, 120, 121; *American Lawyer*, 12 (February 1904), 52-53. See also Lloyd W. Bowers, "The Lawyer of To-Day," *American Law Review*, 38 (November-December 1904), 823-35; Alfred Hemenway, "The American Lawyer," *American Law Review*, 39 (September-October 1905), 641-57; Peter A. Laubie, "Presidential Address," Ohio State Bar Association, *Proc.*, 21 (1900), 97-142; Eugene B. Gary, "The Lawyer," South Carolina Bar Association, *Transactions*, 14 (1907), 83-94.

64. "Words of Advice and Encouragement to the Graduating Classes of 1914 by the Deans of Ten Law Schools," *Case and Comment*, 20 (May 1914), 896.

65. Julius Henry Cohen, *The Law: Business or Profession?* (New York, 1916), 211-12.

66. William Miller, "The Business Elite in Business Bureaucracies," in Miller (ed.), *Men in Business: Essays on the Historical Role of the Entrepreneur* (New York, 1962), 288. Public servants with corporate law firm experience systematized government recruitment practices to conform to the firm model. See Stimson to James Barr Ames, June 30, 1906, Stimson MSS.

CHAPTER TWO

1. ABA *Reports*, 28 (1905), 383-84. Tucker's personal papers are silent on the issue of legal ethics, but they do suggest that Tucker and Roosevelt had been in political opposition since the 1904 presidential campaign, when the Virginia Democrat had campaigned for Alton Parker. See Tucker to William Elliott, February 23, 1905, Folder 173, Tucker MSS.

2. Report of Committee on Code of Professional Ethics, ABA *Reports*, 29 (1906), 600-603.

3. ABA *Reports*, 30 (1907), 678, 680; George Sharswood, *An Essay on Professional Ethics* in ABA *Reports*, 32 (1907), 55, 132, 147-48, 168-69. See Joseph Katz, "The Legal Profession, 1890-1915" (M.A. thesis, Columbia University, 1953).

4. See Barlow F. Christensen, *Lawyers for People of Moderate Means* (Chicago, 1970), 129-30.

5. James F. Brennan, "The Bugaboo 'Ambulance Chasing,' " *California State Bar Journal*, 6 (February 1931), 37-42; John H. Wilbur, "Advertising, Solicitation and Legal Ethics," *Vanderbilt Law Review*, 7 (June 1954), 677-94; "A Critical Analysis of Rules Against Solicitation By Lawyers," *University of Chicago Law Review*, 25 (Summer 1958), 674-85; Philip Schuchman, "Ethics and Legal Ethics," *George Washington Law Review*, 37 (1968-69), 244-69.

6. Crystal Eastman, *Work-Accidents and the Law* (New York, 1910), 119-24, 152. For inadequacies of compensation elsewhere, and for a perceptive overview of the changing law of industrial accidents, see Lawrence M. Friedman and Jack Ladinsky, "Social Change and the Law of Industrial Accidents," *Columbia Law Review*, 67 (1967), 50-82.

7. Moorfield Storey to (?), March 24, 1896, Storey MSS, Massachusetts Historical Society.

8. *Stanton v. Embry*, 93 U.S. 548 (1877).

9. Sharswood, *Professional Ethics*, 160-61. For later expressions of hostility see Cohen, *The Law: Business or Profession?*, 209; Carter, *Ethics*, 57.

10. Joseph H. Choate, "The English Bar," New York State Bar Association, *Proc.*, 21 (1906), 72. See Moorfield Storey to George T. Page, September 16, 1918, Storey MSS, Massachusetts Historical Society.

11. ABA *Reports*, 33 (1908), 61, 63-74.

12. *Ibid.*, 76-80. Italics added.

13. Jessup, *A Study of Legal Ethics*, 168-69. See also New York State Bar Association, *Proc.*, 23 (1908), 99-102, 203; Christensen, *Lawyers for People of Moderate Means*, 144-45. During the 1950's and 1960's, civil rights and poverty lawyers would encounter similar charges—which, again, had less to do with stirring litigation *per se* than with who stirred what for whom.

14. See Benjamin R. Twiss, *Lawyers and the Constitution* (Princeton, 1942) 129-40; Paul, *Conservative Crisis and the Rule of Law*, 15-18.

15. James W. Bollingen, "Upward Tendencies in our Proposed Reforms," Iowa State Bar Association, *Proc.*, 15 (1909), 46; Stiles Burr, "Extortionate Fees," *ABAJ*, 3 (January 1917), 54; New York State Bar Association, *Reports*, 31 (1908), 105-7, 120-23.

16. F. C. Wilkinson, "The Contingent Fee," *Central Law Journal*, 72 (1911), 339-40; Robert H. Patton, "Contingent Fees," *Chicago Legal News* (1904), 127.

17. Reginald Heber Smith, *Justice and the Poor* (New York, 1919), 86; see "Elimination of the Evils of Contingent Fees," *ABAJ*, 5 (January 1919), 61-75; Ashley Cockrill, "The Shyster Lawyer," *Yale Law Journal*, 21 (March 1912), 383-90.

18. Isidor J. Kresel, "Ambulance Chasing, Its Evils and Remedies Therefor," New York State Bar Association, *Proc.*, 52 (1929), 337-39. Kresel was himself a successful Jewish immigrant lawyer who described America as "my passion and my religion." He wrote an unpublished autobiography ("to prove that America is still a land of opportunity") filled with invective against radicals and immigrants less assimilated than he was. Isidor J. Kresel, Unpublished Autobiography (1955), loaned to the author by Professor Stephen Botein. See also Supreme Court of the State of New York, *Report to Appellate Division, First Judicial Department* by Mr. Justice Wasservogel (September 26, 1928), 4, 11, 15, 21.

19. Max Radin, "Maintenance By Champerty," *California Law Review*, 24 (1935-36), 71-72, 74. See Barbara C. Steidle, " 'Reasonable' Reform: The Attitude of Bar and Bench Toward Liability Law and Workmen's Compensation," in Jerry Israel (ed.), *Building the Organizational Society* (New York, 1972), 31-32.

20. Radin, "Maintenance By Champerty," 75; Radin, "Contingent Fees in California," *California Law Review*, 28 (1939-40), 588-89, 598.

21. Jerome E. Carlin, *Lawyers' Ethics* (New York, 1966), 6-7; Schuchman, "Ethics and Legal Ethics," 245-46, 254-55, 258, 266.

22. Carlin, *Lawyers' Ethics*, 177. See also Howard S. Becker, *Outsiders: Studies in the Sociology of Deviance* (New York, 1966), 9, 163.

23. ABA *Reports*, 33 (1908), 58.

24. Final Report of the Committee on Code of Professional Ethics, ABA *Reports*, 33 (1908), 584.

25. David J. Brewer, "The Ideal Lawyer," *Atlantic Monthly*, 98 (November 1906), 596.

26. Comments from *NY Globe and Commercial Advertiser*, Boston *Pilot*, Chicago *Tribune*, and Philadelphia *Press* in "Editorial Comment on the Proposed American Bar Association Canons of Ethics," *Albany Law Journal*, 70 (September 1908), 266, 268, 271, 272. See the typical words of praise in "Professional Ethics," *Outlook*, 89 (June 27, 1908), 408-9; "Codes of Professional Ethics," *Nation*, 83 (November 8, 1906), 388. The *Nation* article called attention to the efforts of other professional groups to formulate ethical codes, commenting: "It is with voluntary acts as these that the defender of American business and professional life can best reply to the critic, whether native or foreign, who would have it that our ideals are perishing in an age of materialism."

27. "Lawyers' Codes of Ethics," *Bench and Bar*, 13 (May 1908), 44.

28. New York State Bar Association, *Proc.*, 3 (1879), 67.

29. Geoffrey C. Hazard, Jr., Foreword, Carlin, *Lawyers' Ethics*, xxiii.

30. William Howard Taft, Preface, Reginald Heber Smith and John S. Bradway, *Growth of Legal Aid Work in the United States*, U.S. Department of Labor, Bureau of Labor Statistics, No. 398 (1926), iv; John S. Bradway, "Legal Aid Clinics and the Bar," *Southern California Law Review*, 3 (1930), 384; Albert F. Bigelow, *Twenty-Five Years of Legal Aid In Boston 1900-1925* (Boston, 1926), 5; Harrison Tweed, *The Legal Aid Society, New York City: 1876-1951* (New York, 1954), 5.

31. Smith, *Justice and the Poor*, 134-39; John M. Maguire, *The Lance of Justice: A Semi-Centennial History of the Legal Aid Society 1876-1926* (Cambridge, Mass., 1928), 19; Emery A. Brownell, *Legal Aid in the United States* (Rochester, N.Y., 1951), 7-8; Financial Statement (1908), Arthur von Briesen MSS, Box 21; Mauro Cappelletti, "Legal Aid: The Emergence of a Modern Theme," *Stanford Law Review*, 24 (January 1972), 357.

32. Maguire, *Lance of Justice*, 50-53; Tweed, *Legal Aid Society*, 7-8.

33. See John G. Sproat, *"The Best Men": Liberal Reformers in the Gilded Age* (New York, 1968); G. Edward White, "The Social Values of the Progressives: Some New Perspectives," *South Atlantic Quarterly*, 70 (Winter 1971), 62-76.

34. Von Briesen quoted in "Legal Aid for the Poor," *Annals*, 17 (January 1901), 165-66; Maguire, *Lance of Justice*, 55. Von Briesen's commitment to

legal aid as an instrument of social stability was entirely consistent with his posture toward other public policy issues of the Progressive era. He opposed women's suffrage; he believed that the New Freedom of Woodrow Wilson was destructive of business enterprise; he referred to Emma Goldman and other radicals as "abnormal" and "degenerated." See von Briesen to Lucy L. Schroder, July 30, 1913, Box 8; von Briesen to Leonard McGee, July 9, 1914, Box 10; von Briesen to William Randolph Hearst, February 10, 1919, Box 17, von Briesen MSS.

35. Joseph W. Errant, "Justice for the Friendless and the Poor" (Chicago Legal Aid Society, 1888), 3, 6, 8, 12.

36. Abbott quoted in Brownell, *Legal Aid,* xiii; Roosevelt quoted in Tweed, *Legal Aid,* 10.

37. Von Briesen to Editor, *New York Times,* June 26, 1911, Box 4; von Briesen to Mrs. William E. Boyes, May 19, 1911, Box 4; von Briesen to National Employment Exchange, October 14, 1910, Box 22; von Briesen to Grace Hamilton (1919), Box 17, von Briesen MSS; von Briesen, "Legal Aid Society," Box 20, *ibid.*

38. Von Briesen to Eleanor E. Nichols, November 15, 1913, Box 8, *ibid.;* Reginald Heber Smith to Louis W. Robey, March 22, 1916, Smith MSS, National Legal Aid and Defender Association MSS; Smith and Bradway, *Growth of Legal Aid,* 77; Waddill Catchings, "The Work of the New York Legal Aid Society," *Green Bag,* 15 (July 1903), 313-14; Smith, *Justice and the Poor,* 156-57.

39. Cappelletti, "Legal Aid," 358-59.

40. Smith and Bradway, *Growth of Legal Aid,* 84-85; Maguire, *Lance of Justice,* 72; Smith, *Justice and the Poor,* 174, 238; Smith, "A Lawyers' Legal Aid Society," *Case and Comment,* 23 (May 1917), 1008-9; Legal Aid Society of Chicago, *Ninth Annual Report* (1913), 11-12.

41. Von Briesen to Oswald Garrison Villard, March 8, 1905, Box 2; von Briesen to President, Knickerbocker Trust Co., November 10, 1909, Box 21; von Briesen to William J. Schieffelin, September 6, 1918, Box 16, von Briesen MSS.

42. 32nd Annual Report of the President of the Legal Aid Society (1907), Box 21, *ibid.;* von Briesen, "Legal Aid Society."

43. Louis Stoiber to Professor Racca, December 25, 1912, Box 7, von Briesen MSS.

44. Smith and Bradway, *Growth of Legal Aid,* 74.

45. Smith, *Justice and the Poor,* 17-34, 37, 182; Smith, "Justice and the Poor," prepared for Dr. Pritchett, August 12, 1920, Smith MSS.

46. *Ibid.;* Smith, *Justice and the Poor,* 6-7, 8, 11, 217.

47. *Ibid.,* ix.

48. Charles Evans Hughes, "Legal Aid Societies, Their Function and Necessity," ABA *Reports,* 43 (1920), 227, 231-32, 235.

49. Maguire, *Lance of Justice*, viii.

50. Smith to T. R. Williamson, December 9, 1921; Smith to Taft, August 2, 1922; Smith to Edgar B. Tolman, April 27, 1926, Smith MSS.

51. See Wiebe, *The Search for Order*, ch. 5; Hurst, *Growth of American Law*, 285-86; George Martin, *Causes and Conflicts: The Centennial History of the Association of the Bar of the City of New York, 1870-1970* (Boston, 1970); Edson R. Sunderland, *History of the American Bar Association and Its Work* (n.p., 1953), 4-6, 38.

52. Baldwin to Hampton Carson, November 13, 1919, Box 178, Baldwin MSS.

53. Lowell S. Nicholson, *The Organized Bar in Massachusetts*, Prepared for the Survey of the Legal Profession (1952), 17; George W. Gale, *The Organized Bar in Chicago*, Survey of the Legal Profession Report (1950), 303, 310; Martin, *Causes and Conflicts*, 179.

54. James G. Rogers, *American Bar Leaders* (Chicago, 1932), iv, vii, *passim*.

55. Peter W. Meldrum, "Address of the President," *ABAJ*, 1 (October 1915), 480-81; Charlton G. Ogburn, "The Lawyer and Democracy," *American Law Review*, 49 (September-October 1915), 739-43.

56. Stephen S. Gregory, "The American Bar Association," *Green Bag*, 24 (June 1912), 281. See also Henry D. Ashley, "The Effect on American Jurisprudence of the Doctrine of Judicial Precedent," *American Law Review*, 39 (September-October 1905), 704.

57. Lawrence M. Friedman, "Law Reform in Historical Perspective," *St. Louis University Law Journal*, 13 (Spring 1969), 356, 358, 370-71.

58. ABA *Reports*, 37 (1912), 93, 95.

59. Wickersham to Oswald G. Villard, March 9, 1912, Box C-192, NAACP MSS; George Whitelock to Wickersham, February 8, 1912, April 6, 1912, *ibid*.

60. Storey to Charles F. Libby, March 1, 1912; Storey to Stephen S. Gregory, April 13, 1912, Storey MSS, Massachusetts Historical Society.

61. Charles J. O'Connor to Villard, June 13, 1912, *ibid.*; ABA *Reports*, 37 (1912), 12-16; Henry St. George Tucker to William Howard Taft, November 22, 1913, Folder 331, Tucker MSS; Storey to James C. Crosby, October 24, 1919; Storey to Gregory, October 29, 1914, Storey MSS, MHS. For comments favorable to exclusion see "The American Bar Association and the Attorney General," *Bench and Bar*, 28 (March 1912), 90-92; *American Law Review*, 46 (March-April 1912), 309-10.

62. Alpheus T. Mason, *Brandeis: A Free Man's Life* (New York, 1946), 506.

63. See Edward F. McClennen, "Louis D. Brandeis as a Lawyer," *Massachusetts Law Quarterly*, 33 (September 1948), 10-11, 15-16; Brandeis to

Charles W. Clifford, July 11, 1912, Brandeis MSS, copy in possession of Professor Melvin Urofsky.

64. Arthur D. Hill to George W. Anderson, February 24, 1916, in U.S. Congress, Senate, *Hearings . . . on the Nomination of Louis D. Brandeis . . .* , 64th Cong., 1st Sess., Document No. 409 (1916), I, 620. The most extended treatments of the confirmation struggle are in Mason, *Brandeis*, chs. 30-31; A. L. Todd, *Justice on Trial: The Case of Louis D. Brandeis* (New York, 1964).

65. Allison Dunham and Philip B. Kurland (eds.), *Mr. Justice* (Chicago, 1964), 177.

66. Brandeis to Henry Morgenthau, Sr., December 5, 1906, in Melvin I. Urofsky and David W. Levy (eds.), *Letters of Louis D. Brandeis* (New York, 1971), I, 507.

67. Brandeis to Charles Nagel, July 12, 1879, *ibid.*, 37.

68. Brandeis to Warren, May 30, 1879, *ibid.*, 35.

69. Brandeis to William H. Dunbar, August 19, 1896, *ibid.*, 125-26.

70. Brandeis to Benjamin Lindsey, April 25, 1912; Brandeis to Charles W. Clifford, July 11, 1912, *ibid.*, II, 609, 647.

71. Brandeis, "The Opportunity in the Law," *American Law Review*, 555, 557-62.

72. Quoted in Mason, *Brandeis*, 105-6.

73. Brandeis *Hearings*, I, 271, 611; "Brief on Behalf of the Opposition," Brandeis *Hearings*, II, 5-6, 14.

74. *Ibid.*, 6-7. George Wickersham, who had fought Brandeis in the Ballinger-Pinchot conservation dispute, gathered the signatures. According to Todd, nine refused to sign. Stephen S. Gregory, ABA president from 1911 to 1912, testified regarding Brandeis' reputation: "I think it is excellent as a lawyer of ability and character." Brandeis *Hearings*, I, 711; Todd, *Justice on Trial*, 159-62. See Storey to R. C. Bolling, February 2, 1916, Storey MSS, MHS; Irving Katz, "Henry Lee Higginson vs. Louis Dembitz Brandeis: A Collision Between Tradition and Reform," *New England Quarterly*, 41 (March 1968), 77. Joel B. Grossman cites the ABA response to the nomination as another "defense of the dominant ideological values" held by ABA leaders at a time when the association "was still primarily a closely knit group of successful lawyers." *Lawyers and Judges: The ABA and the Politics of Judicial Selection* (New York, 1965), 54-55; see also Joseph P. Harris, *The Advice and Consent of the Senate* (Berkeley, Calif., 1953), ch. 7.

75. Taft to Gus Karger, January 31, 1916; Taft to Henry W. Taft, January 31, 1916, Vol. 43; Taft to Wickersham, February 7, 1916, Vol. 44, Letterbooks, Taft MSS.

76. Storey to John Sharp Williams, March 1, 1916; Storey to William A. Ketchum, April 14, 1916, Storey MSS, MHS.

77. See Richard M. Abrams, "Brandeis and the Ascendancy of Corporate Capitalism," Intro. to Louis D. Brandeis, *Other People's Money and How the Bankers Use It* (New York, 1967), xxi-xxiii.

CHAPTER THREE

1. Robert Stevens, "Two Cheers for 1870: The American Law School," in Fleming and Bailyn (eds.), *Law in American History*, 405-548.
2. Quoted in Arthur E. Sutherland, *The Law at Harvard: A History of Ideas and Men, 1817-1967* (Cambridge, Mass., 1967), 184. See Franklin G. Fessenden, "The Rebirth of the Harvard Law School," *Harvard Law Review*, 33 (February 1920), 511.
3. Hurst, *Growth*, 260-65; William R. Vance, "The Ultimate Function of the Teacher of Law," Association of American Law Schools, *Proceedings* (1911), 28-43; *The Centennial History of the Harvard Law School 1817-1917* (Cambridge, Mass., 1918), 70-71. In 1898, a Detroit attorney on the Michigan faculty resigned, stating, "I find that the work of teacher in the Law Department in addition to my work as a practicing lawyer, is more than I can properly do." Quoted in Elizabeth G. Brown, *Legal Education at Michigan, 1889-1959* (Ann Arbor, 1959), 80. In 1920 the Committee on the Status of the Law Teacher of the AALS described the transition from practitioner-teachers to full-time professors as "nothing less than revolutionary." AALS, *Proc.* (1920), 166.
4. James Barr Ames, "The Vocation of the Law Professor," in *Lectures on Legal History and Miscellaneous Legal Essays* (Cambridge, Mass., 1913), 360ff. Ames delivered this address in 1901 at the opening of a new law building at the University of Pennsylvania.
5. Quoted in Sutherland, *The Law at Harvard*, 175.
6. James Bradley Thayer, "Address," ABA *Reports*, 18 (1895), 415-16.
7. Ames, "Vocation," 360. For figures on law school enrollments, see American Bar Foundation, Research Memorandum Series, No. 15: *Compilation of Published Statistics on Law School Enrollments and Admissions to the Bar, 1889-1957* (preliminary draft, 1958).
8. Ames, "Vocation," 361-62.
9. For an earlier expression of instrumentalism, see Morton J. Horwitz, "The Emergence of an Instrumental Conception of American Law, 1780-1820," in Fleming and Bailyn (eds.), *Law in American History*, 287-326.
10. Roscoe Pound, "The Need of a Sociological Jurisprudence," *Green Bag*, 19 (October 1907), 610-15. For parallel developments in England, see Reba N. Soffer, "The Revolution in English Social Thought, 1880-1914," *American Historical Review*, 75 (December 1970), 1938, 1941, 1962.

11. Floyd R. Mechem, "The Opportunities and Responsibilities of American Law Schools," ABA *Reports*, 30 (1906), Pt. 2, 177, 181.

12. *Felix Frankfurter Reminisces*, 19.

13. Felix Frankfurter to Henry L. Stimson, June 26, 1913, Box 101, Stimson MSS.

14. Frankfurter to Stimson, July 7, 1913, Box 102, *ibid*.

15. Felix Frankfurter, "The Law and the Law Schools," *ABAJ*, 1 (October 1915), 534, 538-39. On Frankfurter's appointment to Harvard, see "A Young Teacher of Live Law," *Independent*, 81 (March 22, 1915), 419. The scientific metaphor remained a favorite of Frankfurter's. To a law school colleague he wrote: "For those of us who see and feel that the only solution to the present rut of things is the application of the scientific spirit to our social problems, the task is temperate diagnosis of social difficulties, [and] the persistent . . . gathering of data as a basis of formulating and encouraging change." Frankfurter to Zechariah Chafee, Jr., January 13, 1918, Box 4, File 10, Chafee MSS.

16. Richard Hofstadter, *The Age of Reform* (New York, 1956), 158.

17. Calvin Woodard, "The Limits of Legal Realism: An Historical Perspective," *University of Virginia Law Review*, 54 (May 1968), 710-11.

18. Brian Abel-Smith and Robert Stevens, *In Search of Justice* (London, 1968), 337; Abel-Smith and Stevens, *Lawyers and the Courts* (Cambridge, Eng., 1967), 165-68.

19. A. V. Dicey, "The Teaching of English Law at Harvard," *Harvard Law Review*, 13 (1900), 3; Ames to William R. Harper, March 31, 1902; Joseph H. Beale to Harper, April 2, 1902; Beale to Ernst Freund, April 7, 1902, quoted in Edwin B. Firmage, "Ernst Freund—Pioneer of Administrative Law," *University of Chicago Law Review*, 29 (1962), 764, 766-68; Stevens, "Two Cheers for 1870," 436-37, 446-47.

20. This generalization is based upon my examination of the following: ABA *Reports* (1893-1918); *American Law School Review* (1902-18); AALS *Proceedings* (1901-18); the *Proceedings* of the bar associations of New York, Pennsylvania, Ohio, Illinois, Wisconsin and California (1900-18); and a sampling of legal periodicals, including *American Law Review* (1900-18); *Green Bag* (1900-14); *Bench and Bar* (1905-20); *Central Law Journal* (1900-18); and *Law Student's Helper* (1900-15).

21. AALS, *Proc.* (1906), 9-11.

22. William Draper Lewis, "Legal Education and the Failure of the Bar to Perform its Public Duties," AALS, *Proc.* (1906), 32-49.

23. One listener declared emphatically: "The cradle of American jurisprudence, the cradle of American law, is being rocked today by professors of law in the schools of America." AALS, *Proc.* (1906), 5.

24. Mechem, "Opportunities," 168. See also George W. Kirchwey, "American Law and the American Law School," AALS, *Proc.* (1908), 10-24.

25. Louis D. Brandeis to Chafee, June 5, 1921, Box 4, File 10, Chafee MSS.

26. Pound, "Sociological Jurisprudence," 611-12; Pound, "Law in Books and Law in Action," *ALR,* 44 (January-February 1910), 12-36.

27. Vance, "Ultimate Function," *ALSR,* 3 (Fall 1911), 2, 4-7.

28. Henry M. Bates, "Address of the President," AALS *Proc.* (1913), 29; Frankfurter, "The Law and the Law Schools," 371; Samuel Williston, "The Necessity of Idealism in Teaching Law," ABA *Reports,* 33 (1908), 782, 784, 787, 793. Such optimism was not confined to law teachers. Louis B. Wehle, a 1904 graduate of Harvard Law School (and Brandeis' nephew), spoke of the potential of the law school as "a mighty agent of progress." Wehle, "Social Justice and Legal Education," *ALR,* 51 (January-February 1917), 3.

29. Bates, "Address of the President," 30; Frankfurter to Learned Hand, June 28, 1913, Learned Hand MSS; Erwin A. Meyers to Roscoe Pound, May 18, 1914, Box 28, Roscoe Pound MSS.

30. Armistead M. Dobie, "Reminiscences of a Teacher of Law," Virginia State Bar Association, *Reports,* 44 (1932), 316. As Dean Kirchwey of Columbia told a young protégé: "When you turned your back on the practice of law you turned your back also on prospects of wealth and other worldly rewards . . . If you don't believe that . . . I don't know why you abandoned Wall Street for the arduous path of the academic life." George W. Kirchwey to William Underhill Moore, April 27, 1910, William Underhill Moore MSS.

31. Pound to Ezra R. Thayer, August 26, 1915, Box 30, Pound MSS.

32. See G. Edward White, "The Social Values of the Progressives: Some New Perspectives," *South Atlantic Quarterly,* 70 (Winter 1971), 68-69, 72-73; Brandeis to William Draper Lewis, November 13, 1912; Brandeis to Norman Hapgood, July 30, 1912, Brandeis MSS, copies in possession of Professor Melvin Urofsky.

33. Professor Charles Rosenberg offered this point in his critique of my paper, "Twentieth-Century American Lawyers," American Historical Association (1970).

34. Harlan F. Stone, "The Importance of Actual Experience at the Bar as a Preparation for Law Teaching," *ALSR,* 3 (Fall 1912), 208-9; Clarence D. Ashley, "Legal Education and Preparation Therefor," *Albany Law Journal,* 63 (October 1901), 392.

35. Ames, "Vocation," 137. See Williston to Jerome Frank, June 14, 1933, Box 3, Jerome Frank MSS; Williston, *Life and Law: An Autobiography* (Boston, 1940), 269; Foundation for Research in Legal History, *A History of the School of Law, Columbia University* (New York, 1955), 263; Frederick C. Hicks, *Yale Law School: 1895-1915,* Yale Law Library Publications No. 7 (New Haven, 1938), 45; Taft to James Bradley Thayer, May 4, 1896, May 15, 1896; Wigmore to Thayer, December 30, 1894, Thayer MSS.

36. Harry S. Richards, "Progress in Legal Education," AALS *Proc.* (1915),

69; William Draper Lewis, "The Law Teaching Branch of the Profession," *ALSR*, 5 (March 1925), 447. See Josef Redlich, *The Common Law and the Case Method in American University Law Schools* (New York, 1915), 5. There were, of course, those who preferred to retain ties between teaching and practice, in part perhaps as a check on the reform propensities of the teachers. See Stone, "The Importance of Actual Experience," ABA *Reports*, 37 (1912), 747-60, with comment by Samuel Williston (p. 712); J. Newton Fiero, "Teaching Law Without Experience at the Bar," *ALSR*, 3 (Fall 1914), 558-60; Edward H. Warren, *Spartan Education* (Boston, 1942), 13.

37. Harry S. Richards, "Neglected Phases of Legal Education," ABA *Reports*, 34 (1909), 777-79; ABA *Reports*, 16 (1893), 367-68; Edson R. Sunderland, *History of the American Bar Association and Its Work* (n.p., 1953), 28-33.

38. Harlan F. Stone, "Address of the President," AALS, *Proc.*, 17 (1917), 108; Warren A. Seavey, "The Association of American Law Schools in Retrospect," *Journal of Legal Education*, 3 (Winter 1950), 153-57; ABA *Reports*, 22 (1899), 565; ABA *Reports*, 23 (1900), 447; AALS, "Articles of Association" (adopted August 28, 1900); Edson R. Sunderland, "The Law Schools and the Legal Profession," *Tulane Law Review*, 5 (April 1931), 337.

39. Harlan F. Stone, "Address of the President," *ALSR*, 4 (April 1920), 483. Between 1901 and 1916 the number of students in AALS schools increased by 24.9 percent, but the number in non-member schools increased by more than 100 percent. ABA *Reports*, 23 (1900), 448-52.

40. George L. Reinhard, "American Law Schools and the Teaching of Law," *Green Bag*, 16 (March 1901), 169; W. Jethro Brown, "The American Law School," *Law Quarterly Review*, 21 (January 1905), 71.

41. ABA *Reports*, 27 (1904), 502; James Barr Ames, "Chairman's Address," Section of Legal Education, ABA *Reports*, 27 (1904), 507; Albert J. Harno, *Legal Education in the United States* (San Francisco, 1953), 98-99. For a general survey, see Alfred Z. Reed, *Present-Day Law Schools in the United States and Canada* (New York, 1928), 21-35.

42. Henry M. Bates to Roscoe Pound, April 24, 1915, Pound MSS; Walter Wheeler Cook, "The Improvement of Legal Education and Standards for Admission to the Bar," ABA *Reports*, 42 (1917), 556-57; Eugene A. Gilmore, "Address of the President," AALS, *Proc.* (1920), 40-43; AALS *Proc.* (1926), 30-31. Richards had previously proposed abolition of the ABA's Section of Legal Education and its replacement by the AALS. "It may seem unnatural for the child of the Section . . . to demand the death of its parent, but the efficiency expert cannot allow himself the luxury of the softer emotions." Harry S. Richards, "Progress in Legal Education," AALS *Proc.* (1915), 61-62.

43. Thomas F. Bergin, "The Law Teacher: A Man Divided Against Himself," *Virginia Law Review*, 54 (May 1968), 638-39.

44. ABA *Reports,* 18 (1895), 370; John F. Dillon, "The True Professional Ideal," ABA *Reports,* 17 (1894), 413-14. See also George Wharton Pepper, "This Profession of Ours," State Bar of California, *Proc.* (1936), 102. Pepper referred to the "amiable contempt" directed against teachers by turn-of-the-century practitioners. "Practicing Lawyers as Law School Lecturers," *Law Notes,* 14 (February 1911), 203; D. D. Murphy, "The Law School and its Duty to the State," Iowa Bar Association, *Proc.* (1913), 94-95; Harry S. Richards, "Neglected Phases of Legal Education," ABA *Reports,* 34 (1909), 781; AALS, *Proc.* (1911), 5; Walter Probert and Louis M. Brown, "Theories and Practices in the Legal Profession," *University of Florida Law Review,* 19 (Winter 1966-67), 477-85.

45. ABA *Reports,* 37 (1912), 712.

46. Vance, "Ultimate Function," *ALSR,* 3 (Fall 1911), 7.

47. Thomas Reed Powell, "Law as a Cultural Study," ABA *Reports,* 40 (1917), 575, 578, 582; Arthur D. Hill to Frankfurter, May 5, 1932, Frankfurter MSS, LC.

48. Guthrie to Stone, January 14, 1915, November 29, 1916, in *A History of the School of Law, Columbia University,* 477, n. 77. There is, however, evidence to suggest that the school was to some extent dependent on the good will of Wall Street lawyers vis-à-vis faculty appointments. See William Underhill Moore to Walter Wheeler Cook, February 21, 1917, Moore MSS.

49. Taft to Charles Warren Fairbanks, March 7, 1913, Vol. 1; Taft to J. D. Brannan, April 4, 1913, Vol. 2; Taft to Elihu Root, May 1, 1913, Vol. 3, Letterbook Series 8, Taft MSS.

50. Frankfurter to Chafee, October 9, 1915, Box 4, File 10, Chafee MSS.

51. Cuthbert W. Pound, "Legal Education and the Education of the Lawyer," New York State Bar Association, *Proc.* (1930), 231-32.

52. Elihu Root, "Address," New York State Bar Association, *Reports,* 39 (1916), 479. Louis Marshall accurately perceived the anti-Semitic cast of Root's address. See Marshall to Editor, *New York Times,* January 17, 1916; Marshall to Isadore M. Levy, January 22, 1916, Marshall MSS.

53. Henry Wade Rogers, "Legal Education in the United States," *ALSR,* 1 (Fall 1902), 14; Edwin G. Dexter, "The Educational Status of the Legal Profession," *Green Bag,* 15 (May 1903), 217; Report of the Committee on Legal Education and Admission to the Bar, ABA *Reports,* 24 (1901), 399-401; Eldon R. James, "The Law School and the Practicing Lawyer," ABA *Reports,* 39 (1916), 678.

54. Henry Wade Rogers, "Address," ABA *Reports,* 17 (1894), 394.

55. *Ibid.,* 397; ABA *Reports,* 22 (1899), 531; AALS, *Proc.,* 9 (1909), 11; Hurst, *Growth,* 273.

56. Department of Commerce and Labor, Bureau of the Census, *12th Census of the United States,* Special Reports, *Occupations* (Washington, D.C., 1904); *13th Census of the United States,* Vol. IV, *Population* (Wash-

ington, D.C., 1914); *14th Census of the United States,* Vol. IV, *Population,* 1920 (Washington, D.C., 1923).

57. ABA *Reports,* 23 (1900), 424 ff.

58. AALS, *Proc.,* 3 (1903), 7.

59. AALS, *Proc.,* 4 (1904), 12; AALS, *Proc.,* 6 (1906), 107; ABA *Reports,* 36 (1911), 660-61.

60. Edward T. Lee, "The Evening Law School," *Chicago Legal News,* 37 (May 27, 1905), 332; Paul L. Martin, "Night Law Schools," *ALSR,* 3 (Winter 1914), 454. For an updated version of the argument, see Lon Fuller, *Legal Education in Pennsylvania* (n.p., 1951), 55. Night schools, Fuller maintained, keep open "the gateways of opportunity" and provide "an inspiring example of sheer American grit." For the history of night law schools, see Joseph T. Tinnelly, *Part-Time Legal Education: A Study of the Problems of Evening Law Schools* (Brooklyn, N.Y., 1957), 5-17.

61. F. M. Finch, "Legal Education," *Columbia Law Review,* 1 (February 1901), 102-3; Brown, *Legal Education,* 276; Franklin M. Danaher, "Some Suggestions for Standard Rules for Admission to the Bar," ABA *Reports,* 34 (1909), 785; Bates, "Address," AALS, *Proc.,* 13 (1913), 44. See also George W. Kirchwey, "American Law and the American Law School," AALS, *Proc.,* 8 (1908), 10-11.

62. William Draper Lewis, "Legal Education and the Failure of the Bar to Perform its Public Duties," AALS, *Proc.,* 6 (1906), 42; ABA *Reports,* 36 (1911), 645-46; Thomas T. Sherman to Root, May 10, 1915, Box 129, Root MSS; Richards, "Progress in Legal Education," AALS, *Proc.,* 15 (1915), 63; William Underhill Moore to E. W. Hinton, November 10, 1916, Moore MSS; ABA *Reports,* 40 (1915), 719.

63. Kirchwey, "American Law," 12, 17; William Howard Taft, "The Social Importance of Proper Standards for Admission to the Bar," *ALSR,* 3 (Fall 1913), 326, 333.

CHAPTER FOUR

1. Newton D. Baker, "Some Legal Phases of the War," *ABAJ,* 7 (June 1921), 323; James Beck to Samuel Untermeyer, November 7, 1917, Untermeyer MSS; Marvin B. Rosenberry, "Will the Bar Furnish Our Leaders in the Impending World Crisis?" Wisconsin State Bar Association, *Reports* (1916-17-18), 97; Joseph Hartigan, "The Lawyer and Patriotic Service," *Case and Comment,* 24 (March 1918), 804-5; *ABAJ,* 3 (July 1917), 347; Theodore A. Huntley, *The Life of John W. Davis* (New York, 1924), 160; *ABAJ,* 3 (October 1917), 576; Charles E. Hughes, "New Phases of National Development," *ABAJ,* 4 (January 1918), 93, 109.

2. *ABAJ,* 3 (July 1917), 345-47; *ABAJ,* 4 (January 1918), 87-91; *ABAJ,* 5

(April 1919), 178-86; *ABAJ*, 3 (October 1917), 576-77. According to the new president of the ABA, the meeting "presented one dominant idea—*patriotism.*" "A Word from President Smith," *ABAJ*, 3 (October 1917), 578.

3. C. B. Stuart, "The Lawyer's Relation to the World War," Oklahoma State Bar Association, *Proc.*, 11 (1917), 108, 109, 120; *ABAJ*, 3 (October 1917), 639; "Suggestions of Attorney General Gregory to Executive Committee in Relation to the Department of Justice" (April 16, 1918), *ABAJ*, 4 (July 1918), 314.

4. *In re Margolis*, 112 Atl. 478 (1921); *In re Smith*, 233 Pac. 288 (1925); *In re Hofstede*, 173 Pac. 1087 (1918); *In re Arctander*, 188 Pac. 380, 383 (1920); *Lotto v. State*, 208 S.W. 563, 564 (1919).

5. W. G. Haydon, "The Duty of a Lawyer to His Country in War," New Mexico Bar Association, *Minutes*, 31 (1917), 19; New York State Bar Association, *Proc.* (1918), 225-26; "The Lawyer's Duty as to Clients Attacking Governmental Power During the War," *Central Law Journal*, 84 (June 22, 1917), 451. See also "Lawyer and Legal Events," *Bench and Bar*, 12 N.S. (December 1917), 364. The *American Law Review* hailed the "purifying process" involved in the dismissal of a University of Michigan professor who opposed intervention. *ALR*, 51 (November-December 1917), 912; Samuel Untermeyer to Learned Hand, July 17, 1918, Untermeyer MSS; A. E. Bolton, "Individuality of Bench and Bar," California Bar Association, *Proc.* (1917), 19. For a view contrary to mine, see James J. Cavanaugh, *The Lawyer in Society* (New York, 1963). Cavanaugh defends patriotism as the *sine qua non* of the lawyer's work. A lawyer, he argues, "is a patriot before he is a lawyer . . ." (49-50).

6. Lucien Hugh Alexander, "The Legal Profession and the War," *ABAJ*, 5 (October 1919), 682, 688; William D. Guthrie, "The Public Service of the American Bar," New York State Bar Association, *Proc.* (1922), 143-64; *ALR*, 53 (March-April 1919), 299; Henry W. Taft, "The Bar in the War," New York State Bar Association, *Proc.* (1918), 221.

7. Frederic R. Coudert, "The Crisis of the Law and Professional Incompetency," ABA *Reports*, 36 (1911), 683; Strong, *Landmarks of a Lawyer's Lifetime*, 347; William V. Rowe, "Legal Clinics and Better Trained Lawyers —A Necessity," *Illinois Law Review*, 11 (April 1917), 593, 602-3; Stone, *Law and Its Administration*, 178; Stone to Willard Bartlett, February 14, 1913, Ezra Ripley Thayer MSS.

8. ABA *Reports*, 31 (1907), 519; "Overcrowding the Profession," *Law Student's Helper*, 10 (February 1902), 36; "The Business of Law and of Medicine . . . ," *ALR*, 43 (November-December 1909), 917; ABA *Reports*, 34 (1909), 743-44. For prewar trends nationally, see Edward G. Hartmann, *The Movement to Americanize the Immigrant* (New York, 1948), 7-8.

9. Charles W. Moores, "The Career of the Country Lawyer: Abraham Lincoln," *ALR*, 44 (November-December 1910), 886-90; Tocqueville, *De-*

mocracy in America, II, 103, 109-10, 112. For perceptive comments on democracy and legalism, see Judith N. Shklar, *Legalism* (Cambridge, Mass., 1964), 16.

10. Roscoe Pound to William Howard Taft [1913], Ezra Ripley Thayer MSS.

11. For a parallel example, see David B. Tyack, "The Perils of Pluralism: The Background of the Pierce Case," *American Historical Review*, 74 (October 1968), 74-98.

12. "Report of Seventeenth Annual Meeting," *ALSR*, 4 (April 1920), 508, 511. Prior to 1919 the AALS required of its members four years toward a night-school degree, as against three years toward a day-school degree. At its annual meeting a stiffer requirement was adopted, which in effect meant that no member school could offer a degree solely for night-school work. *Ibid.*, 525-33. A night-school teacher protested that the requirement was designed "to kill the night school. . ." (526).

13. ABA *Reports*, 46 (1921), 656-57. For the economic and ethnic elitism of the Flexner report, see Rosemary Stevens, *American Medicine and the Public Interest* (New Haven, Conn., 1971), 66-72.

14. Alfred Z. Reed, *Training for the Public Profession of the Law* (New York, 1921), 3, 60, 64, 215.

15. *Ibid.*, 56-57, 227, 235, 410, 417-18.

16. *Ibid.*, 55, 57, 399, 402, 416, 418, 419. For insightful comment on this continuing problem, see Robert Stevens, "Aging Mistress: The Law School in America," *Change in Higher Education* (January-February 1970), 32-42.

17. *Harvard Law Review*, 35 (November 1921), 98; Arthur L. Corbin, "Democracy and Education for the Bar," *ALSR*, 4 (March 1922), 727; Harlan F. Stone, "Legal Education and Democratic Principle," *ABAJ*, 7 (December 1921), 639-46. Stone assumed (1) that any lawyer admitted to the bar "may aspire to any position at the bar"; and (2) that any person who was healthy and "worth educating" could have a college education. *Columbia Law Review*, 22 (March 1922), 284-92.

18. Although the Reed study was not officially published until August 1921, copies were available at the December 1920 meeting of the AALS. Members of the ABA's Root committee had access to it in the interim. Alfred Z. Reed, "Raising Standards of Legal Education," *ABAJ*, 7 (November 1921), 571. In May 1920, Harlan Stone and William Draper Lewis had conferred with Root in New York, pleading the urgency of ABA involvement in raising educational standards and bar admission requirements. At the meeting plans were formulated for an ABA committee to report on the problem. Root expressed his willingness to serve on it. William Draper Lewis to Pound, June 23, 1920, Box 10, Pound MSS. In 1915, 28 states required no preliminary education for admission to the bar. Walter Wheeler Cook, "The Improvement of Legal Education and of Standards for Admission to the

Bar," ABA *Reports*, 40 (1917), 550; AALS, *Proceedings of the Summer Meeting and of the Eighteenth Annual Meeting* (1920), 24-25, 40-44; George T. Page, "Government," *ABAJ*, 5 (October 1919), 537; ABA *Reports*, 43 (1920), 465-66.

19. ABA *Reports*, 44 (1921), 656, 681; *ALSR*, 4 (November 1921), 673; ABA *Reports*, 44 (1921), 687-88.

20. *ALSR*, 4 (November 1921), 681; ABA *Reports*, 44 (1921), 662, 666, 668, 671, 672, 684.

21. *Ibid.*, 40, 42-43, 47, 675-76.

22. Elihu Root to Presidents of State and Local Bar Associations, October 31, 1921, in Conference of Bar Association Delegates, *Special Session on Legal Education* (Baltimore, 1922), 5, 120-22. The conference report is an extremely revealing document, and one that until recently was virtually ignored. Professor Preble Stolz, with evident surprise at the attitudes expressed by the delegates, refers to it in his "Clinical Experience in American Legal Education: Why Has It Failed?" in Edmund W. Kitch (ed.), *Clinical Education and the Law School of the Future*, University of Chicago Law School Conference, Series No. 20 (1970), 54-76. See also Stolz, "Training for the Public Profession of the Law (1921): A Contemporary Review," AALS, *Proc.* (1971), I, 166-74; Stevens, "Two Cheers for 1870," 456-58. There is an insightful discussion in Barlow F. Christensen, *Specialization* (Chicago, 1967), of the rhetoric of equality amid the reality of stratification. Lawyers, he notes, "*act as if* they believe all lawyers to be equal and seem to want the public to believe likewise." They do this because "abandonment of the egalitarian ethic would appear as a denial of . . . the validity of both the adversary system and the lawyer's role as an advocate" (18, 20).

23. Conference, *Special Session*, 48, 72, 116, 128-30. See Wickersham, "The Moral Character of Candidates for the Bar," *ABAJ*, 9 (October 1923), 617-21; *ABAJ*, 8 (March 1922), 137.

24. *Conference*, 11; Stone to H. S. Pritchett, March 27, 1918; Reginald H. Smith to Stone, January 20, 1920; Frankfurter to Stone, February 17, 1922, Stone MSS.

25. Conference, *Special Session*, 40, 70-72, 75, 131, 167.

26. *Ibid.*, 59, 144; Christopher Jencks and David Riesman, *The Academic Revolution* (New York, 1968), 77 (Table III), 96.

27. Conference, *Special Session*, 65; John Higham, *Strangers in the Land: Patterns of American Nativism 1860-1925* (New Brunswick, N.J., 1955), 278. Of course, if the young man's family needed the income that his working would provide, even free college education might be beyond his means. On the employment problems of *Columbia Law Review* graduates due to prejudice, see Harlan F. Stone to George B. Case, April 19, 1920, Stone MSS.

28. Richard W. Leopold, *Elihu Root and the Conservative Tradition* (Boston, 1954), 171. Editorial responses to the conference reflected these

divisions. The Newark (N.J.) *News* hailed the resolutions as representative of "the trend of modern thought" which, in an ambiguous way, they were. But the Birmingham (Ala.) *News* criticized them as "undemocratic and unwise," while the Rochester (N.Y.) *Daily Record*, in a long and thoughtful editorial, suggested that while New York City might require "the strong arm of legal repression against those presenting themselves at the door of the profession," other parts of the nation did not. "Conference on Legal Education," *ABAJ*, 8 (March 1922), 156-57; "Slamming the Door to Legal Profession in the Face of Brains," *ALR*, 57 (May-June 1923), 354 ff. Cf. "The Public Profession of the Law," *New Republic*, 29 (February 22, 1922), 351-54.

29. Only West Virginia required that bar applicants be law-school graduates. Alfred Z. Reed, "Recent Progress in Legal Education," *ALSR*, 5 (May 1926), 705; A. M. Hendrickson, "Admission to the Bar," *ALSR*, 4 (May 1922), 799-805.

30. Lawrence Stone, "The Ninnyversity," *New York Review of Books*, 16 (January 28, 1971), 22.

31. Pound to Stone, January 2, 1923, Box 4, File 2, Pound MSS. A Chicago attorney complained to Root that the Root report recommendations would "snuff out the night law schools." Robert McMurdy to Elihu Root, August 13, 1921, Box 139, Root MSS. For concern at Harvard and Columbia over the Jewish problem (i.e., too many Jewish students), see Stone to Williston, October 13, 1922, Stone MSS. Stone conceded the existence of the problem, confessed to ignorance about its resolution, and noted that part of the problem was that Jewish students seemed to work harder and to be intellectually more capable than their Anglo-Saxon classmates. The ABA standards remained higher than those demanded by the AALS of its members. Until 1923, the AALS required only high-school training; after September 1, 1923, one year of college was required; after September 1, 1925, two years.

32. American Bar Foundation, Research Memorandum Series, Number 15: *Compilation of Published Statistics on Law School Enrollments and Admissions to the Bar, 1889-1957* (November 1958). In 1927 alone, more students were enrolled in American law schools than in any other single year prior to 1947. Department of Commerce, Bureau of the Census, 14th Census of the U.S., Vol. IV, *Population*, 1920 Occupations (Washington, D.C., 1923); Department of Commerce, Bureau of the Census, 15th Census of the U.S., Vol. IV, *Population* (Washington, D.C., 1933). See, e.g., Will Shafroth, "The Rising Tide of Advocates," *ABAJ*, 16 (July 1930), 451-53; Philip J. Wickser, "Bar Examinations," *ABAJ*, 16 (November 1930), 733-38; Young B. Smith, "The Overcrowding of the Bar and What Can Be Done About It," *ALSR*, 7 (December 1932), 565-72.

33. Dayton McKean, *The Integrated Bar* (Boston, 1963), 21-38.

34. New York State Bar Association, *Proc.* (1921), 235-36; (1924), 243-44;

"Higher Educational Standards Urged for Admission to Law Study in New York," *ABAJ*, 13 (March 1927), 123; New York State Bar Association, *Proc.* (1923), 331, 335-40.

35. New York State Bar Association, *Proc.* (1922), 85-94; William D. Guthrie, "The Proposed Compulsory Organization of the Bar," *New York Law Review*, 4 (May 1926), 179-80. New York State Bar Association, *Proc.* (1926), 272; Louis Marshall to Guthrie, April 9, 1926, Marshall MSS; Guthrie, "Proposed Compulsory Organization," 186, 187, 225, 236. See also New York State Bar Association, *Proc.* (1923), 287. For a brief discussion of Guthrie's role, which misses the import of nativism, see Martin, *Causes and Conflicts*, 218-22.

36. Marshall to Guthrie, April 9, 1926, Marshall MSS.

37. Marshall to Samuel Seabury, October 21, 1926, *ibid.*

38. Philip J. Wickser, "Bar Associations," *Cornell Law Quarterly*, 15 (April 1930), 405-7, 409; New York State Bar Association, *Proc.* (1926), 293; Clarence F. Giles, "An Analysis of the Integrated Bar," *Albany Law Review*, 17 (January 1953), 239, 241, 247. For Guthrie's efforts to avoid association debate on the issue, and to bury favorable recommendations, see Guthrie to Marshall, December 27, 1926, Marshall MSS. See also I. Maurice Wormser, "Fewer Lawyers and Better Ones," New York State Bar Association, *Proc.* (1929), 352-66; Rollin B. Sanford, "Why We Increase the Membership of the Bar," New York State Bar Association, *Proc.* (1930), 357-70; New York State Bar Association, *Proc.* (1932), 73; Association of the Bar of the City of New York, *Lectures on Legal Topics* (New York, 1924), I, 18, 21.

39. Illinois Bar Association, *Proc.* (1917), 108-15; (1922), 173, 291, 298.

40. *Ibid.* (1923), 263, 292, 434, 436-37. For similar debates, see Ohio State Bar Association, *Proc.* (1922), 52-53ff; (1924), 109-18; California Bar Association, *Proc.* (1926), 60-61, 64-65, 71; Thurman Arnold, "Review of the Work of the College of Law," *West Virginia Law Quarterly*, 36 (1929-30), 326.

41. For historical and sociological background on Philadelphia conditions, see Gary B. Nash, "The Philadelphia Bench and Bar, 1800-61," *Comparative Studies in Society and History*, 7 (January 1965), 203-20; E. Digby Baltzell, *Philadelphia Gentlemen: The Makings of a National Upper Class* (Glencoe, Ill., 1958), 145, 273, 288-89, 395-96.

42. Pennsylvania State Bar Association, *Report* (1922), 248, 347-48; Beck to Charles G. Dawes, May 2, 1923, Box 1, Beck MSS; Pennsylvania State Bar Association, *Report* (1925), 172-74, 223, 225, 233.

43. Report of the Committee on Legal Education, Pennsylvania Bar Association, *Report* (1927), 190-203, 214-27; Lon Fuller, "Legal Education in Pennsylvania" (n.p., 1951), 105-6, 109, 110, 123-24. Local examining committees exercised complete discretion, and as a matter of general policy, as the secretary of the Philadelphia board of examiners explained, "not a great deal

of examination" was required of any applicant related to "a reputable member" of the Philadelphia bar. Albert L. Moise, "Practical Operation of the Pennsylvania Plan In Philadelphia County," *Bar Examiner*, 8 (March 1939), 38.

44. Walter C. Douglas, Jr., "Pennsylvania's New Requirements for Bar Admission," *ABAJ*, 14 (December 1928), 669, 670, 672, 673, 674; Proceedings of the Section on Legal Education of the American Bar Association," *ALSR*, 6 (January 1930), 589; New York State Bar Association, *Proc.* (1932), 79. For euphoric praise of the Pennsylvania Plan, years later, see Will Shafroth, "Character Investigation," in Survey of the Legal Profession, *Reports on Bar Examinations and Requirements for Admission to the Bar* (1952), 258, 261. For a more critical assessment, shortly before the plan was scrapped, see "Admission to the Pennsylvania Bar: The Need for Sweeping Change," *University of Pennsylvania Law Review*, 118 (May 1970), 945-82.

CHAPTER FIVE

1. *American Law Review*, 53 (January-February 1919), 66, 99, 105, 115-20; "The Legal Profession, Its Influence, Responsibility, and Opportunity," *Bench and Bar*, 14 (September 1919), 167, 169; "Coddling Worse Than Treason," *Bench and Bar*, 14 (December 1919), 203; "Plain Talk About Socialism," *Bench and Bar*, 15 (February 1920), 48. See also "The Chief Issue for 1920—The Supremacy of Law," *Central Law Journal*, 90 (January 2, 1920), 2.

2. W. A. Blount, "To the American Bar Association," *ABAJ*, 7 (January 1921), 54; Cordenio A. Severance, "The Constitution and Individualism," *ABAJ*, 8 (September 1922), 538, 541-42. See also Severance, "The Attack on American Institutions," *ABAJ*, 7 (December 1921), 633-35.

3. ABA *Reports*, 47 (1922), 416.

4. R. E. L. Saner, "The Constitution and Nationality," *ABAJ*, 9 (July 1923), 455.

5. ABA *Reports*, 48 (1923), 445.

6. *Ibid.*, 446-49; ABA *Reports*, 49 (1924), 271.

7. Saner to John W. Davis, March 31, 1923, Box 32, John W. Davis MSS; R. E. L. Saner, "Governmental Review," *ABAJ*, 10 (August 1924), 541, 544. Charles Evans Hughes' 1925 presidential address was the only one to depart from the standard clichés of saving the United States from radicalism. See Charles Evans Hughes, "Liberty and Law," *ABAJ*, 11 (September 1925), 563-69. By the following year, however, the pendulum had swung back. See Chester I. Long, "The Advance of the American Bar," *ABAJ*, 12 (August 1926), 517-23.

8. W. Marvin Woodall, "Professionalism Against Commercialism in the

Practice of Law," Alabama State Bar Association, *Proc.*, 46 (1923), 218; Carroll A. Nye, "The Legal Profession and Its Relation to Present Day Problems," Minnesota State Bar Association, *Proc.*, 20 (1920), 46; James H. Wilkerson, "The American Bar—The Nation's Great Conservative Force," Iowa State Bar Association, *Proc.*, 30 (1924), 250; A. T. Dumm, "The Country Lawyer," *American Law Review*, 55 (March-April 1921), 304; Ensign N. Brown, "Presidential Address," Ohio State Bar Association, *Proc.* (1919), 65; Thomas R. Marshall, "Altruistic Evil," *American Law Review*, 55 (May-June 1921), 362-63.

9. ABA *Reports*, 54 (1929), 302.

10. Sherman Whipple, "Is Our Profession Becoming Unduly Commercialized?" *American Law Review*, 58 (January-February 1924), 48, 61, 62.

11. "Lawyer's Office Systems," *Central Law Journal*, 91 (November 5, 1920), 340; Roger Sherman, "Business System in Law Offices," *American Law Review*, 58 (July-August 1924), 531-32; Report of the Committee on Office Management, Illinois Bar Association, *Proc.* (1921), 142-43, 146-47.

12. See, e.g., W. R. Nelson, "Business Methods for Lawyers," Alabama State Bar Association, *Report*, 14 (1891), 66-72; Dwight G. McCarty, "Our Changing Law Practice," Michigan State Bar *Journal*, 7 (November 1927), 102, 103, 104, 106-10, 119; Dwight G. McCarty, *Law Office Management* (New York, 1940), Preface to First Edition, ix. See also Wilmer T. Fox, "Business Methods in a Lawyer's Office," Indiana State Bar Association, *Proc.*, 24 (1920), 111-24.

13. See H. F. Lawson, "Business Methods in a Lawyer's Office," Georgia Bar Association, *Reports*, 39 (1922), 119; A. M. Keene, "The Evolution of the Practice of the Law," Bar Association of Kansas, *Proc.*, 36 (1919), 105-6; George S. Jones, "Business Methods in a Lawyer's Office," Georgia Bar Association, *Reports*, 39 (1922), 131-32; Joseph M. Proskauer, "Office System," in Association of the Bar, *Lectures on Legal Topics*, I, 145; Cornelius W. Wickersham, "Office Systems," *ibid.*, 119-31; George B. Rose, "How to Get on at the Bar," *Central Law Journal*, 93 (November 25, 1921), 372; Roger Sherman, "Business System in Law Offices," Wisconsin State Bar Association, *Proc.*, 14 (1922), 201-2; Henry Wollman, "Lawyers: Business Men," *ABAJ*, 9 (December 1923), 811-12; E. O. Howard, "The Responsibility of the Legal Profession to Society," Idaho State Bar, *Proc.*, 3 (1927), 122-23; M. B. Rosenberry, "Some Observations on the Present Status of the Legal Profession," State Bar Association of Wisconsin, *Proc.*, 17 (1927), 23; T. C. Hannah, "A Little More Business in Practicing the Profession of Law," *Mississippi Law Journal*, 2 (August 1929), 59.

14. H. F. Lawson, "Business Methods," 117. Yet the same military metaphor might also be used without opprobrious connotations. Cf. "How Great Law Offices Work," *American Legal News*, 29 (November 1918), 17. W. A. Marin, "The Lawyer and Commerce," South Dakota Bar Association, *Re-*

port, 23 (1922), 183-84, 187; Henry W. Taft, "Some Responsibilities of the American Lawyer," New York State Bar Association, *Proc.* (1920), 190; Omer F. Hershey, "The Passing of the Lawyer's Primacy," Maryland State Bar Association, *Report*, 30 (1925), 58, 60, 62; William L. Ransom, "The Lawyer Today," *ABAJ*, 16 (September 1930), 589. For parallel developments in England, see H. D. Darbishire, *The Legal Profession and the Rotary Movement* (London, 1924), urging the legal profession to commit itself to the service goals of Rotary.

15. Davis to Edward R. Davis, March 12, 1928, Add. Box 4, Davis MSS. See Davis, "On Legal Training," in Huntley, *Life of John W. Davis*, 241; Davis to H. M. Jacoway, November 7, 1932, Add. Box 1; Davis to W. Matthews, November 15, 1935, Add. Box 4; Davis to E. M. Gilkerson, February 16, 1925, Add. Box 4, Davis MSS.

16. John W. Davis Memoir, Columbia Oral History Collection (cited below as COHC), 34, 46-47, 157.

17. Davis to William H. Sawyers, August 4, 1921, Box 20, Davis MSS. See also Davis to Breckenridge Long, October 19, 1922, Box 27; Davis to Willis Van Devanter, October 31, 1922, Box 27; R. P. Resor to Davis, December 15, 1922, Box 29; Davis to John A. Stewart, March 23, 1923, Box 32, *ibid.*; Taft quoted in William H. Harbaugh, *Lawyer's Lawyer: The Life of John W. Davis* (New York, 1973), 193.

18. Harvey F. Smith to Davis, September 23, 1922; Thomas W. Shelton to Davis, September 14, 1924, Box 26, *ibid.*; William R. Vance to Davis, April 30, 1921, Add. Box 1, *ibid.*

19. Harbaugh, *Lawyer's Lawyer*, 199.

20. Frankfurter to Charles C. Burlingham, February 14, 1924, Frankfurter MSS, LC. For an expression of similar views, several years earlier, see Pound to Stone, February 3, 5, 1921, Stone MSS.

21. "Why Mr. Davis Shouldn't Run," *New Republic*, 38 (April 16, 1924), 194-95; "John W. Davis," *New Republic*, 39 (July 23, 1924), 225-26. Furthermore, Frankfurter told Walter Lippmann, since Davis had returned to practice he had demonstrated "such indifference and cowardice in the face of bigotry as to have made him one of bigotry's effective silent allies." Davis' silence while president of the ABA especially angered Frankfurter. Frankfurter to Lippmann, July 1, 1924; Frankfurter to Herbert Croly, April 18, 1924; Frankfurter to Raymond Fosdick, September 27, 1924; Frankfurter to Learned Hand, October 3, 1924; Frankfurter to Charles C. Burlingham, October 29, 1924, Frankfurter MSS, LC; *Felix Frankfurter Reminisces*, 190.

22. William Howard Taft, "Legal Ethics," *Boston University Law Review*, 1 (October 1921), 244; Illinois Bar Association, *Proc.* (1923), 443; Robert S. and Helen M. Lynd, *Middletown* (New York, 1929), 48; Earle, *Mr. Shearman and Mr. Sterling and How They Grew*, 205, 207, 218, 223-24.

23. *The Journals of David E. Lilienthal. Vol. I: The TVA Years 1939-1945*

(New York, 1964), 12, entry for December 25, 1920. Lilienthal added (p. 13): "Definitely, I have no such intentions. . . . There are lawyers in this country who have done the sort of thing I hope to do. They are *not* corporation lawyers." He named Brandeis, Frankfurter, Chafee, Clarence Darrow, and Frank P. Walsh. Remarks of Judge Henry J. Friendly, in Arthur E. Sutherland (ed.), *The Path of the Law from 1967* (Cambridge, Mass., 1967), 114.

24. Acheson to Frankfurter, December 11, 1920, January 17, 1921, Frankfurter MSS, LC. See Acheson to Pound, March 18, 1920, Pound MSS; Lilienthal, *TVA Years,* 13, 16; Frankfurter to Ehrmann, August 28, 1921, Box 1, Ehrmann MSS. Emory Buckner advised Nathan Margold that the next best thing to public service, where opportunities were limited, was in law teaching. Buckner to Pound, January 2, 1928, Box 8, Pound MSS.

25. "Lawyers' Fees," *Fortune,* 3 (January 1931), 67; D. E. Simmons to David A. Simmons, March 31, 1920, B 19/6, Simmons MSS; Henry L. Stimson Diaries, Vol. 6 (October 18, 1921); "Lawyers Looking at You," *Fortune,* 3 (January 1931), 61; Harbaugh, *Lawyer's Lawyer,* 259.

26. For a fuller discussion of the sample, a more detailed breakdown of year-by-year and school-by-school variations, and a comparison with 1930's patterns, see Jerold S. Auerbach and Eugene Bardach, " 'Born to an Era of Insecurity': Career Patterns of Law Review Editors, 1918-1941," *American Journal of Legal History,* 17 (1973), 3-26.

27. *Felix Frankfurter Reminisces,* 190. Frankfurter was also annoyed that bar leaders did not encourage law teaching as a career: "Insofar as the influence of the bar is concerned, insofar as it helps to shape the ambitions and ideals of young men, no influence whatever emanates from it towards the kind of a life that Ames and Grey and Thayer led." Frankfurter to Emory Buckner, November 30, 1926, quoted in Martin Mayer, *Emory Buckner* (New York, 1968), 151; Harlan F. Stone to Pound, April 22, 1924, December 8, 1924; Pound to Stone, December 22, 1924, Box 4, File 2, Pound MSS. In 1919 the Department of Justice employed sixty-seven attorneys; it grew during the 1920's, largely as the result of prohibition. See *Report of the Attorney General* (1919), 9; Albert Langeluttig, *The Department of Justice of the United States* (Baltimore, 1927), vii. See *Felix Frankfurter Reminisces,* 249.

28. Frankfurter to Stimson, March 22, 1921, Box 202, Stimson MSS; Thomas N. Perkins to Frankfurter, November 26, 1919, Frankfurter MSS, LC.

29. Frankfurter to Arthur D. Hill, October 7, 1927, *ibid.;* Louis C. Joughin and Edmund M. Morgan, *The Legacy of Sacco and Vanzetti* (Chicago, 1964), 319.

30. Frankfurter to Learned Hand, April 3, 1927, Box 104, File 14, Learned Hand MSS; Frankfurter to Pound, August 23, 1927, 4/4, Pound MSS; Frankfurter to Richard Hale, November 5, 1926, Frankfurter MSS, LC; Frankfurter

to Hans Zinsser, December 17, 1927, Sacco-Vanzetti Collection, Frankfurter MSS, Harvard Law School; Herbert B. Ehrmann to Arthur W. Blackman, November 30, 1926, Ehrmann MSS.

31. Frankfurter to Pound, August 31, 1927, 4/4, Pound MSS; Frankfurter to Robert H. Neilson, August 23, 1927, Sacco-Vanzetti, Frankfurter MSS; William Renwick Riddell, "The Sacco-Vanzetti Case from a Canadian Viewpoint," *ABAJ*, 13 (December 1927), 683, 694; Charles S. Whitman, "Presidential Address," *ABAJ*, 13 (September 1927), 491; Frankfurter to Pound, August 22, 1927, Frankfurter MSS, LC; Frankfurter to Chafee, April 2, 1927, Box 4, Chafee MSS; Frankfurter to Charles Nagel, March 21, 1927, Sacco-Vanzetti, Frankfurter MSS. See also Frankfurter to Robert L. Hale, October 28, 1926, *ibid.;* Joughin and Morgan, *Legacy,* 145-49; Frankfurter to Edwin H. Hall, April 2, 1927, Frankfurter MSS, LC.

32. Frankfurter to Hans Zinsser, December 24, 1927; Frankfurter to Henry U. Sims, April 26, 1932, *ibid.;* Frankfurter to Karl N. Llewellyn, April 30, 1927, G, II, Karl Llewellyn MSS; Frankfurter to Thomas Nelson Perkins, April 28, 1927, Frankfurter MSS; Frankfurter to Franklin Hichborn, June 20, 1927, *ibid.;* Frankfurter to Francis Rawle, July 19, 1927, *ibid.;* Frankfurter to Newton D. Baker, April 11, 1927, *ibid.;* Frankfurter to Charles L. Slattery, April 11, 1927, Roll 21, Frankfurter MSS, HLS; Frankfurter to Gilson Gardner, April 20, 1927, Sacco-Vanzetti, Frankfurter MSS.

33. James R. Knapp to Frankfurter, October 15, 1927; Thompson to Stoughton Bell, April 11, 1927, Frankfurter MSS, LC. For variations on this theme see Knapp to Frankfurter, September 20, 1927, *ibid.;* Charles Nagel to Frankfurter, March 19, 1927; Oswald Garrison Villard to Frankfurter, May 9, 1927; Stephen S. Wise to Frankfurter, May 13, 1927, Sacco-Vanzetti, Frankfurter MSS; Robert Grant, *Fourscore: An Autobiography* (Boston, 1934), 373; Joughin and Morgan, *Legacy,* 262, 268, 319, 370; Robert L. Hale to A. Lawrence Lowell, August 4, 1927, Sacco-Vanzetti, Frankfurter MSS; Taft to Elihu Root, May 12, 1927, Box 126, Root MSS; Frank W. Grinnell to Frankfurter, May 4, 1927; Spencer B. Montgomery to Pound, May 9, 1927, Sacco-Vanzetti, Frankfurter MSS.

34. Joughin and Morgan, *Legacy,* 255; Forrest Bailey to Frankfurter, April 14, 1927; Karl Llewellyn to Frankfurter, April 27, 1927, Sacco-Vanzetti, Frankfurter MSS; William Twining, *Karl Llewellyn and the Realist Movement* (London, 1972), 348-49; Charles C. Clark to Frankfurter, May 9, 14, 1927, Frankfurter MSS, HLS; T. R. Powell to Karl Llewellyn, May 27, 1931, A, II, 656, Llewellyn MSS; Frankfurter to Richard Hale, November 25, 1926, Felix Frankfurter MSS, LC; Paul Sayre to Frankfurter, June 23, 1927, *ibid.* The anguish of two of Frankfurter's younger colleagues, expressed on the day of the Sacco-Vanzetti executions, alluded to this paradox. Sheldon Glueck, collaborator with Frankfurter on the Boston Crime Survey, asked despondently: "Our crime survey, our findings, our recommendations—what

can they amount to when the apple is rotten at the core? . . . Will [this tragedy] even bring about a fundamental change in that sinister, cynical logic of our craft which will persist in the confusion of means and end, in the raising of law above justice?" And James M. Landis told Frankfurter that "to many of us more things than Sacco and Vanzetti have been murdered. . . . Bench and bar must learn anew humility and the mob mercy." Sheldon Glueck to Frankfurter, August 23, 1927; James M. Landis to Frankfurter, August 23, 1927, Sacco-Vanzetti, Frankfurter MSS.

35. Frankfurter to William Rosenwald, May 13, 1927, Roll 13, Frankfurter MSS, HLS.

36. Pound, "Liberty of Contract," *Yale Law Journal*, 18 (1909), 464.

37. Grant Gilmore, "Legal Realism: Its Cause and Cure," *Yale Law Journal*, 70 (June 1961), 1038; Eugene V. Rostow, "American Legal Realism and the Sense of the Profession," in *The Sovereign Prerogative* (New Haven, Conn., 1962), 10, 22; Wilfrid E. Rumble, Jr., *American Legal Realism* (Ithaca, N.Y., 1968), 1-20, 74-78; Stevens, "Two Cheers for 1870," 470-81.

38. Pound to E. A. Ross, February 14, 1919, Pound MSS; Herman Oliphant, "A Return to Stare Decisis," *ABAJ*, 14 (February 1928), 71; Stone to Pound, March 27, 1917, Pound MSS; Young B. Smith, "The Abuse of Law," *ALR*, 59 (November-December 1925), 932; Robert M. Hutchins, "Connecticut and the Yale Law School," *Conn. Bar J.*, 2 (1928), 169.

39. "Meeting of the Association of American Law Schools—1928," *ALSR*, 6 (March 1929), 449; Thomas Reed Powell to Jerome Frank, November 5, 1930, A-5, Powell MSS; Powell, "The Recruiting of Law Teachers," *ABAJ*, 13 (February 1927), 72; William O. Douglas, *Go East, Young Man* (New York, 1974), 123, 165-66; Frank to James H. Tufts, April 13, 1931, Box I, Frank MSS.

40. *Congressional Record*, 71st Cong., 2d Sess. (1930), Vol. 72, Pt. 3, 3373; Pt. 4, 3450, 3501, 3516.

41. *Ibid.*, 3452, 3509. For Hughes' defense of his own "complete independence," see David J. Danelski and Joseph S. Tulchin (eds.), *The Autobiographical Notes of Charles Evans Hughes* (Cambridge, Mass., 1973), 285, 295.

42. Danelski and Tulchin (eds.), *Autobiographical Notes*, 297.

43. Harbaugh, *Lawyer's Lawyer*, 23, 46, 412.

44. See F. Scott Fitzgerald, *The Great Gatsby* (New York, 1925); Lynds, *Middletown*, 498; John W. Ward, "The Meaning of Lindbergh's Flight," *American Quarterly*, 10 (1958), 3-16; Keith Sward, *The Legend of Henry Ford* (New York, 1968).

45. Harbaugh, *Lawyer's Lawyer*, 35; Townsend Hoopes, *The Devil and John Foster Dulles* (Boston, 1973), 25, 37, 43-45.

CHAPTER SIX

1. Karl N. Llewellyn, review, *Columbia Law Review*, 31 (1931), 1218.

2. James E. Brenner, "A Survey of Employment Conditions Among Young Attorneys in California," State Bar of California, *Proceedings* (1932), 32-38; AALS, *Proc.* (1934), 77; Isidor Lazarus, "The Economic Crisis in the Legal Profession," *National Lawyers Guild Quarterly* (December 1937), 18-19; James P. Gifford, "Lawyers and the Depression," *Nation*, 137 (August 30, 1933), 236; Melvin M. Fagen, "The Status of Jewish Lawyers in New York City," *Jewish Social Studies*, 1 (1939), 74, 79, 81, 86-87, 92, 95, 104; American Bar Association, *The Economics of the Legal Profession* (Chicago, 1938), 47.

3. Swaine, *The Cravath Firm*, II, xiii-xx, 461; Frank E. Holman, *The Life and Career of a Western Lawyer 1886-1961* (privately printed, 1963), 271; Harrison Tweed Memoir, COHC, 66; Dean, *William Nelson Cromwell*, 172-81; S. Pearce Browning, Jr., to James M. Landis, December 2, 1932, Box 4, Landis MSS. See also "Economic Security and the Young Lawyer: Four Views," *Illinois Law Review*, 37 (1937-38), 662.

4. Stone, *Law and Its Administration*, 165-66, 176; quoted in Alpheus T. Mason, *Harlan F. Stone: Pillar of the Law* (New York, 1956), 375, 376-77 n.

5. *Ibid.*, 382; Harlan F. Stone, "The Public Influence of the Bar," *Harvard Law Review*, 48 (1934), 2, 3, 6, 7.

6. Fred Rodell, *Woe Unto You, Lawyers!* (New York, 1939), 249, 272; Calvert Magruder, "What May Society Expect of Our Profession?" Maryland State Bar Association, *Reports* (1932), 102; Adolf A. Berle, "Modern Legal Profession," *Encyclopaedia of the Social Sciences* (New York, 1933), IX, 340, 343-44; *ABAJ*, 19 (January 1935), 5; Arthur A. Ballantine, "The Lawyer's Outlook Today," *ABAJ*, 24 (1938), 1022.

7. Karl N. Llewellyn, "The Bar Specializes—With What Results?," *Annals of the American Academy of Political and Social Science*, 167 (May 1933), 178, 179; Llewellyn, review, *Columbia Law Review*, 1218.

8. Rodell, *Woe Unto You, Lawyers!*, 3; Thurman Arnold to Roy M. Hardy, May 17, 1937, Thurman Arnold MSS; Ferdinand Lundberg, "The Legal Profession: A Social Phenomenon," *Harper's Magazine*, 178 (December 1938), 3, 11.

9. Thomas Reed Powell to Felix Frankfurter, February 13, 1934, A-6, Thomas Reed Powell MSS, referring to Earle W. Evans, "Lawyers and Legal Events," *United States Law Review*, 68 (1934), 107; Zechariah Chafee, Jr., "What's the Matter With the Law?" Address to Worcester Bar Association (1934), Zechariah Chafee, Jr., MSS; ABA, *Reports* (1935), 180; Albert J. Harno, "Social Planning and Perspective Through Law," *ABAJ*, 19

(1933), 201-6, 250; Arnold to Hardy, May 17, 1937, Arnold MSS; Stone to Frankfurter, October 9, 1934, quoted in Mason, *Stone*, 384; John Dickinson, "The Professor, The Practitioner, and the Constitution," *American Law School Review*, 8 (1936), 479-86.
10. Charles E. Clark, "Law Professor, What Now?" AALS, *Proc.* (1933), 15; James C. Bonbright, review, *Columbia Law Review*, 32 (February 1932), 395-96.
11. Esther Lucile Brown, *Lawyers, Law Schools and the Public Service* (New York, 1948), 12.
12. Karl Llewellyn, "Some Realism About Realism—Responding to Dean Pound," *Harvard Law Review*, 44 (June 1931), 1236-37; Jerome Frank to Thomas Reed Powell, December 2, 1930, A-5, Powell MSS; Powell to Max Lerner, May 2, 1933, A-10, *ibid.*; Edward A. Purcell, Jr., "American Jurisprudence Between the Wars: Legal Realism and the Crisis of Democratic Theory," *American Historical Review*, 75 (December 1969), 424-46; Shklar, *Legalism*, 95-98.
13. Davis to Julian S. Gravely, February 25, 1935, Davis MSS; ABA, *Reports* (1935), 156-57; Walter P. Armstrong, "A Practicing Lawyer Looks at Legal Education," *ALSR*, 9 (1940), 776, 780; Joseph Auerbach, *The Bar of Other Days* (New York, 1940), 18; Holman, *Western Lawyer*, 314.
14. Frankfurter to Philip L. Miller, October 31, 1911, Roll 14, Felix Frankfurter MSS, HLS; Frankfurter to Miller, March 11, 1913, Frankfurter MSS, LC; Frankfurter, "The Law and the Law Schools," *ABAJ*, 1 (1915), 538-39; Frankfurter to Charles C. Burlingham, February 14, 1924, Frankfurter MSS, LC; Frankfurter, "John W. Davis," *New Republic*, 39 (July 23, 1924), 225-26; Frankfurter to Learned Hand, October 3, 1924, Frankfurter MSS, LC; *Felix Frankfurter Reminisces*, 190, 248; Mayer, *Buckner*, 141; Joseph N. Welch to Frankfurter, November 27, 1925, Frankfurter MSS, LC.
15. Felix Frankfurter and Nathan Greene, *The Labor Injunction* (New York, 1930), 227; Frankfurter to Stone, April 2, April 4, 1932, Frankfurter MSS, LC.
16. Frankfurter to Learned Hand, March 18, January 30, 1932, 104-22, Learned Hand MSS; Frankfurter, "Democracy and the Expert," *Atlantic Monthly*, 146 (November 1930), 652, 660; Frankfurter to Henry J. Friendly, February 17, 1932, Frankfurter MSS, LC; Frankfurter to David Lilienthal, April 6, 1931, Roll 8, Frankfurter MSS, HLS.
17. Frankfurter to Walter Lippmann, April 17, 1933, Frankfurter MSS, LC; Frankfurter, "Dean James Barr Ames and the Harvard Law School," in Philip Kurland (ed.), *Of Law and Life and Other Things That Matter: Papers and Addresses of Felix Frankfurter 1956-1963* (Cambridge, Mass., 1965), 32; Frankfurter to William O. Douglas, January 16, 1934; Frankfurter to George A. Brownell, January 12, 1934, Frankfurter MSS, LC; Frankfurter to Henry L. Stimson, December 19, 1933, February 20, 1934, Roll 25, Frank-

furter MSS, HLS; Frankfurter to Landis, January 22, 1935, Frankfurter MSS, LC; Frankfurter to James Couzens, December 7, 1933, Frankfurter MSS, LC; Frankfurter to Max Lowenthal, January 23, 1926, Frankfurter MSS, LC; James M. Landis, *The Administrative Process* (New Haven, 1938), 46.

18. Frankfurter to Lippmann, April 17, 1933; Frankfurter to Stone, November 14, 1935, Frankfurter MSS, LC; James Willard Hurst, "Themes in United States Legal History," in Wallace Mendelson (ed.), *Felix Frankfurter: A Tribute* (New York, 1964), 200.

19. Frankfurter to Franklin D. Roosevelt, January 9, 1937, in *Roosevelt and Frankfurter: Their Correspondence 1928-1945*, annotated by Max Freedman (Boston, 1967), 374; Frankfurter to FDR, January 18, 1937, Roll 3, Frankfurter MSS, HLS; Frankfurter to Milton Katz, May 25, 1936, Frankfurter MSS, LC; Frankfurter to Jerome Frank, September 29, 1933, Jerome Frank MSS.

20. Hugh Johnson, "Think Fast, Captain," *Saturday Evening Post*, 208 (October 26, 1935), 85; John Franklin Carter, *The New Dealers* (New York, 1934), 309; Fred Rodell, "Felix Frankfurter, Conservative," *Harper's Magazine*, 183 (October 1941), 453.

21. Frankfurter to Max Lerner, March 17, 1937, Max Lerner MSS, Yale University; Frankfurter to Nathan Margold, November 29, 1932, Vol. 615, American Civil Liberties Union Archives; Frankfurter to Charles Clark, February 18, 1934, Frankfurter MSS, LC; Frankfurter to Charles Robbe, January 10, 1936, Roll 16; Frankfurter to Raymond Moley, Moley to Frankfurter, October 31, 1935, Roll 4, Frankfurter MSS, HLS; Frank to Frankfurter, April 18, 1933, Frank MSS; Frankfurter to Frank, April 24, 1933; Thomas Corcoran to Frankfurter, April 22, 1933, Frankfurter MSS, LC; Frankfurter to Charles E. Wyzanski, Jr., December 30, 1938, 1-5, Charles E. Wyzanski, Jr., MSS.

22. Frankfurter to Henry W. Bickle, January 30, 1936, Roll 14; Frankfurter to Eugene Meyer, May 24, 1938, Roll 15, Frankfurter MSS, HLS; Frankfurter to Grenville Clark, March 6, 1937, Frankfurter MSS, LC; Freedman, *Roosevelt and Frankfurter*, 744. In the later Cold War climate, Frankfurter went to considerable lengths to repudiate both his role and its policy implications. See *Felix Frankfurter Reminisces*, 248-50. For a suggestive assessment, see Sanford V. Levenson, "The Democratic Faith of Felix Frankfurter," *Stanford Law Review*, 25 (1973), 430-48.

23. Samuel Lubell, *The Future of American Politics* (New York, 1965), 43-44; American Bar Foundation, Research Memorandum Series, No. 15: *Compilation of Published Statistics on Law School Enrollments and Admissions to the Bar 1889-1957* (Chicago, 1958); U.S., Department of Commerce, Bureau of the Census, 14th Census, *Population, 1920 Occupations* (1923); U.S., Department of Commerce, Bureau of the Census, 15th Census, *Population* (1933).

24. In the careers of Thurman Arnold and William O. Douglas, the challenge of Western insurgency to the Eastern legal establishment is discernible. Douglas, especially, liked to describe himself as a country boy; after his accession to the Supreme Court he insisted that all his law clerks come from the Northwest, where students had limited opportunities to achieve such positions. His contempt for the "elite," the "Establishment," the "First Families," and the "affluent" is evident in his autobiography, *Go East, Young Man*. Douglas to Llewellyn, May 17, 1934, July 6, 1939, R, IV, 5, Karl Llewellyn MSS.

25. Francis Biddle, *A Casual Past* (New York, 1961), 366; Lloyd Landau to Frankfurter, February 4, 1933, Frankfurter MSS, LC; Leonard Davidow to Frank, May 26, 1933, Frank MSS; Landis to Frankfurter, January 27, 1934, Frankfurter MSS, LC; Corcoran quoted in Joseph Alsop and Robert Kitner, "We Shall Make America Over," *Saturday Evening Post*, 211 (November 16, 1938), 89.

26. *In Memory of Robert Houghwout Jackson* (Washington, D.C., 1955), 25; Warner W. Gardner, "Robert H. Jackson, 1892-1954, Government Attorney," *Columbia Law Review*, 55 (1955), 438.

27. Robert H. Jackson, "An Organized American Bar," *ABAJ*, 18 (1932), 383; Eugene C. Gerhart, *America's Advocate: Robert H. Jackson* (Indianapolis, 1958), 49; Jackson, "The Lawyer: Leader or Mouthpiece?" *Journal of the American Judicature Society*, 18 (1934), 72; Jackson, "The Bar and the New Deal," *ABAJ*, 21 (1935), 93-96.

28. Charles S. Desmond, "The Role of the Country Lawyer in the Organized Bar and the Development of the Law," in Desmond *et al.*, *Mr. Justice Jackson: Four Lectures in His Honor* (New York, 1969), 25; Glendon Schubert, *Dispassionate Justice: A Synthesis of the Judicial Opinions of Robert H. Jackson* (Indianapolis, 1969), 5-6.

29. Jerome N. Frank Memoir, COHC, 1-13; Frank to Powell, November 20, 1930, A-5, Powell MSS; Frank to Pound, November 30, December 2, 1930; Frank to Powell, November 24, 1930, in *ibid.*; Thurman Arnold, "Judge Jerome Frank," *University of Chicago Law Review*, 24 (1957), 635.

30. Frank to Powell, December 2, 1930, A-5, Powell MSS; Frank to Arnold, October 20, 1932; Frank to Adolf A. Berle, November 9, 1932; Frank to Arnold, December 29, 1932; Frank to Julian W. Mack, April 17, 1933, Frank MSS.

31. Jerome Frank, "Realism in Jurisprudence," *ALSR*, 7 (1934), 1066; Frank, "Experimental Jurisprudence and the New Deal," U.S., *Congressional Record*, 73d Cong., 2nd Sess., 78 (1934), 12412-14.

32. Frank, "Realism in Jurisprudence," 1067; Frank to Chester Davis, February 9, 1934; Frank to Frankfurter, November 29, 1935; Frankfurter to Frank, December 2, 1935; Frank to Frankfurter, January 21, 1936, Frank MSS.

33. Frank to Frankfurter, June, 1935, Frankfurter MSS, LC. See G. Edward White, "From Sociological Jurisprudence to Realism: Jurisprudence and Social Change in Early Twentieth-Century America," *Virginia Law Review*, 58 (1972), 999-1028.

34. Charles E. Wyzanski, Jr., Memoir, COHC, 6, 44, 92-93, 101.

35. *Ibid.*, 108, 149, 154, 175, 194, 274.

36. Wyzanski to Frankfurter, March 18, 1935, January 1, 1936, Frankfurter MSS, LC; Wyzanski, "The Lawyer's Relation to Recent Social Legislation," Kentucky Bar Association, *Proc.* (1937), 127; Wyzanski to Frankfurter, June 16, 1937, Frankfurter MSS, LC.

37. William O. Douglas, "An Age Coming to Birth," in *Being an American* (New York, 1948), 72; Karl Llewellyn, "On What Is Wrong With So-Called Legal Education," *Columbia Law Review*, 35 (May 1935), 662.

38. Louis Auchincloss, *Tales of Manhattan* (New York, 1967), 154.

39. These figures are drawn from Jerold S. Auerbach and Eugene Bardach, " 'Born to an Era of Insecurity': The Career Patterns of Law Review Editors, 1918-1941," *American Journal of Legal History*, 17 (January 1973), 3-26. They are based upon examination of the careers of 774 law review editors who were graduated from Harvard, Yale, and Columbia between 1918 and 1941, a 69 percent sample of the 1,125 law review editors who were graduated from the schools during that period.

40. Hurst, "Themes in United States Legal History," in Mendelson (ed.), *Frankfurter*, 200; Frankfurter to Roosevelt, January 18, 1937, Roll 3, Frankfurter MSS, HLS.

41. Abe Fortas to Richard Rovere, October 25, 1946, Group 31 R, Roosevelt Papers, Franklin D. Roosevelt Library; Francis Shea to Frankfurter, October 7, 1933, Frankfurter MSS, LC; David C. Shaw to Landis, March 25, 1937, Box 13, Landis MSS. See also Gordon Dean Memoir, COHC, 46-47, 53, 61.

42. Richberg to Clarence Martin, November 13, 1933, Box 1, Richberg MSS; Arnold to Henry Edgerton, June 11, 1934, Arnold MSS; Cohen to Frankfurter, October 9, 1933, Box 215, Frankfurter MSS, LC; Landis to Frankfurter, March 6, 1934, Box 217, *ibid.*

43. Stevenson to James Hamilton Lewis, July 17, 1933, in Walter Johnson (ed.), *The Papers of Adlai E. Stevenson* (Boston, 1972), I, 247, 248.

44. Nathan Witt to Frankfurter, December 7, 1931, Frankfurter MSS, LC; Lee Pressman Memoir, COHC, 5, 8; Pressman to Frank, April 12, 1933, Frank MSS.

45. Thomas I. Emerson Memoir, COHC, 1, 172-74, 177-79, 181, 196, 221-27, 265-66, 1759.

46. Arthur Schlesinger, Jr., *The Politics of Upheaval* (Boston, 1960), 96.

47. I am grateful to Professor John Murrin of Princeton University for calling some of these points to my attention in his comments on my paper,

"The American Legal Profession: Social Structure and Conservatism," Conference on American Legal History, Harvard Law School (1971).

48. Frankfurter to Samuel Becker, January 28, 1937, Roll 14, Frankfurter MSS, HLS; Herbert B. Ehrmann to Frankfurter, January 4, 1932; Joseph N. Welch to Frankfurter, December 30, 1931, 1-5, Herbert B. Ehrmann MSS; Buckner to Frankfurter, February 27, 1934; Buckner to Edmund Morgan, January 31, 1930, Frankfurter MSS, LC; Charles E. Clark to Frank, March 26, 1931, Frank MSS; interview with Raoul Berger (1971), for Chicago conditions; Arnold to J. G. Driscoll, June 12, 1934, Arnold MSS.

49. Edmund Morgan to Frankfurter, June 1, 1932, Frankfurter MSS, LC; Mary (?) to Frankfurter [April 1936], Frankfurter MSS, LC; Thurman Arnold to David L. Landy, January 29, 1936; Arnold to Edward J. Dimock, December 4, 1935; George B. Case to Arnold, April 4, 1934; Dimock to Arnold, November 25, 1935; Horace G. Hitchcock to Arnold, December 31, 1935, Arnold MSS; Frankfurter to LaRue Brown, January 13, 1937, Roll 14, Frankfurter MSS, HLS; Ernst to Stone, January 4, 1939, Ernst MSS.

50. M. T. Van Hecke to Landis, December 2, 1932, Box 8, Landis MSS; Landis to Albert J. Harno, February 9, 1931, Box 6, Landis MSS; Leon Green to Arnold, February 28, 1936, Arnold MSS.

51. Malcolm A. Hoffmann, *Government Lawyer* (New York, 1956), 237. For a strikingly similar statement by a young Jewish economist, see Studs Terkel, *Hard Times: An Oral History of the Great Depression* (New York, 1970), 266; FDR to Burlingham, February 6, 1936, Roll 3, Frankfurter MSS, HLS.

52. Nathan Margold to Frankfurter, March 27, 1933, Roll 21, Frankfurter MSS, HLS; Frank to Mack, April 17, 1933; Frank to Frankfurter, April 18, 1933, Frank MSS; Stevenson to Ellen Stevenson [July 1933], in Johnson (ed.), *Papers of Adlai E. Stevenson*, I, 249; Harry Sagotsky to Frankfurter, November 6, 1935; Benjamin Pollack to Frankfurter, September 24, 1935; Asher W. Schwartz to Frankfurter, October 16, 1935; Benjamin J. Levin to Frankfurter, January 29, 1936; Allan Rosenberg to Frankfurter, June 8, 1936; Jerome R. Hellerstein to Frankfurter, April 19, 1933; Jacob Salzman to Frankfurter, April 15, 1936; Vartak Bulbankian to Frankfurter, February 25, 1936, Frankfurter MSS, LC; Charles W. Quick to Pound, March 24, 1939, 6-4, Pound MSS; Nelson H. Nichols, Jr., to Charles Houston, February 24, 1934, C-81, NAACP MSS.

CHAPTER SEVEN

1. "In Time of Stress," *ABAJ*, 17 (1931), 666.
2. Report of the Special Committee on Administrative Law, ABA *Reports*,

57 (1934), 542-43, 549; Jacob M. Lashley, "Administrative Law and the Bar," *Virginia Law Review*, 25 (1939), 645, 648. See Walter Gellhorn, *Federal Administrative Proceedings* (Baltimore, 1941), 1-3.

3. William Ransom, "Address," State Bar Association of North Dakota, *Bar Briefs*, 13 (December 1936), 51; Ransom, "The Profession of Law: Its Present and Future," in *Law: A Century of Progress, 1835-1935* (New York, 1937), 152; Walter Tuller to Davis, May 2, 1935, Davis MSS.

4. Guy H. Thompson, "Address," New Jersey State Bar Association, *Yearbook* (1932-33), 117-28; Clarence E. Martin, "The Law in Retrospect and Prospect," *ABAJ*, 19 (1933), 137-41; Newton D. Baker, "The Lawyer's Function in Modern Society," *ABAJ*, 19 (1933), 261-64; Earle W. Evans, "Responsibility and Leadership," *ABAJ*, 20 (1934), 589-93; William L. Ransom, "Government and Lawyers," Iowa State Bar Association, *Proceedings* (1936), 88; Adrian Raymond, "Revolutions and the Profession," *ABAJ*, 18 (1932), 864; Jacob K. Javits, "The Lawyer's Place in the Coordination of Government and Business," *Commercial Law Journal* (March 1938), 71; John A. Garver to James Beck, February 2, 1933, Box 2, Beck MSS. See also ABA *Reports*, 61 (1936), 2.

5. Quoted in Francis Biddle, *A Casual Past* (New York, 1961), 358.

6. Swaine, *Cravath Firm*, II, 451n; Henry W. Taft to James Beck, November 20, 1935, Box 3, Beck MSS; Henry P. Lawther to David A. Simmons, March 26, 1935, B 19/21, Simmons MSS; Pa. Bar Association, *Reports* (1935), 68-69.

7. George Wolfskill, *The Revolt of the Conservatives* (Boston, 1962), 71-72; James M. Beck, "The Duty of the Lawyer in the Present Crisis," American Liberty League Document No. 69 (1935), 7.

8. Wolfskill, *Revolt of the Conservatives*, 72.

9. Beck, "The Duty of the Lawyer," 3; Sveinbjorn Johnson, "The Fifty-Eight Lawyers," *USLR*, 70 (1936), 24, 26; "The Fifty-Eight Lawyers," *USLR*, 69 (1935), 505. Johnson's critique was not from a radical perspective; he was appalled that lawyers should "appeal . . . to the mob" on a matter within the province of the judiciary (p. 29).

10. Wolfskill, *Revolt of the Conservatives*, 75-76; Beck to Raoul Desvernine, November 7, 1935, Box 1, Beck MSS; ABA Committee on Professional Ethics, Opinion 148 (November 16, 1935), copy in Vol. 763, ACLU Archives; Desvernine to Beck, November 22, 1935, Box 2, Beck MSS.

11. Stone to Frankfurter, October 28, 1935, Box 372, Frankfurter MSS, LC; James G. Rogers to David A. Simmons, February 8, 1935, B 19/8, Simmons MSS.

12. D. A. Beckman to Roosevelt, March 25, 1937, OF 41, Box 89, FDRL.

13. Davis to Thomas M. Bell, March 22, 1937, Davis MSS; William L. Ransom to Frederick H. Stinchfield, March 5, 1937, copy in Simmons MSS; Burlingham to Frankfurter, February 9, 1937, Box 306, Frankfurter MSS, LC.

14. *ABAJ*, 23 (May 1937), 316; Sterling McNees to Roosevelt, March 6, 1937; F. M. Sercombe to Roosevelt, March 8, 1937; Charles H. Strong to Roosevelt, February 26, 1937, OF 41, Box 108, FDRL.

15. Ransom to Stinchfield, March 30, 1937, Box 24, Newton Baker MSS.

16. Stinchfield to Arthur T. Vanderbilt, November 26, 1937, Simmons MSS; Stinchfield, "The Attack on the Legal Profession," *Vital Speeches*, 4 (October 15, 1937), 28.

17. Newton Baker to Stinchfield, March 2, 1937, Box 24, Baker MSS; H. D. Kissinger to Roosevelt, February 12, 1937, OF 41, Box 101, FDRL; New York State Bar Ass'n, *Proc.* (1938), 109-10.

18. "Lawyers and Legal Events," *USLR*, 68 (February 1934), 107; T. R. Powell to Frankfurter, February 13, 1934, A-6, Powell MSS; Chafee to James C. Collins, April 18, 1936, 2-23, Chafee MSS; Pound to Chafee, September 8, 1934, 2-22, Chafee MSS; Collins to Chafee, April 16, 1936, 2-23, Chafee MSS. Jerome Frank had earlier described Evans as "a small town lawyer with a good heart and little education or experience to wake him from his dogmatic slumbers," Frank to Frankfurter, February 29, 1932, Box II, Frank MSS; H. Milton Colvin to FDR, September 4, 1933, OF 3260, FDRL; James G. Rogers to Frankfurter, January 6, 1933, Frankfurter MSS.

19. Arnold to Roy M. Hardy, May 17, 1937, Arnold MSS; ABA *Reports*, 60 (1935), 178-80.

20. The Court fight also encouraged lawyers at the lower strata of professional life to voice their grievances against the professional elite. The overwhelming preponderance of favorable responses to the judicial reform bill came from solo practitioners, country or urban, who did not share the values of corporate partners or bar association members. Urban Jewish lawyers were most enthusiastically in favor of the Court plan and used it to voice their displeasure with corporate lawyers and professional associations. One New Jersey attorney reminded Roosevelt that opposition to his plan arose solely from "moneyed lawyers," while another urged the President to disregard the opinions of lawyers entirely because "as a class they were never on the side of progress." A New York City lawyer dismissed bar association polls because they ignored the sentiments of the professional rank-and-file. It was, another lawyer complained, "the old men of the Bar" who were as disproportionately represented in professional activities as they were on the Supreme Court itself. See Stanley Cohen to Roosevelt, March 11, 1937, OF 41A; Louis Josephson to FDR, March 8, 1937; Elias H. Avram to FDR, February 27, 1937; Saul Klein to FDR, March 5, 1937; Oscar Anderson to FDR, March 7, 1937, all in OF 41, FDRL.

21. Thomas I. Emerson, "The Role of the Guild in the Coming Year," *Lawyers Guild Review*, 10 (1950), 1; National Lawyers Guild constitutional preamble quoted in Esther Lucile Brown, *Lawyers and the Promotion of Justice* (New York, 1938), 147.

22. Morris Ernst, *A Love Affair With the Law* (New York, 1968), 10-12; Ernst, "The Future of the Small Law Office Under the New Deal," May 5, 1934, copy in Ernst MSS; Henry Muller to Ernst, May 5, 1934; Ernst to Muller, May 8, 1934, *ibid.*; Ernst to Charles Houston, November 10, 1936, C-81, NAACP MSS.

23. Mortimer Riemer, "Report of the Executive Secretary," February 19, 1937, National Lawyers Guild MSS; Ernst to Houston, November 10, 1936, C-81, NAACP MSS; Karl Llewellyn to Norbert Brockman, April 18, 1961, R XIV.4, Llewellyn MSS; Thomas Emerson Memoir, COHC, 521-22; "A Call to American Lawyers" [Winter, 1936-37], Walsh MSS; John P. Devany, "The Quarterly," *NLGQ*, 1 (December 1937), 1-2; Ernst to FDR, January 18, 1937, Ernst MSS.

24. NLG Press Release (1936); Walsh to Erving C. Bland, December 18, 1936; Minutes of Executive Board Meeting, June 26, 1937; Walsh to Victor Hemphill, January 2, 1937; NLG Membership Report, May 31, 1937; J. Francis Reilly to Walsh, December 18, 1936; Walsh to William G. Rice, Jr., January 4, 1937; Raymond Pace Alexander to Walsh, January 12, 1937; C. B. McCullar to Walsh, January 6, 1937, Walsh MSS; William Hastie to Walter White, February 15, 1939, C-64, NAACP MSS.

25. Membership Report, June 30, 1938, Walsh MSS; Minutes of Executive Board meeting, June 26, 1937, NLG MSS; Ernst to Devany, October 19, 1937, Ernst MSS; Herbert M. Brune, Jr., to Frankfurter, February 8, 1937, Roll 14, Frankfurter MSS, HLS; "The Function of Committee Reports of the National Lawyers Guild: A Symposium," *NLGQ*, 1 (1938), 411-18.

26. Ernst to Walsh, December 31, 1936; W. B. Rubin to Walsh, August 25, 1937, Walsh MSS; M. L. Sovern to Ernst, November 19, 1937; Ernst to Thurman Arnold, February 17, 1938, Ernst MSS; Ernst to Ferdinand Pecora, February 28, 1938, Box 15, Frank MSS.

27. Minutes of Executive Board, February 10, 1939, NLG MSS.

28. *Ibid.*

29. Ernst to Frederic R. Coudert, Jr., March 3, 1939; Ernst to M. L. Sovern, May 19, 1939; Emil Schlesinger to George Quilici, May 26, 1939; Ernst to Robert Jackson, February 14, 1940, Ernst MSS.

30. Frank to Robert Jackson, March 3, 1939, enclosing letter to John Gutknecht, Thurman Arnold MSS; Frank to Gutknecht, March 8, 1939, Box 12, Frank MSS.

31. Ernst was more successful in the American Civil Liberties Union, which adopted his anti-totalitarian resolution, purged a member of the Communist Party from its board of directors, and thereby maintained its organizational life at the sacrifice of libertarian principle. See Jerold S. Auerbach, "The Depression Decade," in Alan Reitman (ed.), *The Pulse of Freedom: American Liberties, 1920-1970's* (New York, 1975), 65-104.

32. Frank Walsh to Ernst, May 2, 1940, Ernst MSS.

33. Llewellyn, "The Bar's Troubles, and Poultices—and Cures?" *Law and Contemporary Problems*, 5 (Winter 1938), 109.

34. Quoted in Stephen Love *et al.*, "Economic Security and the Young Lawyer: Four Views," *Illinois Law Review*, 32 (1937-38), 666, 668.

35. *Ibid.*, 667-68; Llewellyn, "The Bar's Troubles," 109-10, 112, 114-17; AALS, *Proc.*, 31 (1933), 64-65.

36. Llewellyn, "The Bar Specializes," 191-92; AALS, *Proc.*, 31 (1933), 66.

37. Llewellyn to Edward T. Lee, October 30, 1933, A, II, 57; Llewellyn to Harrison Tweed, November 25, 1935, H, III, 2, Llewellyn MSS; Llewellyn to Jerome Frank, November 3, 1933, Box 3, Frank MSS; Llewellyn, "The Bar's Troubles," 124-26.

38. Charles E. Clark and Emma Corstvet, "The Lawyer and the Public: An A.A.L.S. Survey," *Yale Law J.*, 47 (1937-38), 1293; Clark to W. B. Petermann, January 8, 1937, H, III, 2, Llewellyn MSS; Love *et al.*, "Economic Security," 670; Reginald Heber Smith, "The Bar Association Law Office for Persons of Moderate Means (Designed for a Large City)," *B.U. Law R.*, 19 (April 1939), 227-28, 242; John S. Bradway, "Low Cost Legal Service Bureaus," *North Carolina Law R.*, 17 (1938-39), 114.

39. Arthur A. Brooks, Jr., to Llewellyn, January 27, 1940; Elliott Goldstein to Llewellyn, February 16, 1940; James O. McCulloch to Llewellyn, July 1, 1938; Sharl B. Bass to Llewellyn, April 20, 1937, H, III, 2, Llewellyn MSS.

40. Herman Steerman to Llewellyn, August 20, 1937, H, III, 2, Llewellyn MSS; Isidor Lazarus, "The Economic Crisis in the Legal Profession," *NLGQ*, 1 (1937-38), 23-26; Garrison, "The Legal Profession and the Public," *NLGQ*, 1 (1937-38), 128, 131.

41. "Proposal for a Legal Service Bureau for the Metropolitan Area of Chicago," *NLGQ*, 1 (1937-38), 149-54.

42. Robert D. Abrahams, "Law Offices to Serve Householders in the Lower Income Group," *Dickinson Law R.*, 42 (1937-38), 133-39; Abrahams, "The Neighborhood Law Office Experiment," *U. Chicago Law R.*, 9 (1942), 410-11, 412, 414, 416-17; Abrahams, "The New Philadelphia Lawyer," *Atlantic Monthly*, 185 (April 1950), 72; Abrahams, "The Neighborhood Law Office Plan," *U. Wis. Law R.* (1949), 634-40, 643, 646.

43. "Report of the Committee on Professional Economics on the Plan for Neighborhood Law Offices Authorized by the National Lawyers Guild" (n.d.), 4, 6-8.

44. Names identified from NLG letter head, February 1938, A, II, 36, Llewellyn MSS.

45. The guild even had an impact upon the American Bar Association. In the summer preceding the guild's birth the ABA, irritated with criticism that it was merely a self-appointed group of self-interested corporate lawyers, had transformed itself into a more inclusive, federalized association. Al-

though its leaders described the guild as, at best, an "affront," and, in less charitable moments, as a "viper," they began to appreciate the need to further modify their own image. In 1938 association president Arthur T. Vanderbilt read a newspaper account of an address by Grenville Clark, senior partner in a prominent Wall Street firm, in which Clark had observed that the defense of civil liberties had been permitted "to drift largely into the hands of elements of 'the Left' " while conservatives remained "unfortunately quiescent" in the face of intrusions on the Bill of Rights. Clark urged bar associations to reverse direction. Vanderbilt arranged a meeting with Clark and Frank J. Hogan, the association's president elect, to implement the proposal. In his inaugural address Hogan urged ABA members to "make it clear to the whole country that . . . the organized bar and lawyers as citizens . . . think in terms of human welfare, the rights, the security, and the happiness of the average man." Hogan called for the establishment of a new committee vested with authority "to take a staunch and militant stand" in defense of the Bill of Rights. Clark, its first chairman, observed what a "new departure" this was for the association. As the committee declared in its *amicus* brief in *Hague v. C.I.O.*, an important freedom of assembly case, ABA participation marked "a significant step forward in the history of the Association and its relationship to public affairs." It was a step prompted by the challenge of the National Lawyers Guild's expanded definition of professional responsibility, a definition that had thrust the guild forward as the only professional association committed to a vigorous defense of the Bill of Rights. Sunderland, *History of the ABA*, 173-79; William Ransom to Edgar B. Tolman, March 11, 1937; Arthur T. Vanderbilt to Tolman, March 11, 1937, Box 24, Newton Baker MSS; Grenville Clark, "Conservatism and Civil Liberty," *ABAJ*, 24 (August 1938), 641, 644; Irving Dilliard, "Grenville Clark: Public Citizen," *American Scholar*, 33 (1963-64), 98-101; Frank J. Hogan, "Lawyers and the Rights of Citizens," *ABAJ*, 24 (August 1938), 615-17; Clark to Chafee, August 13, 1938, Bill of Rights Committee, Chafee MSS, "Association's Committee Intervenes to Defend Rights of Public Assembly," *ABAJ*, 25 (January 1939), 7.

46. Quoted in Harvard Law School *Bulletin*, 22 (February 1971), 13; Charles Houston, "Tentative Findings Re Negro Lawyers," 3/3, Pound MSS; Carter G. Woodson, *The Negro Professional Man and the Community* (Washington, D.C., 1934), 194-95, 228.

47. Houston, "Survey of Howard University Law Students" (1927), 3/3, Pound MSS; Houston, "Tentative Findings"; Woodson, *Negro Professional Man*, 191-96, 221; William Henri Hale, "The Career Development of the Negro Lawyer in Chicago" (unpublished Ph.D. thesis, University of Chicago, 1949), 71, 79, 150.

48. Houston, "Tentative Findings"; National Bar Association, *Proc.*, 6

(1930), viii, 22; Raymond Pace Alexander to James Weldon Johnson, September 21, 1929, D-4, NAACP MSS; Alexander, "The National Bar Association—Its Aims and Purposes," *National Bar J.*, 1 (1941), 4.

49. G. James Fleming, "A Philadelphia Lawyer," *Crisis*, 46 (November 1939), 329-30, 347; Frankfurter to Ernst, June 5, 1930, Box 324, Frankfurter MSS, LC; John P. Davis to Julian W. Mack, March 17, 1928; Mack to Pound, March 20, 1928, 2/4, Pound MSS.

50. Alexander, "The Negro Lawyer: His Duty in a Rapidly Changing Social, Economic and Political World," National Bar Association, *Proc.*, 6 (1930), 4-5, 7; Houston, "Tentative Findings"; William T. Andrews to Alexander, September 25, 1929, D-4, NAACP MSS; Woodson, *Negro Professional Man*, 224.

51. Clement E. Vose, *Caucasians Only: The Supreme Court, the NAACP and the Restrictive Covenant Cases* (Berkeley, Calif., 1967), 40; Memo from Committee on Negro Work to Directors, American Fund for Public Service, October 18, 1929, C-196, NAACP MSS; Walter White to American Fund for Public Service, June 21, 1930, AFPS MSS; *Twenty-First Annual Report of the NAACP* (1930), 17.

52. Frankfurter to Ernst, June 5, 1930, Box 324, Frankfurter MSS, LC; Nathan Margold biographical information; Agreement between NAACP and Margold, May 26, 1930; Margold to Walter White, March 31, 1931, all in C-196; Margold, "Preliminary Report," C-200, NAACP MSS.

53. Walter White, *A Man Called White* (New York, 1948), 142.

54. Houston, "Survey"; Houston to Walter White, May 21, 1935, C-64, NAACP MSS; William H. Hastie, "Toward an Equalitarian Legal Order, 1930-1950," *Annals*, 407 (May 1973), 21-22.

55. Jack Greenberg, *Race Relations and American Law* (New York, 1959), 22; Walter White to Ralph Harlow, December 19, 1933, C-104; White to Joel Spingarn, January 4, 1933, D-63; James Weldon Johnson to Ira Jayne, April 9, 1926, D-86; Houston to NAACP, 26th Annual Conference, June 24, 1935, C-87, NAACP MSS; Oliver Allen, "Chief Counsel for Equality," *Life*, 38 (June 13, 1955), 141; Loren Miller, *The Petitioners* (Cleveland, 1966), 260-61, 446-47.

56. Woodson, *Negro Professional Man*, 224; Stenographer's Minutes, *U.S. v. William Z. Foster et al.* (1949), 9378-79, 9385, 9392, 9395-96.

57. Grenville Clark to Walter White, February 2, 1939, C-192, NAACP MSS; Henry U. Sims to J. Glenn Tucker, October 23, 1943; A. G. Powell to David A. Simmons, November 3, 1943; James L. Shepherd, Jr., to J. T. Powell, November 13, 1943; Sims to Simmons, November 15, 1943, Simmons MSS.

58. *The Autobiography of Malcolm X* (New York, 1964), 36.

59. Harold A. Katz, "Labor Lawyers," *Va. Law Weekly Dicta: Labor Law*, 3 (1950-51), 83; Brainerd Currie, "The Materials of Law Study," *J.*

Legal Ed., 8 (1955), 4-5n; Pound to S. E. Turner, March 24, 1921, Box 32, Pound MSS.

60. U.S. Comm. on Industrial Relations, *Final Report and Testimony*, XI, 64th Cong., 1st Sess., Doc. 415 (1916), 10775; Darrow to Miss S. (n.d.), Box 1, Darrow MSS.

61. Lowell S. Hawley and Ralph Bushnell Potts, *Counsel for the Damned: A Biography of George Francis Vanderveer* (Philadelphia, 1953), 181-91, 222-28, 312, 320; "The Lawyers of Labor," *Fortune*, 63 (March 1961), 213.

62. Louis Waldman, *Labor Lawyer* (New York, 1944), 17-25, 128; Harry S. Kessler to Roger Baldwin, September 29, 1921, Vol. 181; Harry M. McKee to Baldwin, August 8, 1922, Vol. 213, ACLU Archives; *New York Times*, January 15, 1922; Jacob Margolis to Baldwin, February 19, 1921, Vol. 185, ACLU Archives.

63. Sidney Fine, *Sit-Down: The General Motors Strike of 1936-1937* (Ann Arbor, Mich., 1969), 164, 194; Malcolm A. Hoffmann, *Government Lawyer* (New York, 1956), 14; Francis J. McTernan quoted in Ann Fagan Ginger (ed.), *The Relevant Lawyers* (New York, 1972), 144.

64. Edward Lamb, *No Lamb for Slaughter* (New York, 1963), 37-38, 49, 52; "The Lamb Disbarment Proceedings," *Yale Law Journal*, 47 (1938), 1386, 1389n; *The Attempted Disbarment of Edward Lamb* (National Lawyers Guild, 1938).

65. Charles Garry and Allan Brotsky quoted in Ginger, *Relevant Lawyers*, 85, 98.

66. Waldman, *Labor Lawyer*, 144-45; Robert M. Segal, "Labor Union Lawyers: Professional Services of Lawyers to Organized Labor," *Industrial and Labor Relations Review*, 5 (April 1952), 343n, 347, 349, 363.

67. Archibald Cox, "The Labor Law Courses," Harvard Law School *Bulletin*, 2 (April 1951), 6; "Lawyers of Labor," 214.

68. Bernard Schwartz, "Administrative Law," *International Encyclopedia of the Social Sciences*, I, 81; Charles Evans Hughes, "Important Work of Uncle Sam's Lawyers," *ABAJ*, 17 (April 1931), 237-39; James G. Rogers, "Forces Remolding the Lawyer's Life," *ABAJ*, 17 (October 1931), 640-41; Robert M. Cooper, "Administrative Justice and the Role of Discretion," *Yale Law J.*, 47 (February 1938), 600-601.

69. James M. Landis, *The Administrative Process* (New Haven, Conn., 1938), 46; Louis L. Jaffe, "James Landis and the Administrative Process," *Harvard Law R.*, 78 (1964), 320; Frankfurter, "Foreword," *Yale Law J.*, 47 (1938), 517-18; Frankfurter to Thomas D. Thacher, December 3, 1934, Frankfurter MSS, HLS. See also Gellhorn, *Federal Administrative Proceedings*, 116-22.

70. Dulles to Frankfurter, October 24, 1938; Frankfurter to Dulles, December 31, 1938; Dulles, "Administrative Law," address at Harvard Law School, January 14, 1939, ABA 1938-52, John Foster Dulles MSS.

71. Douglas to Dulles, February 28, 1939, *ibid.*

72. Charles A. Horsky, *The Washington Lawyer* (Boston, 1952), 10; Karl Krastin, "The Lawyer in Society—A Value Analysis," *Western Reserve Law R.*, 8 (1957), 422.

73. For Frankfurter's concern about lawyers capitalizing upon government expertise in the service of private clients, see Frankfurter to Landis, January 11, 1936, Box 10, Landis MSS.

74. "The Federal Legal Service," *Federal Bar Association J.*, 2 (1934), 99-101; Harold M. Stephens to Frankfurter, September 16, 1933, Frankfurter MSS, HLS; Richard A. Solomon, "Practice of Law in the Federal Government—Career or Training Ground," *George Washington Law R.*, 38 (1969-70), 754; Peyton Ford, *The Government Lawyer* (New York, 1952), 5-6.

75. Charles Gordon, "The Lawyer and the Civil Service," *NLGQ*, 1 (1938), 294-96; Paul V. McNutt, "The Lawyer in Government," *Federal Bar Association J.*, 3 (November 1939), 359-404.

76. Gerard D. Reilly, "Founding a Career System for Government Lawyers," *Lawyers Guild R.*, 1 (1941), 1-2; *Report of President's Committee on Civil Service Improvement*, 77th Cong., 1st Sess., House Document No. 118 (1941), 12, 29-31; Robert H. Jackson, "Government Counsel and Their Opportunity," *ABAJ*, 26 (May 1940), 412.

77. *Felix Frankfurter Reminisces*, 248.

78. See Shklar, *Legalism*, 8-10. She writes: "The dislike of vague generalities, the preference for case-by-case treatment of all social issues, the structuring of all possible human relations into the form of claims and counterclaims under established rules, and the belief that the rules are 'there'—these combine to make up legalism as a social outlook. . . . It is an openly, intrinsically, and quite specifically conservative view . . ." (p. 10). A somewhat different expression of legalism appears in Henry M. Hart, Jr., and Albert M. Sacks, "The Legal Process: Basic Problems in the Making and Application of Law," mimeographed (Cambridge, Mass., 1958). The authors refer to "the central idea of law" as "the principle of institutional settlement [which] expresses the judgment that decisions which are the duly arrived at result of duly established procedures . . . ought to be accepted as binding upon the whole society unless and until they are duly changed." As the authors concede—revealing the core of legalism—"we . . . brush aside further discussion of what it [law] 'ought' to be" (p. 4).

79. Charles A. Reich, *The Greening of America* (New York, 1970), 51.

80. Geraint Perry, *Political Elites* (New York, 1969), 32-33.

81. See Christopher Lasch, *The Agony of the American Left* (New York, 1969), 137.

CHAPTER EIGHT

1. Joseph C. Goulden, *The Superlawyers* (New York, 1972), 156-61; Thurman Arnold, *Fair Fights and Foul* (New York, 1965), 190, 204; Abe Fortas, "Thurman Arnold and the Theatre of the Law," *Yale Law J.*, 79 (1970), 990.

2. Harbaugh, *Lawyer's Lawyer*, 254.

3. N.Y. State Bar Ass'n, *Proc.*, 71 (1948), 109. See the discussion at 93 ff.

4. Tom C. Clark, "Civil Rights: The Boundless Responsibility of Lawyers," *ABAJ*, 32 (August 1946), 456-57.

5. "To Insure Domestic Tranquility," *ABAJ*, 34 (January 1948), 44; "The Present Crisis," *ABAJ*, 34 (April 1948), 300; Frank B. Ober, "Communism v. the Constitution: The Power to Protect Our Free Institutions," *ABAJ*, 34 (August 1948), 777.

6. ABA *Reports*, 73 (1948), 85, 145; Ross Malone to ABA President Frank E. Holman, Oct. 7, 1948, ABA Committees–A, Malone MSS.

7. U.S. Cong., House, Committee on Un-American Activities, *Report on the National Lawyers Guild*, 81st Cong., 2d Sess., Report No. 3123 (1950), 1, 5, 21.

8. Thomas I. Emerson, "Answer to the Report of the House Committee on Un-American Activities on the National Lawyers Guild," *Lawyers Guild R.*, 10 (1950), 45-46; "The National Lawyers Guild: Legal Bulwark of Democracy," *Lawyers Guild R.*, 10 (1950), 93-110; *Lawyers Guild R.*, 14 (1954), 17.

9. ABA *Reports*, 78 (1953), 340.

10. Jessica Davidson, "Report of the Administrative Secretary," *Lawyers Guild R.*, 14 (1954-55), 168.

11. *Ibid.*; *Lawyers Guild R.*, 14 (1954), 15-17; *Lawyers Guild R.*, 18 (1958), 135.

12. Royal W. France, "The Reasons for My Present Predicament," *Lawyers Guild R.*, 16 (1956), 20.

13. Alex Elson, "Extending Legal Service to the Low and Moderate Income Groups," *Lawyers Guild R.*, 8 (1948), 295-99; "The Availability of Legal Services and Judicial Processes to the Low and Moderate Income Groups," *Lawyers Guild R.*, 10 (1950), 9, 12, 21. For the Rushcliffe recommendations, see Brian Abel-Smith and Robert Stevens, *In Search of Justice* (London, 1968), 32.

14. Robert G. Storey, "The Legal Profession versus Regimentation," State Bar of California, *Proc.*, 23 (1950), 22, 25, 27; Harold J. Gallagher, "Annual Address," Pa. Bar Ass'n., *Report*, 55 (1950), 107.

15. U.S. Cong., Senate, Special Subcommittee on Investigation of Committee on Government Operations, 83rd Cong., 2d Sess. (1954), 2426-30.

16. ABA *Reports,* 75 (1950), 154.
17. ABA *Reports,* 76 (1951), 591-92, 600.
18. *ABAJ,* 37 (1951), 312-13.
19. ABA *Reports,* 77 (1952), 178-79.
20. *Ibid.,* 282; ABA *Reports,* 78 (1953), 132, 291-93. By this time some committee members had second thoughts. See pp. 133, 304-8.
21. *ABAJ,* 36 (1950), 972.
22. Chafee to Abe Fortas, January 10, 1951, 85-6, Chafee MSS. For Chafee's early career, see Jerold S. Auerbach, "The Patrician as Libertarian: Zechariah Chafee, Jr. and Freedom of Speech," *New England Q.,* 42 (1969), 511-31.
23. Chafee, "Purge Trials Are for Russian Lawyers, Not American Lawyers," *N.J. Law Journal,* 74 (May 24, 1951), 169-71; Chafee to Grenville Clark, October 31, 1950, 85-8; Chafee to Ernest Angell, October 20, 1950, 84-9; Chafee to Carl L. Shipley, November 16, 1950, 85-7, Chafee MSS.
24. Chafee to Angell, October 31, 1950, 85-8; Angell to Chafee, November 3, 1950, 85-10; copy of statement, December 8, 1950, 85-5, Chafee MSS.
25. Nelson Trottman to Chafee, February 15, 1951, 85-3; Charles H. Woodward to Chafee, February 3, 1951, 85-1; Carl L. Shipley to Chafee, December 7, 1950, 85-7, Chafee MSS.
26. For general studies of postwar loyalty issues, see Eleanor Bontecou, *The Federal Loyalty-Security Program* (Ithaca, N.Y., 1953); Alan D. Harper, *The Politics of Loyalty* (Westport, Conn., 1969).
27. Vern Countryman, "Loyalty Tests for Lawyers," *Lawyers Guild R.,* 13 (1953), 149.
28. Chafee, *The Blessings of Liberty* (Philadelphia, 1956), 159-60, 337-38n; Chafee to Monte M. Lemann, January 26, 1953, 85-2, Chafee MSS; Ralph S. Brown, Jr., and John D. Fassett, "Loyalty Tests for Admission to the Bar," *Chicago Law R.,* 20 (1952-53), 497-98; Martin, *Causes and Conflicts,* 274-85.
29. "The Independence of the Bar," *Lawyers Guild R.,* 13 (1953), 163; Fowler Harper, "Loyalty and Lawyers," *Lawyers Guild R.,* 11 (1951), 206.
30. Milton R. Konvitz, *Expanding Liberties* (New York, 1966), 130.
31. Stenographer's Minutes, *U.S. v. William Z. Foster et al.* (1948-49), 16, 130, copy in Medina MSS.
32. Report of the Association of the Bar of the City of New York, Special Committee on Courtroom Conduct, *Disorder in the Court* (New York, 1973), 50.
33. Medina to Spencer Gordon, January 20, 1949; J. Edgar Hoover and J. Howard McGrath to Medina, October 25, 1949; McCarthy to Medina, October 31, 1949, Medina MSS.
34. George W. Alger to Medina, March 11, 1949, Medina MSS; Chafee to Harry Nims, June 22, 1951, 34-10, Chafee MSS.
35. *U.S. v. Sacher,* 182 F. 2d 416, 430 (1950). Medina's belief that defense lawyers were spreading Communist propaganda and intending to abort the

trial and wear him down physically and mentally, is presented in Hawthorne Daniel, *Judge Medina* (New York, 1952), 231-32.

36. Harold R. Medina, "Courage and Independence at the Bar," *Ohio Bar*, 25 (May 26, 1952), 387; Vern Countryman, "The Bigots and the Professionals," *Nation*, 174 (June 28, 1952), 641.

37. This account is drawn from Marvin Schick, *Learned Hand's Court* (Baltimore, 1970), 290-91. Schick's version is based upon research in the Frank and Clark MSS.

38. *Ibid.*, 291.

39. *U.S. v. Sacher*, 182 F. 2d 416, 420, 428-29 (1950).

40. *Ibid.*, at 430-31.

41. *Ibid.*, at 455.

42. *Ibid.*, at 420.

43. The importance of Frank's switch for *Dennis* was implicitly acknowledged in the opinion in that case, written by Learned Hand, who observed: "It is not irrelevant that this court decided that [defense counsel] so far exceeded the bounds of professional propriety as to deserve a sentence for criminal contempt." *U.S. v. Dennis*, 183 F. 2d 201, 225 (1950).

44. *U.S. v. Sacher* at 454.

45. Memorandum for Conference from Justice Robert H. Jackson, on petition for rehearing of *Sacher v. U.S.*, October 9, 1951, Frankfurter MSS, LC.

46. *Sacher et al. v. U.S.*, 343 US 1, 11 (1952).

47. *Ibid.* at 18, 37-38.

48. *Association of the Bar of the City of New York v. Sacher*, 206 F. 2d 358, 360-62, 366 (1953).

49. *Sacher v. Association of the Bar of the City of New York*, 347 US 388, 389 (1954).

50. *In re Isserman*, 9 NJ 269, 279 (1952).

51. *Re Disbarment of Abraham J. Isserman*, 345 US 286, 293 (1953). In another instance, not mentioned, John W. Davis had been cited for contempt for hitting an attorney in court; subsequently he threw an inkwell. Harbaugh, *Lawyer's Lawyer*, 60.

52. *Re Disbarment of Abraham J. Isserman*, 348 US 1 (1954).

53. *Association of the Bar of the City of New York v. Isserman*, 271 F. 2d 784, 785 (1959).

54. *In re Isserman*, 35 NJ 198, 204 (1961). For preliminary and interim comment, see Fowler Harper and David Haber, "Lawyer Troubles in Political Trials," *Yale Law J.*, 60 (1951), 1-56; David L. Weissman, "Sacher and Isserman in the Courts," *Lawyers Guild R.*, 12 (1952), 39-47; *Lawyers Guild R.*, 14 (1954), 19-31.

55. "The Nature and Consequences of Forensic Misconduct in the Prosecution of a Criminal Case," *Columbia Law R.*, 54 (1954), 980.

56. William O. Douglas, "The Black Silence of Fear," *N.Y. Times Magazine* (January 13, 1952), 7, 37-38.

57. *Hallinan v. U.S.*, 182 F. 2d 88 (1950); *MacInnis v. U.S.*, 191 F. 2d 157, 161 (1951). For Hallinan's account see his *A Lion in Court* (New York, 1963), 244 ff.

58. *Schlesinger v. Musmanno*, 367 Pa. 476 (1951).

59. *Schlesinger Appeal*, 404 Pa. 584, 590 (1961).

60. *Sheiner v. State*, 82 So. 2d 657, 662 (1955).

61. *Braverman v. Bar Association of Baltimore*, 209 Md. 328, 336 (1956); cert. denied 352 US 830 (1956).

62. *New York Times*, October 8, 1973, December 13, 1973.

63. *In re Sawyer*, 260 F. 2d 189, 192, 200 (1958); *In re Sawyer*, 360 US 622, 631 (1959).

64. "The Independence of the Bar," *Lawyers Guild R.*, 13 (1953), 161-62.

65. ABA *Reports*, 78 (1953), 307.

66. Hannah Bloom, "The Law's Diminishing Returns," *Nation*, 173 (December 29, 1951), 556.

67. Douglas, "The Black Silence of Fear," 38. See also Samuel M. Koenigsberg and Morton Stavis, "Test Oaths: Henry VIII to the American Bar Association," *Lawyers Guild R.*, 11 (1951), 126; Morris L. Ernst and Alan U. Schwartz, "The Right to Counsel and the 'Unpopular Cause,'" *University of Pittsburgh Law R.*, 20 (1959), 727-29; Charles Alan Wright, "Right to Counsel and Counsel's Rights," *Nation*, 177 (November 21, 1953), 427; Leon Jaworski, "The Unpopular Cause," *ABAJ*, 47 (July 1961), 714.

68. Fowler Harper, "Loyalty and Lawyers," *Lawyers Guild R.*, 11 (1951), 205; *New York Times*, August 9, 1951; "Independence of the Bar," *Lawyers Guild R.*, 13 (1953), 165; Joseph E. Downs and Alvin L. Goldman, "The Obligation of Lawyers to Represent Unpopular Defendants," *Howard Law J.*, 9 (1963), 49, 55-56. For a contrary view, blaming Communists for their predicament, see the address by Whitney North Seymour in N.Y. State Bar Association, *Proc.*, 75 (1952), 54-55.

69. Steve Nelson, *The 13th Juror* (New York, 1955), 118; "Independence of the Bar," 164-65.

70. *Application of Levy*, 214 F. 2d 331 (1954); 348 US 978 (1955).

71. *Schware v. Board of Bar Examiners of the State of New Mexico*, 355 US 232, 236-39, 244, 246 (1957).

72. *Konigsberg v. State Bar of California*, 353 US 252, 262 (1957).

73. John T. McTernan, "Schware, Konigsberg and Independence of the Bar: The Return to Reason," *Lawyers Guild R.*, 17 (1957), 48-53. For a gloomier assessment, see William H. Rehnquist, "The Bar Admission Cases: A Strange Judicial Aberration," *ABAJ*, 44 (March 1958), 229-32.

74. *In re Patterson*, 210 Ore. 495 (1956); 353 US 952 (1957).

75. *Application of Patterson,* 213 Ore. 398 (1958); 356 US 947 (1958).
76. *Konigsberg v. State Bar of California,* 366 US 36 (1961).
77. *Ibid.,* 80.
78. *In re George Anastaplo,* 121 N.E. 2d 826 (1954); 348 US 946 (1955).
79. *In re George Anastaplo,* 163 N.E. 2d 429, 439 (1959).
80. *In re George Anastaplo,* 366 US 82, 89 (1961).
81. *Ibid.,* 114, 115-16.
82. Malcolm P. Sharp, "The Conservative Fellow Traveler," *Chicago Law R.,* 30 (1963), 717; for additional comment on the case see "The Illinois Bar and Individual Freedom," *Northwestern Law R.,* 50 (1955), 94-104; Harry Kalven, Jr., and Roscoe T. Steffen, "The Bar Admission Cases: An Unfinished Debate Between Justice Harlan and Justice Black," *Law in Transition,* 21 (1961-62), 155-96; Osmond K. Fraenkel, "Heresy and the Illinois Bar," *Lawyers Guild R.,* 12 (1952), 163-78.
83. Arthur F. Canfield, "The Bar Must Repudiate the Communist Lawyer," *N.J. Law J.,* 74 (June 14, 1951), 6.
84. Quoted in Harbaugh, *Lawyer's Lawyer,* 437.
85. Frankfurter to Charles E. Wyzanski, Jr., February 22, 1951, Wyzanski MSS.
86. The examples, but not the interpretation, are drawn from Harbaugh, *Lawyer's Lawyer,* 437-61. According to Harbaugh (p. 437), Davis "set a high example of devotion to the lawyer's ideal."
87. *Ibid.,* 443, 445.
88. Clark to Ross Malone, April 9, 1952, ABA Committees I, Malone MSS.
89. Martin, *Causes and Conflicts,* 277.
90. Quoted in Daniel H. Pollitt, "Counsel for the Unpopular Cause: The 'Hazard of Being Undone,'" *N. Ca. Law R.,* 43 (1964-65), 18. See also Milnor Alexander, "The Right to Counsel for the Politically Unpopular," *Law in Transition,* 22 (Spring 1962), 41-42.
91. Jack H. Olender, "Let Us Admit Impediments," *U. Pittsburgh Law R.,* 20 (1959), 752.
92. Benjamin Dreyfus and Doris Brin Walker, "Grounds and Procedure for Discipline of Attorneys," *Lawyers Guild R.* (Summer 1958), 69; Alexander, "Right to Counsel," 41-42; Koenigsberg and Stavis, "Test Oaths," 126n.
93. Quoted in Robert Shogan, *A Question of Judgment* (Indianapolis, 1972), 59. See Owen Lattimore, *Ordeal by Slander* (Boston, 1950), 27-28, 214-15.
94. Jonathan D. Casper, "Lawyers and Loyalty-Security Litigation," *Law & Society R.,* 3 (May 1969), 579-82, 584.
95. Statement by George Crockett, Stenographer's Minutes, pp. 16, 144, Medina MSS.

96. Royal W. France, *My Native Grounds* (New York, 1957), 61, 79, 94-98, 100-101; France to Frankfurter, January 4, 1952, Frankfurter MSS, LC; France, "The Reasons for My Present Predicament," 19-20.

97. Even the American Civil Liberties Union had a dismal Cold War record. See Mary S. McAuliffe, "The Politics of Civil Liberties: The American Civil Liberties Union During the McCarthy Years," in Robert Griffith and Athan Theoharis (eds.), *The Specter: Original Essays on the Cold War and the Origins of McCarthyism* (New York, 1974), 154-70. See Otto Kirchheimer, *Political Justice* (Princeton, N.J., 1961), 6, 233-34, and chs. 3-4, 6 for a theoretical framework for considerations of political justice.

98. "The Privilege to Practice Law *versus* the Fifth Amendment Privilege to Remain Silent," *Northwestern Law R.*, 56 (1961), 644; Lloyd Wright, "The Lawyer's Responsibility and the Fifth Amendment," *Neb. Law R.*, 34 (1954-55), 584; Ralph Brown, "Lawyers and the Fifth Amendment: A Dissent," *ABAJ*, 40 (May 1954), 407.

99. For the abuses of disciplinary procedures, see "The Imposition of Disciplinary Measures for the Misconduct of Attorneys," *Columbia Law R.*, 52 (1952), 1039, 1047-48, 1053.

100. See Walter Murphy, *Congress and the Court* (Chicago, 1962).

101. Learned Hand, *The Bill of Rights* (New York, 1964), 38-39, 67, 71-72.

102. Herbert Wechsler, "Toward Neutral Principles of Constitutional Law," *Harvard Law R.*, 73 (1959), 11, 12, 15, 17, 19.

103. Henry M. Hart, Jr., "Foreword: The Time Chart of the Justices, The Supreme Court, 1958 Term," *Harvard Law R.*, 73 (1959), 99, 101, 125; G. Edward White, "The Evolution of Reasoned Elaboration: Jurisprudential Criticism and Social Change," *Va. Law R.*, 59 (1973), 287, 289-90.

104. Paul L. Murphy, *The Constitution in Crisis Times 1918-1969* (New York, 1972), 471. The most outspoken contemporary advocate of this school has been Alexander M. Bickel, whose *The Least Dangerous Branch* (Indianapolis, 1962) and *The Supreme Court and the Idea of Progress* (New York, 1970) bracketed a decade which exposed Bickel's "passive virtues" as a legalistic smokescreen for conservative politics. The Court, Bickel wrote in *The Least Dangerous Branch*, "must pronounce only those principles which can gain 'widespread acceptance' " (p. 239). For a telling critique of the implications of such passivity for civil liberties, see Martin Shapiro, *Freedom of Speech: The Supreme Court and Judicial Review* (Englewood Cliffs, N.J., 1966). An effective rebuttal to Bickel appears in J. Skelly Wright, "Professor Bickel, The Scholarly Tradition, and the Supreme Court," *Harvard Law Review*, 84 (1971), 769-805.

105. Thurman Arnold, "Professor Hart's Theology," *Harvard Law Review*, 73 (1960), 1310; Hart, "Foreword," 101.

106. Arnold, "Professor Hart's Theology," 1310, 1314, 1316-17. For ABA opposition to the Warren Court, see John Nolan, "The Supreme Court

versus the ABA," *Commonweal*, 70 (May 15, 1959), 179; Murphy, *Congress and the Court*, 95-96, 118-19, 164-65, 225-27. In 1959 Chief Justice Warren resigned from the ABA. *New York Times*, February 21, 1959.

107. See Stephen C. Yeazell, "The Ideology of Legal Method: 1880-1925" (Third-year paper, Harvard Law School, 1974), 8, 43-44.

108. White, "Reasoned Elaboration," 289-90.

109. Lon Fuller, "American Legal Philosophy at Mid-Century," *Journal of Legal Education*, 6 (1954), 464.

CHAPTER NINE

1. Marvin E. Frankel, "*The Alabama Lawyer*, 1954-64: Has the Official Organ Atrophied?" *Columbia Law Review*, 64 (1964), 1254-55; August Meier and Elliott Rudwick, *CORE: A Study in the Civil Rights Movement 1942-1968* (New York, 1973), 270-71; Chauncey M. Depuy to David E. Maxwell, October 18, 1956, Committees I #3, Ross Malone MSS; Daniel H. Pollitt, "Timid Lawyers and Neglected Clients," *Harper's*, 229 (August 1964), 81-86; *New York Times*, July 4, 1963; *In re Brown*, 346 F. 2d 903 (1965); William M. Kunstler and Arthur Kinoy, "Southern Justice: Lawyers Walk in Fear," *Nation*, 198 (June 8, 1964), 576-80.

2. Gunnar Myrdal, *An American Dilemma* (New York, 1944), 325-26; G. Franklin Edwards, *The Negro Professional Class* (Glencoe, Ill., 1959), 23, 135-38; William H. Hale, "The Negro Lawyer and His Clients," *Phylon*, 13 (1952), 57, 59; Earl L. Carl and Kenneth R. Callahan, "Negroes and the Law," *Journal of Legal Education*, 17 (1964-65), 254; Kenneth S. Tollett, "Black Lawyers, Their Education, and the Black Community," *Howard Law J.*, 17 (1972), 326-57; Ernest Gellhorn, "The Law Schools and the Negro," *Duke Law J.* (1968-69), 1070; W. Willard Wirtz, "Address," *AALS Proc.* (1963), 100.

3. "Summer Project: Mississippi, 1964," *National Lawyers Guild Practitioner*, 24 (1965), 32-43; *New York Times*, August 14, 1963, June 28, 1964; John Honnold, "The Bourgeois Bar and the Mississippi Movement," *ABAJ*, 52 (March 1966), 228-29.

4. U.S. District Court, Eastern District of Louisiana, New Orleans Division, Civil Action No. 67-243, *Sobol et al. v. Perez et al.*, Brief of the United States (1968), loaned to the author by Professor Owen Fiss; *NAACP v. Button*, 371 US 415 (1963); *Sobol v. Perez*, 289 F. Supp. 392 (1968); *New York Times*, April 17, 1967.

5. Victor S. Navasky, *Kennedy Justice* (New York, 1971), 184-85, 205, 207, 219-21.

6. William M. Kunstler, *Deep In My Heart* (New York, 1966), 36, 358 ff.; Charles Morgan, Jr., *A Time to Speak* (New York, 1964), 41, 83; Mark

Howe, "Foreword: Our Splendid Bauble," in Leon Friedman (ed.), *Southern Justice* (New York, 1965), vii.

7. Harrison Tweed to Charles Evans Hughes, April 27, 1937, Hughes MSS; Report of Standing Committee on Lawyer Referral Service, ABA *Reports*, 77 (1952), 235-36.

8. "Neighborhood Law Offices: The New Wave in Legal Services for the Poor," *Harvard Law R.*, 80 (1967), 805-6; Harry P. Stumpf, "Law and Poverty: A Political Perspective," *Wisconsin Law R.* (1968), 695 ff.; Edgar S. and Jean C. Cahn, "The War on Poverty: A Civilian Perspective," *Yale Law J.*, 73 (1964), 1317-52; Patricia M. Wald, *Law and Poverty: 1965* (Washingington, D.C., 1965), 44-49.

9. Memo from F. William McCalpin, December 29, 1964, ABA Correspondence, NLADA MSS; Cahn and Cahn, "The War on Poverty," 1334; Richard Pious, "Congress, The Organized Bar, and the Legal Services Program," *Wisconsin Law R.* (1972), 418-19.

10. Cahn and Cahn, "The War on Poverty," 1335; National Conference on Law and Poverty, *Proceedings* (Washington, D.C., 1965), v, 64-65; "Neighborhood Law Offices," 805; Earl Johnson, Jr., "The O.E.O. Legal Services Program," *Catholic Lawyer*, 14 (Spring 1968), 100.

11. Memo from F. William McCalpin, December 29, 1964; Lewis Powell, Jr. to Bert H. Early, January 24, 1965, ABA Correspondence, NLADA MSS; *ABAJ*, 51 (June 1965), 548; Pious, "Congress, The Organized Bar, and the Legal Services Program," 420.

12. Lewis F. Powell, Jr., "Guidelines for Legal Service Proposals," April 3, 1965, ABA Correspondence, NLADA MSS. The contents of this memo contradict claims that the ABA played "a central role" in the creation of OEO legal services, or that Powell displayed "open-mindedness to new ideas" on the subject of federal legal services. See F. Raymond Marks, *Lawyers, The Public, and Professional Responsibility* (Chicago, 1972), 186; Wald, *Law and Poverty*, 98.

13. Pious, "Congress, The Organized Bar, and the Legal Services Program," 420-22.

14. See Harry P. Stumpf *et al.*, "The Legal Profession and Legal Services: Explorations in Local Bar Politics," *Law & Society R.*, 6 (August 1971), 47-48, 59-64; Jerry A. Green and Ellen S. Green, "The Legal Profession and the Process of Social Change: Legal Services in England and the United States," *Hastings Law J.*, 21 (1970), 583-84, 592; Stumpf, "Law and Poverty," 703-4; Norman Dorsen (ed.), "Poverty, Civil Liberties, and Civil Rights: A Symposium," *NYU Law R.*, 41 (1966), 336-37, 346.

15. Statement by Sargent Shriver, April 26, 1966, ABA Correspondence, NLADA MSS; Shriver, "Legal Services and the War on Poverty," *Catholic Lawyer*, 14 (1968), 95. See Gary G. Bellow, "The Extension of Legal Serv-

ices to the Poor: New Approaches to the Bar's Responsibility," in Arthur E. Sutherland (ed.), *The Path of the Law from 1967* (Cambridge, Mass., 1968), 122-23.

16. Dorsen, "Poverty, Civil Liberties, and Civil Rights," 328; Alan F. Westin in *New York Times Book Review* (November 24, 1968).

17. Johnson, "The O.E.O. Legal Services Program," 101-4; Edward V. Sparer, "The Role of the Welfare Client's Lawyer," *UCLA Law R.,* 12 (1965), 375; Louis H. Masotti and Jerome R. Corsi, "Legal Assistance for the Poor: An Analysis and Evaluation of Two Programs," *J. Urban Law,* 44 (1967), 483-502; E. Clinton Bamberger, Jr., "The Legal Services Program of the Office of Economic Opportunity," *Notre Dame Lawyer,* 41 (1965-66), 847-48.

18. "Neighborhood Legal Services and the War Against Poverty," *Welfare Law Bulletin,* No. 6 (December 1966), 1-2.

19. *New York Times,* June 26, 1965; Theodore Vorhees to Powell, June 16, 1965; Powell to Bert Early, September 24, 1965, ABA Correspondence, NLADA MSS.

20. "Neighborhood Law Offices," 843-44; Stumpf, "Law and Poverty," 716; A. Kenneth Pye and Raymond F. Garraty, Jr., "The Involvement of the Bar in the War Against Poverty," *Notre Dame Lawyer,* 41 (1965-66), 864-66; Billie Bethel and Robert Kirk Walker, "Et Tu, Brute!" *Tenn. Bar J.,* 1 (August 1965), 13, 18, 25.

21. Jerome Carlin, "Will Poverty Program Impoverish Lawyers?" *NLG Practitioner,* 24 (1965), 132-33; Stumpf, "Law and Poverty," 708-11.

22. Harry P. Stumpf and Robert J. Janowitz, "Judges and the Poor: Bench Responses to Federally Financed Legal Services," *Stanford Law R.* (May 1969), 1074-76; Frederic R. Merrill, "Utilization of Legal Services By the Poor and the Private Practicing Lawyer" (Chicago, 1969), 18-19, 41.

23. *Troutman v. Shriver,* 417 F. 2d 171 (1969); *Matter of Community Action for Legal Services, Inc.,* 26 App. Div. 2d 354 (1966); James J. Graham, "The Ghetto Lawyer," *Catholic Lawyer,* 14 (Winter 1968), 37-45, 87; Michael Botein, "The Constitutionality of Restrictions on Poverty Law Firms: A New York Case Study," *NYU Law R.,* 46 (1971), 749-61, 765-66. The strength of professional values within OEO also limited its potential for social change. For one example of the conflict between the interests of OEO lawyers and their clients, see Chester Hartman, *Yerba Buena: Land Grab and Community Resistance in San Francisco* (San Francisco, 1974), 137-40.

24. John D. Robb, "Controversial Cases and the Legal Services Program," *ABAJ,* 56 (April 1970), 329-31; Stumpf, *et al.,* "Legal Profession and Legal Services," 65 n. 7; Fred J. Hiestand, "The Politics of Poverty Law," in Bruce Wasserstein and Mark J. Green (eds.), *With Justice for Some* (Boston, 1970), 160-89; Ginger, *Relevant Lawyers,* 245.

25. "With the Editors," *Harvard Law R.*, 71 (March 1958), vii-viii; Spencer Klaw, "The Wall Street Lawyers," *Fortune*, 57 (February 1958), 140-44, 192-202.

26. Ralph Nader, "Law Schools and Law Firms," *Case and Comment* (May-June 1970), 34.

27. Alan A. Stone, "Legal Education on the Couch," *Harvard Law R.*, 85 (1971), 393-94.

28. Neil M. Levy, review in *Michigan Law R.*, 72 (1974), 916-17; Paul N. Savoy, "Toward a New Politics of Legal Education," *Yale Law J.*, 79 (January 1970), 455 n; Nader, "Law Schools and Law Firms," 30-31, 34; Edgar S. and Jean Camper Cahn, "Power to the People or the Profession?—The Public Interest in Public Interest Law," *Yale Law J.*, 79 (May 1970), 1025, 1027.

29. Savoy, "Toward a New Politics of Legal Education," 502; Arthur Kinoy, "The Present Crisis in American Legal Education," *Rutgers Law R.*, 24 (Fall 1969), 2-3, 5-6; Stone, "Legal Education on the Couch," 393-94.

30. Duncan Kennedy, "How the Law School Fails: A Polemic," *Yale Review of Law and Social Action*, 1 (Spring 1970), 80; Dorsen, "The Role of the Lawyer," 61.

31. Robert Lefcourt, "Introduction," Lefcourt (ed.), *Law Against the People* (New York, 1971), 4; Coffin quoted in Jessica Mitford, *The Trial of Dr. Spock* (New York, 1969), 77; *Time* (May 4, 1971), 52.

32. Karl Mannheim, "The Problem of Generations," in *Essays on the Sociology of Knowledge* (New York, 1952), 293-94, 301; Jerry J. Berman and Edgar S. Cahn, "Bargaining for Justice: The Law Students' Challenge to Law Firms," *Harvard Civil Rights–Civil Liberties Law R.*, 5 (January 1970), 16-17; Ronald Dworkin, "There Oughta Be a Law," *New York Review of Books* (March 14, 1968), 18; Richard D. Schwartz, "From the Editors," *Law & Society R.*, 3 (August 1968), 3.

33. *New York Times*, February 15, 17, 1968; Michael Garrett and Jean Zimmer Pennington, "Will They Enter Private Practice?" *ABAJ*, 57 (July 1971), 663-66; Mark Green, "Law Graduates: The New Breed," *Nation*, 210 (June 1, 1970), 660.

34. Berman and Cahn, "Bargaining for Justice," 22-25; John E. Robson, "Private Lawyers and Public Interest," *ABAJ*, 56 (April 1970), 332-34; "With the Editors," *Harvard Law R.*, 83 (1970), vii; *Time* (April 18, 1969), 77; Derek C. Bok, "A Different Way of Looking at the World," Harvard Law School *Bulletin*, 20 (March-April 1969), 6.

35. Abe Fortas, "Thurman Arnold and the Theatre of the Law," *Yale Law J.*, 79 (May 1970), 991, 997.

36. Robert T. Swaine, "Impact of Big Business on the Profession: An Answer to Critics of the Modern Bar," *ABAJ*, 35 (1949), 89, 170-71.

37. *New York Times*, May 24, 1969.

38. Fortas, "Thurman Arnold and the Theatre of the Law," 995-96, 998.

39. Lloyd N. Cutler, review in *Harvard Law Review*, 83 (1970), 1746, 1750; Mimeographed statement from Wilmer, Cutler & Pickering, November 1969, copy in author's possession courtesy of Mr. Cutler, who graciously expanded upon the points raised in his review in private correspondence.

40. Barlow F. Christensen, *Specialization* (Chicago, 1967), 20.

41. "The New Public Interest Lawyers," *Yale Law J.*, 79 (1970), especially 1070-71 n.

42. Marks, *The Lawyer, The Public, and Professional Responsibility*, 10-29; Dietrich Rueschemeyer, *Lawyers and Their Society* (Cambridge, Mass., 1973), 112-15; David P. Riley, "The Challenge of the New Lawyers: Public Interest and Private Clients," *George Washington Law R.*, 38 (1969-70), 549-50.

43. David Mellinhoff, *The Conscience of a Lawyer* (St. Paul, Minn., 1973), 9.

44. Riley, "The Challenge of the New Lawyers," 549-50, 551, 553.

45. Marks, *The Lawyer, The Public, and Professional Responsibility*, 89, 135. For the *pro bono* activities of one firm, see Peter S. Smith and John E. Kratz, Jr., "Legal Services for the Poor—Meeting the Ethical Commitment," *Harvard Civil Rights–Civil Liberties Law R.*, 7 (May 1972), 509-19. The final assessment is in Marks, pp. 246, 275.

46. Ford Foundation, *The Public Interest Law Firm: New Voices for New Constituencies* (New York, 1973), 5, 9, 31-32, 34; Peter E. Sitkin and J. Anthony Kline, "Financing Public Interest Litigation," *Arizona Law R.*, 13 (1971), 823-27; Lenney Hegland, "Beyond Enthusiasm and Commitment," *Arizona Law R.*, 13 (1971), 805-17.

47. Robert Lefcourt, "The First Law Commune," *Evergreen Review*, 15 (October 1971), 29-31, 55-57; "An Interview with Gerald Lefcourt," in Jonathan Black (ed.), *Radical Lawyers* (New York, 1971), 311; James Douglas, "Organization, Ego and the Practice of Alternative Law," *Yale Review of Law and Social Action*, 2 (Fall 1971), 88-89, 92; *New York Times*, September 5, 1971.

48. Riley, "The Challenge of New Lawyers," 552; "Public Interest Lawyers," 1119-20, 1146-47; Peter Vanderwicken, "The Angry Young Lawyers," *Fortune*, 84 (September 1971), 74-77, 125-27.

49. Marlise James, *The People's Lawyers* (New York, 1972), 17-19, 97-98, 101; *New York Times*, December 16, 1969; Ginger, *Relevant Lawyers*, 73; Victor S. Navasky, "Right On! With Lawyer William Kunstler," *New York Times Magazine* (April 19, 1970), 30-31, 88-93; Kunstler, *Deep In My Heart*, 358.

50. Ginger, *Relevant Lawyers*, 26; Paul Harris, "You Don't Have to Love the Law to be a Lawyer," *NLG Practitioner*, 28 (Fall 1969), 97-100; Henry

di Suvero, "The Movement and the Legal System," in Black, *Radical Lawyers*, 57; Kunstler, "Open Resistance: In Defense of the Movement," in Lefcourt, *Law Against the People*, 270; James, *People's Lawyers*, 225, 229.

51. Rueschemeyer, *Lawyers and their Society*, 115.

52. *Brotherhood of Railroad Trainmen v. Va.*, 377 US 1 (1964); *United Mine Workers of America v. Illinois State Bar Association*, 389 US 217 (1967). See *United Transportation Union v. State Bar of Michigan*, 401 US 576, 585 (1971), for the assertion that "collective activity undertaken to obtain meaningful access to the courts is a fundamental right within the protection of the First Amendment."

53. Peter L. Zimroth, "Group Legal Services and the Constitution," *Yale Law J.*, 76 (1966-67), 966-68, 984; *Code of Professional Responsibility* (1969), EC 1-1, EC 2-7, EC 2-25; Eugene L. Smith, "Canon 2: 'A Lawyer Should Assist the Legal Profession in Fulfilling its Duty to Make Legal Counsel Available,'" *Texas Law R.*, 48 (1969-70), 286-88.

54. R. W. Nahstoll, "Limitations on Group Legal Services Arrangements Under the Code of Professional Responsibility, DR 2-103 (D)(5): Stale Wine in New Bottles," *Texas Law R.*, 48 (1969-70), 334, 341, 345; Smith, "Canon 2," 289, 309; Harold Brown, "ABA Code of Professional Responsibility: In Defense of Mediocrity," *Valparaiso Law R.*, 5 (1970), 106.

55. John F. Sutton, Jr., "The American Bar Association Code of Professional Responsibility: An Introduction," *Texas Law R.*, 48 (1969-70), 260-62; James H. Burnley IV, "Solicitation by the Second Oldest Profession: Attorneys and Advertising," *Harvard Civil Rights–Civil Liberties Law R.*, 8 (1973), 80-84, 93-94, 96.

56. "Legal Ethics and Professionalism," *Yale Law J.*, 79 (1970), 1182, 1197; Smith, "Canon 2," 287-88; Donald T. Weckstein, "Maintaining the Integrity and Competence of the Legal Profession," *Texas Law R.*, 48 (1969-70), 268-69.

57. Report of the Association of the Bar of the City of New York, Special Committee on Courtroom Disorder, *Disorder in the Court* (New York, 1973), xiii-xiv; *New York Times*, May 19, 1971, March 16, 1974; ABA *Reports*, 96 (1971), 495; ABA, Special Committee on Evaluation of Disciplinary Enforcement, *Problems and Recommendations in Disciplinary Enforcement* (1970), xvii; William A. Stannmeyer, "The New Left and the Old Law," *ABAJ*, 55 (April 1969), 319-23.

58. Mark L. Levine, George C. McNamee, Daniel Greenberg (eds.), *The Tales of Hoffman* (New York, 1970), 40, 225-26; "A Lawyer for Hire," *ABAJ*, 56 (June 1970), 552.

59. Navasky, "Right On!" 92-93; *New York Times*, May 19, 1973, February 21, 1974.

60. *Hallinan v. Committee of Bar Examiners*, 65 Cal. 2d, 447 (1966); *Baird v. State Bar of Arizona*, 401 US 1, 9, 22 (1971); *Application of Stolar*, 401

US 23 (1971); Jerold S. Auerbach, "The Xenophobic Bar," *Nation*, 214 (June 19, 1972), 784-86.

61. Report of the Association of the Bar of the City of New York, *Disorder in the Court*, xiii-xiv.

62. David N. Rockwell, "Controlling Lawyers by Bar Associations and Courts," *Harvard Civil Rights–Civil Liberties Law R.*, 5 (1970), 301-3, 309, 312-14; Christopher S. Lyman, "State Bar Discipline and the Activist Lawyer," *Harvard Civil Rights–Civil Liberties Law Review*, 8 (1973), 385, 392.

63. Seymour Warkov with Joseph Zelan, *Lawyers in the Making* (Chicago, 1965), xviii; Harry T. Edwards, "New Role for the Black Law Graduate—A Reality or an Illusion," *Michigan Law R.*, 69 (1971), 1407-10; "Symposium on the Black Lawyer in America Today," *Harvard Law School Bulletin*, 22 (February 1971), 46-47; Earl L. Carl, "The Shortage of Negro Lawyers: Pluralistic Legal Education and Legal Services for the Poor," *J. Legal Education*, 20 (1967-68), 23; *New York Times*, April 7, 1974.

64. Robert Stevens, "Law Schools and Law Students," *Virginia Law R.*, 59 (1973), 572-73, 598-600; Robert M. O'Neil, "Preferential Admissions: Equalizing Access to Legal Education," *University of Toledo Law R.* (1970), 281, 294-95, 298, 309. For a recent example of elite opposition to "lowering" of standards, see Robert W. Meserve, "The Quality of Intellectual Competition," *J. Legal Education*, 25 (1973), 378-85. Meserve (then ABA President), like Felix Frankfurter a generation earlier, viewed law school grading as a "realistic objective standard of evaluation" which established a true meritocracy.

65. *New York Times*, December 22, 1970, April 7, 1974; Wayne Green, "Taking the Bar Exam to Court," *Juris Doctor* (January 1974), 27-29; "Southern Lawyers: Blacks and the Bar," *Civil Liberties* (March 1974), 1-2.

66. Charles Anderson, "Black Lawyers in the 20 Largest Firms: It's Better than Before and Worse than Ever," *Juris Doctor* (January 1973), 6-7, 12-13; Edwards, "A New Role for the Black Law Graduate," 1428, 1440; Marion S. Goldman, *A Portrait of the Black Attorney in Chicago* (Chicago, 1972), 49; "Symposium on Black Lawyers," 44.

67. O'Neil, "Preferential Admissions," 296; Arthur J. Paone and Robert Ira Reis, "Effective Enforcement of Federal Nondiscrimination Provisions in the Hiring of Lawyers," *Southern California Law R.*, 40 (1966-67), 618 n; Harlan F. Stone to Charles C. Burlingham, March 12, 1917, Stone MSS; James J. White, "Women in the Law," *Michigan Law R.*, 65 (1967), 1051-1122; Cynthia Fuchs Epstein, "Women and Professional Careers: The Case of the Woman Lawyer" (Unpublished Ph.D. thesis, Columbia University, 1968), 4, 194-95, 220; Janette Barnes, "Women and Entrance to the Legal Profession," *J. Legal Education*, 23 (1971), 293. For a prematurely optimistic assessment, see Erwin O. Smigel, "The Wall Street Lawyer Reconsidered," *New York*, 2 (August 18, 1969), 41.

68. Mark J. Green, "The Young Lawyers 1972: Goodbye to Pro Bono," *New York,* 5 (February 21, 1972), 29, 33-34; *Harvard Law Record,* 58 (March 1, 1974), 5; Theodore M. Becker and Peter R. Meyers, "A Survey of Chicago Law Student Opinions and Career Expectations," *Northwestern Law R.,* 67 (1972-73), 628-42. For a different view, based on a narrow sample and a form of tabulation that conceals changes within a decade, see Rita J. Simon *et al.,* "Have There Been Significant Changes in the Career Aspirations and Occupational Choices of Law School Graduates in the 1960's?" *Law & Society R.,* 8 (1973), 95-108.

69. Anthony J. Mohr and Kathryn J. Rodgers, "Legal Education: Some Student Reflections," *J. Legal Education,* 25 (1973), 403-26; Herbert L. Packer and Thomas Ehrlich, *New Directions in Legal Education* (New York, 1972), 21, 55.

70. Letter from Don Solomon, *Harvard Law Record,* 58 (March 1, 1974), 11.

71. J. Skelly Wright, "Professor Bickel, the Scholarly Tradition, and the Supreme Court," *Harvard Law R.,* 84 (1971), 770-72, 777, 780, 803, 805. Wright's targets, beyond Bickel, included Learned Hand, Herbert Wechsler, Erwin Griswold, Henry Hart, and Philip Kurland.

72. U.S. Senate, Subcommittee on Employment, Manpower, and Poverty of Committee on Labor and Public Welfare, *Hearings on S. 1305 and S. 2007,* 92nd Cong., 1st Sess. (1971), 1580; Stephen Wexler, "Practicing Law for Poor People," *Yale Law J.,* 79 (1970), 1049, 1059; Arlie Schardt, "Saving Legal Services," *Civil Liberties* (May 1973), 3-4; *New York Times,* July 6, 1972.

73. *Hearings on S. 2007,* 1569-70; *New York Times,* February 14, 19, 1973, June 13, 1973, July 12, 19, 1974. In 1974 the Supreme Court severely restricted class action suits by requiring that all potential parties to the suit must be notified—at no small expense to the litigant.

74. Statements by Jacob Fuchsberg of the American Trial Lawyers Association (also a member of the Legal Services National Advisory Commission) and Edward L. Wright, ABA President in 1971, *Hearings on S. 2007,* 1461, 1478; Don Kates of CRLA, quoted in Ginger, *Revelant Lawyers,* 249; Lois G. Forer, *"No One Will Lissen": How Our Legal System Brutalizes the Youthful Poor* (New York, 1970), 19; Wexler, "Poor People," 1067. A *New York Times* study of sentencing practices in state and federal courts in New York disclosed wide disparities that correlated with economic status, race, and geography of defendants. *New York Times,* July 5, 1973.

75. Morris L. Ernst, "Free Speech and Civil Disobedience," *American Criminal Law Q.,* 3 (1964), 15; Eugene V. Rostow (ed.), *Is Law Dead?* (New York, 1971), 7, 11. See Robert Paul Wolff (ed.), *The Rule of Law* (New York, 1971), 8.

76. *New York Times,* May 3, 1973; Anthony Lewis, "And You Are A Lawyer?" *New York Times,* March 28, 1974.

77. "An Awful Lot of Lawyers Involved," *Time* (July 9, 1973).

78. *New York Times,* July 16, 1973, May 29, 1974; Editorial, "Lawyers' Watergate," *New York Times,* June 11, 1974.

79. Former Supreme Court Justice Tom Clark had warned that public dissatisfaction might generate pressure for public regulation. ABA Special Committee on Evaluation of Disciplinary Enforcement, *Problems and Recommendations in Disciplinary Enforcement* (1970), 2. By 1973, Michigan and Minneosta had experimented with lay involvement in professional discipline; Whitney North Seymour, Jr., added his influential voice to the idea of quasi-public enforcement. "Lawyer Discipline: The Public Is Banging at the Door," *Juris Doctor* (October 1973), 21.

80. *New York Times,* May 29, 1974.

81. Alexander M. Bickel, "Watergate and the Legal Order," *Commentary,* 57 (January 1974), 19-25. For another attempt, before Watergate, to equate the New Left with official lawlessness, see the comment of Harris Wofford, Jr., in Rostow (ed.), *Is Law Dead?,* 33. A more recent professional apology for Watergate lawyers appeared in Joseph W. Bishop, Jr., "Lawyers at the Bar," *Commentary,* 58 (August 1974), 48-53. Bishop argued that (1) because lawyers are always criticized, post-Watergate criticism need not be taken too seriously; (2) given the predominance of lawyers in public life, only their absence from Watergate crimes would deserve notice; (3) law students who are eager to use the legal process for reform bear "an odd resemblance" to Nixon administration lawyers, who also used the legal system for political ends. Bishop here displayed the lawyer's characteristic inability, or refusal, to distinguish between competing sets of values. The issue is not whether law promotes political ends, but whether the political ends that law inevitably promotes are democratic, fair, and just.

82. Tom Wicker, "A Double Standard for Mr. Agnew," *New York Times,* December 28, 1973; Richard Harris, "Reflections: The Watergate Prosecutions," *New Yorker* (June 10, 1974), 46-52, 63.

83. Anthony Lewis, "On the Evidence?" *New York Times,* July 18, 1974; Richard Reeves, "The Trouble With Lawyers: The Case of James St. Clair," *New York* (July 29, 1974), 27-28, 30-31.

AFTERWORD

1. Karl N. Llewellyn, *The Bramble Bush: Our Law and Its Study* (New York, 1930), 144-45.

2. U.S. Congress, Senate, *Hearings . . . on the Nomination of Louis D. Brandeis,* 64th Cong., 1st Sess. (1916), I, 344.

Bibliographical Essay

Abundant annotation makes superfluous anything but a brief list of those sources that I found especially useful. Although I approached every manuscript collection with great expectations, the private papers of practicing lawyers were generally disappointing. Their legal correspondence remained with their law firms; their personal correspondence rarely was reflective about professional issues. The more revealing collections included the following: John W. Davis (Yale) for the difficult transformation from country to corporate lawyer; Morris Ernst (University of Texas, Austin), especially for the National Lawyers Guild; Jerome Frank (Yale), for legal Realism and professional life during the New Deal; James M. Landis (Library of Congress), for academic life and New Deal lawyering; Reginald Heber Smith (National Legal Aid and Defender Association, Chicago), for legal aid; Henry Stimson (Yale), for the merger of corporate practice and public service; William Howard Taft (Library of Congress), for hostility to every major current of legal reform during the Progressive era; and Arthur von Briesen (Princeton), indispensable for the history and ideology of the legal aid movement. I read copies of the Louis D. Brandeis correspondence in the possession of Melvin I. Urofsky, whose excellent edition of the *Letters* incorporates the important items. For the internal politics of the American Bar Association after the New Deal, the papers of David A. Simmons

(University of Texas, Austin) and Ross L. Malone (Texas Technological University) were helpful. Other collections, which yielded specific items or general impressions, included: Newton Baker (Library of Congress); Simeon E. Baldwin (Yale); James Beck (Princeton); Charles J. Bonaparte (Library of Congress); Louis Boudin (Columbia); Clarence Darrow (Library of Congress); John Foster Dulles (Princeton); Herbert Ehrmann (Harvard Law School); Charles Fahy (Library of Congress); Learned Hand (Harvard Law School); Arthur Garfield Hays (Princeton); Morris Hillquit (microfilm edition, State Historical Society of Wisconsin); Charles Evans Hughes (Library of Congress); Robert W. Kenny (University of California, Berkeley); Louis Marshall (American Jewish Archives, Hebrew Union College); Harold R. Medina (Princeton); Frank Polk (Yale); Joseph M. Proskauer (private collection); Donald Richberg (Library of Congress); James G. Rogers (Colorado Historical Society); Elihu Root (Library of Congress); Francis Lynde Stetson (Williams College); Moorfield Storey (Library of Congress and Massachusetts Historical Society); Henry St. George Tucker (Southern Collection, University of North Carolina, Chapel Hill); Samuel Untermeyer (American Jewish Archives, Hebrew Union College); Frank Walsh (New York Public Library); Harry Weinberger (Yale); and Charles E. Wyzanski, Jr. (Harvard Law School).

The papers of law teachers were more illuminating; indeed, it is possible to reconstruct virtually the entire teaching profession through its first two generations from relatively few collections. This task is facilitated by the dominance of Harvard Law School in early legal education and by the splendid archives at that school, a relatively new repository but already the single most important one for modern legal historians. Included are the papers of Zechariah Chafee, Jr., Roscoe Pound, Thomas Reed Powell, Ezra Ripley Thayer, and James Bradley Thayer. The Felix Frankfurter Papers have been divided (unwisely and badly) between the school and the Library of Congress; they comprise a basic source for comprehending the relationship between law schools and public life and the nagging tension between legal education and law practice. Other important collections, which offer perspectives on different schools and educational issues, are: Thurman Arnold (University of Wyoming, Laramie), for the University of West Virginia, Yale, and legal Realism; Charles Clark (Yale Law School), for Yale; Robert M. Hutchins (University of Chicago), for Yale and Chicago; Karl N. Llewellyn (University of Chicago Law School), for Chicago and professional issues during the Depression; William Underhill Moore and Harlan F. Stone (Columbia), for Columbia.

Several organizational collections were helpful: American Civil Liberties Union (Princeton University), for the special problems of civil liberties lawyers; American Fund for Public Service (New York Public Library), for the origins of civil rights litigation strategy; Carnegie Foundation—Correspondence Relating to the Bar (Harvard Law School), for the Reed Report; National Association for the Advancement of Colored People (Library of Congress); National Lawyers Guild (Alexander Meiklejohn Civil Liberties Library, Berkeley); National Legal Aid and Defender Association (American Bar Foundation, Chicago). For the 1930's, the Franklin D. Roosevelt Library (Hyde Park) is essential, especially for lawyers' responses to the Court plan.

Oral history memoirs offer the advantages and distortions of a self-conscious backward look for posterity. Among those of value in the Columbia Oral History Collection were: George W. Alger; Henry Breckinridge; Charles C. Burlingham; Frederic R. Coudert; John W. Davis; Gordon Dean; Thomas I. Emerson; Jerome Frank; Learned Hand; James M. Landis; Chester T. Lane; Lee Pressman; Thomas D. Thacher; Harrison Tweed; and Charles E. Wyzanski, Jr. I conducted interviews with Raoul Berger, Herbert Ehrmann, Paul Freund, Morton J. Horwitz, and Louis Jaffe for the Wiener Oral History Library of the American Jewish Committee. Regrettably, the finances and politics of institutionalized oral history, which require that attention be lavished upon prominent persons, usually leave untouched the far more representative (and often far more valuable) recollections of relatively anonymous individuals who, like Faulkner's Dilsey, just endured.

Legal periodicals must be used cautiously. Although these printed pages have their documentary allure, they represent the voices of a few speaking to many. For the pre–New Deal period, however, before the legal profession became vast and fragmented, some journals reflected the values of a relatively homogeneous elite defending its interest in professional stability. The following were read seriatim for the dates indicated: *Albany Law Journal* (1900-1908); *American Bar Association Journal* (1915-74); *American Law Review* (1900-29); *American Law School Review* (1902-41); *American Lawyer* (1900-1908); *American Legal News* (1916-25); *Bench and Bar* (1905-20); *Case and Comment* (1908-42); *Central Law Journal* (1900-27); *Green Bag* (1900-14); *Law Student's Helper* (1900-15); National Lawyers Guild *Quarterly* (1937-40) and *Review* (1941-54); *United States Law Review* (1929-39); *Women Lawyers' Journal* (1926-40). The *Reports* of the American Bar Association (1890-1941) and the *Proceedings* of the Association of American Law Schools (1901-41) were very revealing. For a sampling of activity at other levels I read the following annual state bar association proceedings: California (1910-41); Illinois (1900-39); New York (1900-41); Ohio (1900-27); Pennsylvania (1900-36); Virginia (1900-14, 1919-21, 1931-36); and Wisconsin (1900-18). Otherwise I relied upon the *Index to Legal Periodicals* for appropriate citations to the voluminous law review literature.

Among lawyers' autobiographies and collections of published letters, the following were useful: Dean Acheson, *Morning and Noon* (Boston, 1965); Thurman Arnold, *Fair Fights and Foul* (New York, 1965); Joseph S. Auerbach, *The Bar of Other Days* (New York, 1940); John T. Barker, *Missouri Lawyer* (Philadelphia, 1949); Francis Biddle, *A Casual Past* (New York, 1961); Melvin I. Urofsky and David W. Levy (eds.), *Letters of Louis D. Brandeis* (Vols. 1-3; Albany, N.Y., 1971-74); John R. Dos Passos, *The American Lawyer* (New York, 1907); William O. Douglas, *Go East Young Man: The Early Years* (New York, 1974); *Felix Frankfurter Reminisces*, Recorded in Talks with Harlan B. Phillips (New York, 1960); *Roosevelt and Frankfurter: Their Correspondence, 1928-1945*, Annotated by Max Freedman (Boston, 1967); Morris Gisnet, *A Lawyer Tells the Truth* (New York, 1931) (see the review by Karl Llewellyn in *Columbia Law Review*, 31 (1931), 1215-20); Vincent Hallinan, *A Lion in Court* (New York, 1963); Arthur Garfield Hays, *City Lawyer* (New York, 1942); Malcolm A. Hoffmann, *Government Lawyer* (New York, 1956); Frank E. Holman, *The Life and Career of a Western Lawyer 1886-1961* (Privately printed, 1963); David J. Danelski and Joseph S. Tulchin (eds.), *The Autobiographical Notes of Charles Evans Hughes* (Cambridge, Mass., 1973); Isidor J. Kresel, unpublished autobiography (1955) (in possession of Stephen Botein); William M. Kunstler, *Deep In My Heart* (New York, 1966); Edward Lamb, *No Lamb for Slaughter* (New York, 1963); Charles Reznikoff (ed.), *Louis Marshall, Champion of Liberty* (2 vols.; Philadelphia, 1957); George Wharton Pepper, *Philadelphia Lawyer* (Philadelphia, 1944); Charles P. Sherman, *Academic Adventures* (New Haven, Conn., 1944); Theron G. Strong, *Landmarks of a Lawyer's Lifetime* (New York, 1914); Henry W. Taft, *A Century and a Half at the New York Bar* (New York, 1938); Louis Waldman, *Labor Lawyer* (New York, 1944); Edward H. Warren, *Spartan Education* (Boston, 1942); Harry Weinberger, "A Rebel's Interrupted Autobiography," *American Journal of Economics and Sociology*, 2 (October 1942), 111-22; Samuel Williston, *Life and Law: An Autobiography* (Boston, 1940); F. Lyman Windolph, *The Country Lawyer* (Philadelphia, 1938). Revealing insights into the legal profession just after the turn of the century appear in Louis D. Brandeis, *Business—A Profession* (Boston, 1914) and Julius Henry Cohen, *The Law: Business Or Profession?* (New York, 1916).

Biographies of lawyers and lawyers turned judges or statesmen abound; good ones are sparse. Invariably, biographers reflect professional norms uncritically, or relegate lawyering activities to the periphery so that they can concentrate on more conspicuous public events. Among those with imaginative perspectives or useful information were Alpheus T. Mason, *Brandeis: A Free Man's Life* (New York, 1946); David W. Levy, "The Lawyer as Judge: Brandeis' View of the Legal Profession," *Oklahoma Law Review,* 22 (November 1969), 374-95; Martin Mayer, *Emory Buckner* (New York, 1968); Otto E. Koegel, *Walter S. Carter: Collector of Young Masters* (New York, 1953); Arthur H. Dean, *William Nelson Cromwell, 1854-1948* (New York, 1957); William Twining, *Karl Llewellyn and the Realist Movement* (London, 1972); Philip C. Jessup, *Elihu Root* (2 vols.; New York, 1938); William B. Hixson, Jr., *Moorfield Storey and the Abolitionist Tradition* (New York, 1972); Alpheus T. Mason, *Harlan Fiske Stone: Pillar of the Law* (New York, 1956). In a class of excellence by itself, as a biography which displays critical sensitivity to the interplay of professional and cultural issues, is William H. Harbaugh, *Lawyer's Lawyer: The Life of John W. Davis* (New York, 1973). For exploration of some methodological problems that beset historians who write about lawyers, see the Review by Jerold S. Auerbach in *Harvard Law Review,* 87 (March 1974), 1100-11.

Histories of law firms, invariably written by a partner, are inevitably celebratory. The most comprehensive of these is Robert T. Swaine, *The Cravath Firm and Its Predecessors 1819-1947* (2 vols.; New York, 1946). (See the reviews by Eugene V. Rostow in *Yale Law Journal,* 58 (1948-49), 650-55, and Mark De Wolfe Howe in *Harvard Law Review,* 60 (May 1947), 838-42.) Other useful firm histories include *Davis Polk Wardwell Gardiner & Reed: Some of the Antecedents* (Privately printed, 1935); Thomas B. Gay, *The Hunton Williams Firm and Its Predecessors 1877-1954* (Richmond, Va., 1971); Timothy N. Pfeiffer and George W. Jaques, *Millbank, Tweed, Hadley & McCloy: Law Practice in a Turbulent World* (New York, 1965); Albert Boyden, *Ropes-Gray 1865-1940* (Boston, 1942); Walter K. Earle, *Mr. Shearman and Mr. Sterling and How They Grew* (Privately printed, 1963).

Until quite recently, the history, sociology, and psychology of legal education was *terra incognita*. Histories of law schools resemble histories of law firms in their ponderous institutionalism. Among the better ones are Foundation for Research in Legal History (under the direction of Julius Goebel, Jr.), *A History of the School of Law, Columbia University* (New York, 1955); Arthur E. Sutherland, *The Law at Harvard: A History of Ideas and Men, 1817-1967* (Cambridge, Mass., 1967); Elizabeth G. Brown, *Legal Education at Michigan, 1859-1959* (Ann Arbor, 1959). The pioneer explorers, whose work has not yet been superseded, were Josef Redlich, *The Common Law and the Case Method in American University Law Schools*, Carnegie Foundation for the Advancement of Teaching, Bulletin No. 8 (New York, 1915); and Alfred Z. Reed, *Training for the Public Profession of the Law*, Carnegie Foundation for the Advancement of Teaching, Bulletin No. 15 (New York, 1921) and *Present-Day Law Schools in the United States and Canada*, Carnegie Foundation for the Advancement of Teaching, Bulletin No. 21 (New York, 1928). There is valuable information in Lon Fuller, "Legal Education in Pennsylvania" (Unpublished ms., 1951) and in Joseph T. Tinnelly, *Part-Time Legal Education: A Study of the Problems of Evening Law Schools* (Brooklyn, N.Y., 1957). Reed's books are far more venturesome than the recent Carnegie Commission Report by Herbert L. Packer and Thomas Ehrlich, *New Directions in Legal Education* (New York, 1972). For an excellent assessment of the Reed Report, see Preble Stolz, "Training for the Public Profession of the Law (1921): A Contemporary Review," AALS *Proceedings* (1971), I, 142-83. There is important information in *The AALS Study of Part-Time Legal Education*, Final Report, Charles D. Kelso (Study Director), AALS *Proceedings* (1972). By far the best historical survey is Robert Stevens, "Two Cheers for 1870: The American Law School," in Donald Fleming and Bernard Bailyn (eds.), *Law in American History* (Boston, 1971). For the professionalization of law teaching in historical context, see Jerold S. Auerbach, "Enmity and Amity: Law Teachers and Practitioners, 1900-1922," also in *Law and American History*. Within the past decade, critical explorations of legal education have appeared in profusion. Among the best of these, either for the mood they reflect or for their qualities of insight and originality, are Thomas F. Bergin, "The Law Teacher: A Man Divided Against Himself," *Virginia Law Review*, 54 (May 1968), 637-57; Duncan Kennedy, "How the Law School Fails: A Polemic," *Yale Review of Law and Social Action*, 1 (Spring 1970), 71-90; Anthony J. Mohr and Kathryn J. Rodgers, "Legal Education: Some Student Reflections," *Journal of Legal Education*, 25 (1973), 403-26; Paul N. Savoy, "Toward a New Politics of Legal Education," *Yale Law Journal*, 79 (January 1970), 444-504; Robert Stevens, "Law Schools and Law Students," *Virginia Law Review*, 59 (1973), 551-685; Preble Stolz, "Clinical Experience in American Legal Education: Why Has It Failed?" in Edmund W. Kitch (ed.), *Clinical Education and the Law School of the Future*, University of Chicago Law School Conference, Series No. 20 (1970), 54-76; Alan A. Stone, "Legal Education on the Couch," *Harvard Law Review*, 85 (December 1971), 392-441.

The history of the American legal profession is virtually unwritten. With a few conspicuous exceptions, most historians who have attempted this task, even within narrow chronological boundaries, have so enclosed themselves in professional norms and pieties that their work resembles the standard self-congratulatory accounts that punctuate bar association proceedings. The best overview remains James Willard Hurst, *The Growth of American Law: The Law Makers* (Boston, 1950), Chs. 12-13. Some of his assumptions, however, invite reconsideration, especially his belief in the passivity of American law and in the absence of professional responsibility for community values. A superficial and uncritical account of the early history of the bar appears in Anton-Hermann Chroust, *The Rise of the Legal Profession in America* (2 vols.; Norman, Okla., 1965), supplemented by "The American Legal Profession: Its Agony and Ecstasy (1776-1840)," *Notre Dame Lawyer*, 46 (Spring 1971), 487-525. Chroust's account is one among many which reflect the conventional (and contradictory) wisdom about the nineteenth-century antebellum professional culture: that it simultaneously represented the "formative era" of American law (as Roscoe Pound insisted) and the nadir of the legal profession. For a broad and telling critique of the assumptions on which these and related conclusions rest, see Morton J. Horwitz, "The Conservative Tradition in the Writing of American Legal History," *American Journal of Legal History*, 17 (1973), 275-94. See also R. Kent Newmyer, "Daniel Webster as Tocqueville's Lawyer: The *Dartmouth College* Case Again," *American Journal of Legal History*, 11 (1967), 127-47; Perry Miller, *The Life of the Mind in America: From the Revolution to the Civil War* (New York, 1965). There are several chapters on the legal profession in Lawrence M. Friedman, *A History of American Law* (New York, 1973). Among books and articles that examine important slices of professional history, the following were especially helpful: Gary B. Nash, "The Philadelphia Bench and Bar, 1800-61," *Comparative Studies in Society and History*, 7 (January 1965), 203-20; William Miller, "American Lawyers in Business and in Politics," *Yale Law Journal*, 60 (January 1951), 66-76; Arnold M. Paul, *Conservative Crisis and the Rule of Law: Attitudes of Bar and Bench, 1887-1895* (New York, 1960); Benjamin R. Twiss, *Lawyers and the Constitution* (Princeton, N.J., 1942); Richard M. Brown, "Legal and Behavioral Perspectives on American Vigilantism," in Fleming and Bailyn (eds.), *Law in American History;* Lawrence M. Friedman, "Law Reform in Historical Perspective," *St. Louis University Law Journal*, 13 (Spring 1969), 351-72; Edward A. Purcell, Jr., "American Jurisprudence Between the Wars: Legal Realism and the Crisis of Democratic Theory," *American Historical Review*, 75 (December 1969), 424-46; G. Edward White, "The Evolution of Reasoned Elaboration: Jurisprudential Criticism and Social Change," *Virginia Law Review*, 59 (1973), 279-302. The best bar association history is George Martin, *Causes and Conflicts: The Centennial History of the Association of the Bar of the City of New York 1870-1970* (Boston, 1970). There is no adequate history of the American Bar Association. For a dated, but still useful, history of legal aid, see John M. Maguire, *The Lance of Justice: A Semi-Centennial History of the Legal Aid Society 1876-1926* (Cambridge, Mass., 1928). Also valuable is Mauro Cappelletti, "Legal

Aid: The Emergence of a Modern Theme," *Stanford Law Review*, 24 (1972), 347-86. Surely the time has come for professional histories that are unrelated to anniversary celebrations. The sociology of law is an exciting new discipline. Sociologists, unlike historians, have been willing to explore aspects of the professional culture that lawyers prefer to ignore. Until the 1950's, the most searching scrutiny came from academic lawyers: Adolf A. Berle, "Modern Legal Profession," *Encyclopaedia of the Social Sciences* (New York, 1933), IX, 340-45; Karl N. Llewellyn, "The Bar Specializes—With What Results?" *Annals of the American Academy of Political and Social Science*, 167 (May 1933), 177-92; Llewellyn, "The Bar's Troubles, and Poultices—And Cures?" *Law and Contemporary Problems*, 5 (Winter 1938), 104-34; Llewellyn, *The Bramble Bush: Our Law and Its Study* (New York, 1930); Fred Rodell, *Woe Unto You, Lawyers!* (New York, 1939). For the sociology of law and the legal profession the following were especially helpful: David Riesman, "Toward an Anthropological Science of Law and the Legal Profession," in *Individualism Reconsidered* (Glencoe, Ill., 1954); Riesman, "Law and Sociology: Recruitment, Training and Colleagueship," *Stanford Law Review*, 9 (July 1957), 643-73; Karl Krastin, "The Lawyer in Society—A Value Analysis," *Western Reserve Law Review*, 8 (September 1957), 409-55; F. James Davis *et al.*, *Society and the Law* (New York, 1962); William Evan (ed.), *Law and Sociology* (Glencoe, Ill., 1962); Edwin M. Schur, *Law and Society: A Sociological View* (New York, 1968); Philip Selznick, "The Sociology of Law," *International Encyclopedia of the Social Sciences* (New York, 1968), IX, 50-58; Jerome H. Skolnick, "The Sociology of Law in America: Overview and Trends," Supplement to *Social Problems*, 13 (Summer 1965), 4-39; Philippe Nonet and Jerome E. Carlin, "The Legal Profession," *International Encyclopedia of the Social Sciences*, IX, 66-72; Walter Probert and Louis M. Brown, "Theories and Practices in the Legal Profession," *University of Florida Law Review*, 19 (Winter 1966-67), 447-85.

Professional differentiation and stratification is, by now, a well-worked subject. Early studies include Charles A. Horsky, *The Washington Lawyer* (Boston, 1952) and Beryl Harold Levy, *Corporation Lawyer: Saint or Sinner?* (Philadelphia, 1961). The pathbreaking books were Jerome Carlin, *Lawyers On Their Own* (New Brunswick, N.J., 1962) and Erwin O. Smigel, *The Wall Street Lawyer* (New York, 1964). (For a sampling of critical reviews, see Arthur H. Dean, *California Law Review*, 52 (1964), 1062-67; Detlev F. Vagts, *Harvard Law Review*, 78 (1964), 491-96; Geoffrey C. Hazard, Jr., "Reflections on Four Studies of the Legal Profession," Supplement to *Social Problems*, 13 (Summer 1965), 46-54; Monroe H. Freedman, *American University Law Review*, 16 (1966), 177-82.) For other types of practice, see Joel F. Handler, *The Lawyer and His Community: The Practicing Bar in a Middle-Sized City* (Madison, Wis., 1967); Thomas Reuben Bell, "Law Practice in a Small Town" (Unpublished JSD thesis, Harvard Law School, 1969); Jack Ladinsky, "The Social Profile of a Metropolitan Bar," Michigan State Bar *Journal*, 43 (February 1964), 12-24. Varieties of practice are examined in John D. Donnell, "The Corporate Counsel: A Role Study" (Unpublished thesis, Harvard Graduate School of Business Administration, 1966); Jonathan D. Casper, *Lawyers Before the Warren Court; Civil Liberties and Civil Rights, 1957-66* (Urbana, Ill., 1972); Cynthia Fuchs Epstein, "Women and Professional Careers: The Case of the Woman Lawyer" (Unpublished Ph.D. thesis, Columbia University, 1968); William Henri Hale, "The Career Development of the Negro Lawyer in Chicago" (Unpublished Ph.D. thesis, University of Chicago, 1949).

There are several excellent studies of professional socialization: Jack Ladinsky, "Careers of Lawyers, Law Practice, and Legal Institutions," *American Sociological Review*, 28 (February 1963), 47-54; Ladinsky, "The Impact of Social Backgrounds of Lawyers on Law Practice and the Law," *Journal of Legal Education*, 16 (1963), 127-44; Dan C. Lortie, "Laymen to Lawmen: Law School, Careers, and Professional Socialization," *Harvard Education Review*, 29 (Fall 1959), 352-69; Seymour Warkov, *Lawyers in the Making* (Chicago, 1965).

The relationship between lawyers' ethics, professional stratification and self-interest has been explored in Max Radin, "Maintenance by Champerty," *California Law Review*, 24 (1935-36), 48-78; Radin, "Contingent Fees in California," *California Law Review*, 28 (1939-40), 587-98; "A Critical Analysis of Rules Against Solicitation by Lawyers," *University of Chicago Law Review*, 25 (Summer 1958), 674-85; Jerome E. Carlin, *Lawyers' Ethics* (New York, 1966); Philip Schuchman, "Ethics and Legal Ethics: The Propriety of the Canons as a Group Moral Code," *George Washington Law Review*, 37 (1968-69), 244-69; James H. Burnley IV, "Solicitation by the Second Oldest Profession: Attorneys and Advertising," *Harvard Civil Rights–Civil Liberties Law Review*, 8 (1973), 77-103. The new Code of Professional Responsibility is explored in a symposium in *Texas Law Review*, 48 (1969-70).

The most comprehensive study of the class bias in American justice and the complicity of lawyers in it remains Reginald Heber Smith, *Justice and the Poor* (New York, 1919). For the trials and tribulations involved in trying to provide legal services based upon need rather than income, see Edgar S. and Jean C. Cahn, "The War on Poverty: A Civilian Perspective," *Yale Law Journal*, 73 (July 1964), 1317-52, and "Power to the People or the Profession?—The Public Interest in Public Interest Law," *Yale Law Journal*, 79 (May 1970), 1005-48; Jerome E. Carlin and Jan Howard, "Legal Representation and Class Justice," *UCLA Law Review*, 12 (January 1965), 381-437; Peter L. Zimroth, "Group Legal Services and the Constitution," *Yale Law Journal*, 76 (1966-67), 966-92; Harry P. Stumpf, "Law and Poverty: A Political Perspective," *Wisconsin Law Review* (1968), 694-733; "Neighborhood Law Offices: The New Wave in Legal Services for the Poor," *Harvard Law Review*, 80 (February 1967), 805-50; Michael Botein, "The Constitutionality of Restrictions on Poverty Law Firms: A New York Case Study," *NYU Law Review*, 46 (October 1971), 748-66; Peter S. Smith and John E. Kratz, Jr., "Legal Services for the Poor—Meeting the Ethical Commitment," *Harvard Civil Rights–Civil Liberties Law Review*, 7 (May 1972), 509-19; Richard Pious, "Congress, the Organized

Bar, and the Legal Services Program," *Wisconsin Law Review* (1972), 418-46. Also useful were Barlow F. Christensen, *Lawyers for People of Moderate Means* (Chicago, 1970), and, for the inadequacies of private firms in this area, F. Raymond Marks, *The Lawyer, the Public, and Professional Responsibility* (Chicago, 1972).

For insights into contemporary generational conflict and alternatives to private practice which compel historical reflection, see Jerry J. Berman and Edgar S. Cahn, "Bargaining for Justice: The Law Students' Challenge to Law Firms," *Harvard Civil Rights–Civil Liberties Law Review*, 5 (January 1970), 16-31; "The New Public Interest Lawyers," *Yale Law Journal*, 79 (May 1970), 1069-1152; "Models of Legal Practice which Enrich the Soul: A Discussion with Four Activist Lawyers," *Yale Review of Law and Social Action*, 1 (Winter 1970), 101-17; James Douglas, "Organization, Ego and the Practice of Alternative Law," *Yale Review of Law and Social Action*, 2 (Fall 1971), 88-92; Robert Lefcourt (ed.), *Law Against the People* (New York, 1971); Jonathan Black (ed.), *Radical Lawyers* (New York, 1971); Ann Fagan Ginger (ed.), *The Relevant Lawyers* (New York, 1972); Marlise James, *The People's Lawyers* (New York, 1972). I have given disproportionate weight in this essay to studies such as these because they are unusually self-conscious and informative about current professional issues whose historical antecedents are explored in this book.

For a sampling of other professional subjects of both historical and contemporary importance, see "Symposium on the Black Lawyer in America Today," *Harvard Law School Bulletin*, 22 (February 1971); Harry T. Edwards, "A New Role for the Black Law Graduate—A Reality or an Illusion," *Michigan Law Review*, 69 (August 1971), 1407-42; "The Jewish Law Student and New York Jobs—Discriminatory Effects in Law Firm Hiring Practices," *Yale Law Journal*, 73 (March 1964), 625-60; David N. Rockwell, "Controlling Lawyers by Bar Associations and Courts," *Harvard Civil Rights–Civil Liberties Law Review*, 5 (April 1970), 301-33; Jerold S. Auerbach, "The Xenophobic Bar," *Nation*, 214

(June 19, 1972), 784-86; Joaquin G. Avila, "Legal Paraprofessionals and Unauthorized Practice," *Harvard Civil Rights–Civil Liberties Law Review*, 8 (1973), 104-27; Barlow F. Christensen, *Specialization* (Chicago, 1967). A defense of traditional styles of practice appears in Abe Fortas, "Thurman Arnold and the Theatre of the Law," *Yale Law Journal*, 79 (May 1970), 988-1004.

Although I have not incorporated a comparative perspective in this book, I have tried to remain aware of professional developments outside the United States. Especially useful were Brian Abel-Smith and Robert Stevens, *Lawyers and the Courts* (Cambridge, Mass., 1967) and *In Search of Justice* (London, 1968); Quintin Johnstone and Dan Hopson, Jr., *Lawyers and Their Work* (Indianapolis, 1967); W. J. Reader, *Professional Men: The Rise of the Professional Classes in Nineteenth-Century England* (New York, 1966); Michael Zander, *Lawyers and the Public Interest* (London, 1968); Dietrich Rueschemeyer, *Lawyers and Their Society: A Comparative Study of the Legal Profession in Germany and in the United States* (Cambridge, Mass., 1973); Marc Galanter, "The Study of the Indian Legal Profession," *Law and Society Review*, 3 (1968-69), 201-17; Richard D. Schwartz, "Lawyers in Developing Societies, Especially India," *Law and Society Review*, 3 (1968-69), 195-99; Lauro Martines, *Lawyers and Statecraft in Renaissance Florence* (Princeton, N.J., 1968); Donald D. Barry and Harold J. Berman, "The Soviet Legal Profession," *Harvard Law Review*, 82 (November 1968), 1-41; Farhat J. Ziadeh, *Lawyers, The Rule of Law and Liberalism in Modern Egypt* (Stanford, Calif., 1968).

Acknowledgments

The happiest task for an author is to express appreciation to those who provided safe passage through the hazards of thinking and writing. Free time and costly research required substantial financial assistance. Grants from the Penrose Fund of the American Philosophical Society, the National Endowment for the Humanities, the Faculty Publications Gift Fund of Wellesley College, and a Faculty Research Grant from the Social Science Research Council were indispensable. I am especially grateful to the John Simon Guggenheim Memorial Foundation for a fellowship that provided a rare and splendid year unfettered by obligations other than the completion of this book.

Among the various institutions that facilitated my research none served in so many ways as the Harvard Law School, where I was a Liberal Arts Fellow in Law and History during 1969-70 and a welcomed squatter thereafter. By now the Liberal Arts program has its own band of loyal alumni among college teachers to testify to the genial persistence of Harold J. Berman, its guiding mentor. The vast library resources of the school were matched only by the tolerance of Morris Cohen, Librarian, and Erika S. Chadbourn, Curator of Manuscripts, for my incessant requests. Friends on the faculty mixed skepticism and encouragement in appropriate measure. The members of the Legal History group offered collegiality that was deeply sustaining and provocative questions that provided essential guidance.

Librarians at the archives listed in the Bibliographical Essay, and at Lincoln's Inn, London, gave generous assistance. My thanks to them collectively for their many individual acts of kindness. Several of my former students at Brandeis University and Wellesley College served valiantly as research assistants: David Englander, Dorothy Siegal, Susan Esserman, and Susan Gordon. David, at the beginning, and Susie, near the end, offered good ideas and good cheer when my own resources were depleted. Nancy Donovan prepared the Index; she and Joy Archambault did superb typing under pressure. Sheldon Meyer of Oxford University Press was patient and helpful during the evolution of an idea into a manuscript. The sensitive editing of Stephanie Golden transformed the manuscript into a book. My colleagues and students at Wellesley College, and (during 1974-75) at Tel Aviv University, provided congenial institutional environments that sustained me for the solitude of research and writing.

So many lawyers, law teachers, scholars, and friends spoke and corresponded with me about the subject of this book that it would require another volume to mention them individually and to acknowledge the fullness of my appreciation. Some must, however, escape anonymity. Eugene Bardach treated me to the rewards and inscrutable mysteries of interdisciplinary collaboration in our mutual exploration of lawyers' career patterns. Lloyd L. Weinreb scrutinized the penultimate version of the manuscript and offered many valuable suggestions. Alfred S. Konefsky, a discerning critic, was always available when needed. So much of the creation of this book was nourished and sustained by Morton J. Horwitz that I cannot imagine its existence without him. The pleasure of authorship, however, is nothing compared to the joy of a friendship that made it, and so much else, possible.

The members of my family offered contributions commensurate with their size, talents, and ingenuity. Jeffrey rescued me from my study with long walks and strenuous hockey games. Pammy made my presence there infinitely more tolerable with her pictures, flowers, and surprise visits. Judy contributed her special blend of editorial expertise and love.

Index